ELEMENTS OF STRENGTH OF MATERIALS

ELEMENTS OF STRENGTH

S. TIMOSHENKO

D. H. YOUNG

Princeton, New Jersey Toronto London Melbourne

OF MATERIALS /*fifth edition*

Professor of Engineering Mechanics, Emeritus, Stanford University

Silas H. Palmer Professor of Civil Engineering, Stanford University

D. VAN NOSTRAND COMPANY, INC.

Van Nostrand Regional Offices: *New York, Chicago, San Francisco*

D. Van Nostrand Company, Ltd., *London*

D. Van Nostrand Company, (Canada), Ltd., *Toronto*

D. Van Nostrand Australia Pty. Ltd., *Melbourne*

Fifth Edition Published January 1968
Reprinted October 1968

PRINTED IN THE UNITED STATES OF AMERICA

PREFACE

This textbook is an outgrowth of Timoshenko's two-volume *Strength of Materials*, first published in 1930. Whereas the two-volume edition presents both elementary and advanced topics, the present volume is considerably abridged and is designed primarily for undergraduate courses in elementary strength of materials in American colleges and engineering schools.

In this fifth edition, *Elements of Strength of Materials* represents a recasting and rewriting of the original abridgment, although an attempt has been made to retain the same general approach to the subject that characterized the original work. This consists primarily in proceeding gradually from the simplest cases to the more complex ones and relying on physical and geometrical considerations of deformation to establish the patterns of stress distribution under various types of loading. This, of course, characterizes the "strength of materials approach" as contrasted with that of the "theory of elasticity." Such an approach may seem old-fashioned to some, but the authors firmly believe that, for the beginner, it represents a sounder pedagogy. We must all learn to walk before we attempt to run.

New examples have been added and new sets of problems have been substituted for the old ones throughout the book. Answers are given to all problems.

In the first chapter, the ideas of stress and strain within the elastic range of behavior are treated thoroughly before introduction of the complications associated with nonlinear stress-strain behavior. The second chapter begins with a discussion of the stress conditions on an oblique section of a bar in tension in order that the complete stress-strain diagram with proportional limit, yield point, ultimate strength, etc., may be better appreciated. This chapter also contains a section on Plastic Analysis or Limit Design.

Chapter 3 begins with a discussion of stresses in thin-walled pressure vessels which serves to introduce the problem of biaxial stress. Analysis of biaxial stress is then developed in detail and Mohr's circle is introduced. This leads logically to a discussion of pure shear, which is essential to a proper treatment of torsion as taken up in Chapter 4. In the next two chapters (5 and 6), the question of bending stresses and shearing stresses

V

in beams is taken up. The first of these contains the fundamentals of bending theory, and the second treats a number of special topics in bending of beams. Chapter 7 deals with the general problem of plane stress and the notion of principal stresses. Applications to principal stresses in beams and stresses due to combined bending and torsion are fully treated. Chapter 7 ends with sections on the analysis of plane strain and the use of strain rosettes.

Chapter 8 is devoted to methods of calculating deflections of beams. These include the differential equation of the elastic line, the moment-area theorems, and the method of superposition. Statically indeterminate beams are discussed in Chapter 9. Since the concept of strain energy has been developed in earlier chapters, it is natural at this point to discuss Castigliano's theorem and its application to statically indeterminate problems. This chapter ends with a section on limit analysis of statically indeterminate beams, using the concept of the plastic hinge.

Chapter 10 deals with the theory of columns and has been written so as to emphasize the rational approach and minimize the attention given to empirical column formulas. The text proper ends with a chapter on the mechanical properties of materials. It is hoped that the inclusion of such material in an undergraduate textbook will serve to give the student a better appreciation of the importance of the experimental side of the subject of Strength of Materials.

S. Timoshenko
D. H. Young

CONTENTS

NOTATION

A	area
a, b, c	dimensions
c	distance from neutral axis to extreme fiber
d	diameter
E	modulus of elasticity
e	eccentricity
F	force
G	shear modulus
g	gravitational acceleration constant
h	height; depth of a beam
hp	horsepower
I	moment of inertia of area
i	radius of gyration
J	polar moment of inertia of area
K	stress concentration factor
k	symbol for $\sqrt{P/EI}$; spring constant; factor
l	length
M	bending moment
N	normal force
n	factor of safety; r.p.m.; number
P	force; load
p	pressure per unit area; pitch
Q	force; statical moment of area
q	load per unit length
R	reaction, radius
r	radius; radius of gyration
S	stress resultant
s	arc length
T	torque; temperature
t	thickness
U	strain energy
u	strain energy per unit volume
V	shearing force; volume
v	velocity
W	weight; total load

w	load per unit length; weight per unit volume
X, Y, Z	forces
x, y, z	coordinates
Z	section modulus
α	temperature coefficient of expansion; angle
β	angle
γ	shearing strain; weight density
δ	deflection; total elongation
ϵ	tensile or compressive strain
θ	slope of elastic line; angle of twist per unit length
μ	Poisson's ratio
ρ	radius of curvature; radial coordinate
σ	normal stress
τ	shearing stress
ϕ	angle of twist; angular coordinate
ω	angular velocity

1

Tension, Compression, and Shear: I

1.1 Introduction

Various structures and machines — bridges, cranes, airplanes, ships, etc. — will be found, upon examination, to consist of numerous parts or *members* connected together in such a way as to perform a useful function and to withstand externally applied loads. Consider, for example, the simple press shown in Fig. 1.1a. The function of this press is to test specimens of various materials in compression. To accomplish this, the specimen is placed on the floor of the base A and the end of the screw is forced down against it by turning the handwheel at the top. This action subjects the specimen as well as the lower portion of the screw to *axial compression* (Fig. 1.1d) and the side members N to *axial tension* (Fig. 1.1b). It will be observed also that the crosshead M is subjected to *bending* (Fig. 1.1c) and the upper part of the screw to twist or *torsion* (Fig. 1.1e). These four basic types of loading of a member are frequently encountered in both structural and machine design problems. They may be said to constitute essentially the principal subject matter of Strength of Materials. In subsequent chapters we will consider them in the order of their complexity: tension,

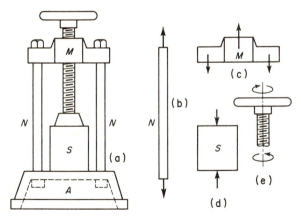

FIG. 1.1

1

compression, torsion, and bending. We will also see later that in many cases a particular member of a structure or machine may be simultaneously subjected to the action of two or more of these basic types of loading in combination. In such cases, the problems of analysis and design of the member can become somewhat more involved.

Analysis and design of any structure or machine like the press in Fig. 1.1 involve two major questions: (a) Is the structure strong enough to withstand the loads applied to it and (b) is it stiff enough to avoid excessive deformations and deflections? In Statics, the members of a structure were treated as *rigid bodies;* but actually all materials are deformable and this property will henceforth be taken into account. Thus Strength of Materials may be regarded as the statics of deformable or *elastic bodies.* For example, it is clear that compression of the specimen in Fig. 1.1a can be increased only by advancing the screw of the press downwards through the crosshead M. This relative displacement between two parts of the machine is partly accounted for by shortening of the specimen and the lower part of the screw and partly by extension of the side bars N as well as some bending deflection of the crosshead M. Thus, the amount of compressive force on the specimen that will correspond to one turn of the handwheel will depend upon the relative stiffness of the various members of the machine.

Both the *strength* and *stiffness* of a structural member are functions of its size and shape and also of certain physical properties of the material from which it is made. These physical properties of materials are largely determined from experimental studies of their behavior in a testing machine. The study of Strength of Materials is aimed at predicting just how these geometric and physical properties of a structure will influence its behavior under service conditions. The applications of the subject are broad in scope and will be found in all branches of engineering. We begin with a study of the simplest type of loading, namely, axial tension or compression of a straight prismatic bar.

1.2   Internal Force; Stress

In Fig. 1.2, a prismatic bar AB is subjected to axial tension by the action of a vertical load P applied at B and acting along the axis AB of the bar, the proper weight of which is neglected. This action on the bar stretches it slightly and also tends to pull it apart, i.e., to produce *rupture.* This tendency to rupture is resisted by internal forces within the bar, i.e., by actions and reactions between its various particles. To visualize these internal forces, imagine that the bar is cut by a section mn perpendicular to its axis and that the lower portion is isolated as a *free body* (Fig. 1.2b). At the lower end of this portion of the bar, the external force P is applied. On the upper end are the internal forces representing the actions of the particles of the upper part of the bar on those of the lower part. These forces are continuously distributed over the cross-section *mn*. In dealing with such distributed forces, the *intensity of force*, i.e., the force per unit area, is of great importance. Visualizing the bar as made up of a bundle of longitudinal fibers, each of which carries its fair share of the load, it appears reasonable to assume, in this case, that the distribution of forces over the cross-section will be *uniform*.* From the condition of equilibrium of the free body (Fig. 1.2b), it is seen that the resultant of this uniform distribution of internal forces must be equal to the external load P. Thus, if A denotes the cross-sectional area of the bar and σ, the force per unit area, we have $S = \sigma A = P$, from which

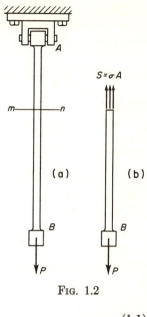

FIG. 1.2

$$\sigma = \frac{P}{A}. \tag{1.1}$$

This force per unit area is called the *stress* in the bar; the total tension $S = \sigma A$ is sometimes called the *stress resultant*. Force is usually measured in pounds and area in square inches so that stress has the dimension of pounds per square inch, denoted by "psi."

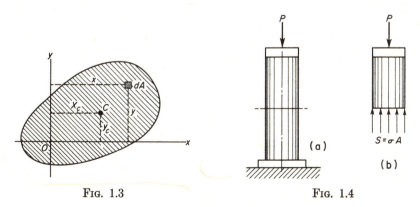

FIG. 1.3   FIG. 1.4

*At cross-sections near the junction points A and B, the distribution may be somewhat non-uniform; but this effect is very localized and will be ignored for the present. For further discussion see Art. 2.5, p. 46.

In order that the applied load P in Fig. 1.2 will actually induce a uniform stress σ over each cross-section of the bar as assumed above, its line of action must pass through the centroid of each cross-section, i.e., P must act along the centroidal axis of the bar. To prove this, consider an arbitrary shape of cross-section as shown in Fig. 1.3 and let dA be any element of area therein. Then for the assumed uniform stress distribution, σ is constant over the cross-section and the element of force acting on dA is σdA, normal to the plane of the section. The resultant of these parallel forces is

$$S = \int \sigma dA = \sigma \int dA = \sigma A, \tag{a}$$

also normal to the section.

The point of application of the stress resultant S can be found from the theorem of moments: namely, the moment of the resultant about either of the coordinate axes x or y must equal the algebraic sum of moments of the elemental forces σdA about the same axis. Thus, denoting by $\bar{x}$ and $\bar{y}$ the coordinates of the point of application of the resultant, we have

$$\left. \begin{aligned} \sigma A \bar{x} &= \int x \sigma dA = \sigma \int x dA = \sigma A x_c, \\ \sigma A \bar{y} &= \int y \sigma dA = \sigma \int y dA = \sigma A y_c, \end{aligned} \right\} \tag{b}$$

where x_c and y_c are the coordinates of the centroid C of the cross-section. From eqs. (b), it is seen that $\bar{x} = x_c$ and $\bar{y} = y_c$. Thus for a uniform stress distribution, the stress resultant S acts through the centroid of the cross-section. Furthermore, it can be seen from Fig. 1.2b that the force S must be collinear with the applied force P. Therefore, P can produce a uniform stress distribution over each cross-section only if it acts through their centroids.*

All of the foregoing discussion applies also to the case of a short post or *strut* subjected to a compressive load P as shown in Fig. 1.4. Here also the

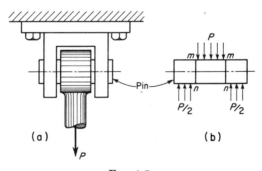

(a)   (b)

FIG. 1.5

*A tensile load P that does not act along the centroidal axis of a bar will produce bending as well as tension of the bar. This case is discussed in Art. 10.1, p. 264.

load P must act along the centroidal axis of the post to produce the uniform compressive stress σ indicated in Fig. 1.4b. In the case of compression members, this condition is sometimes difficult to fulfill, so that the compression of long slender struts or *columns* requires special consideration which will be taken up later in Chapter 10.

Direct Shear. Referring again to Fig. 1.2, let us consider now in some detail the connection between the tension member AB and the ceiling at its upper end. Clearly, in the interests of good design, this connection should be strong enough to develop the full load-carrying capacity of the bar AB itself. An enlarged detail of this connection is shown in Fig. 1.5a, where it is seen that the load P on the tension member must be transmitted to the fork by the horizontal pin connecting the two parts. A free-body diagram of this pin is shown in Fig. 1.5b and it is seen that the pin is primarily in a condition of *shear* which tends to cut it across the sections mn.* Assume now that the internal shearing forces resisting this tendency are uniformly distributed over each of the cross-sections mn. Then denoting by τ_{av} the shear force per unit area, i.e., the *average shear stress*, we see that equilibrium conditions of the middle portion of the pin require that $\tau_{av}A_s = P$, from which

$$\tau_{av} = \frac{P}{A_s},\tag{1.2}$$

where A_s is the total area in shear — in this case, twice the cross-sectional area A of the pin.

Since shearing conditions are never as simple as assumed above, it must be realized that the average shear stress as calculated from eq. (1.2) may be only a rough approximation to the actual stresses that exist in the material. Nevertheless, lacking any more exact knowledge of the true stress distribution, the designer is often forced to use this simple concept of average shear stress as a basis for design.

In dealing with various kinds of machines and structures, the engineer frequently encounters members subjected to simple direct tension, compression, or shear as discussed above. The general problem of design of such members consists in proportioning them so that they can safely and economically withstand the loads that they have to carry. As a basis of doing this, many materials have been tested in the laboratory to establish their strength or resistance to rupture under various types of loading and thereby establish allowable or safe *working stresses*† to be used in design.

---

*There is also some bending of the pin, but if the clearances are small this will be of secondary importance. Only the shearing action will be considered in the present discussion.

†The establishment of working stresses is a very complex question which will not be discussed in any detail at this point. For further discussion, see Art. 2.2, p. 32.

An allowable working stress is usually taken as $1/n$ times the value of the stress at which failure of the test specimen took place. Thus in using such a working stress, the designer has a so-called *factor of safety* n to allow for overloading or other unforeseen adverse effects. Using these somewhat arbitrarily assigned working stresses together with eqs. (1.1) and (1.2), the designer can determine the proper dimensions for the various members of a machine or structure subjected to the action of given loads. Or, if the structure has already been built, he can establish safe values for the allowable loads in a similar manner.

EXAMPLE 1. A vertical load $P = 5000$ lb is supported by two inclined steel wires AC and BC as shown in Fig. 1.6. Determine the required cross-sectional area A of each wire if the allowable working stress in tension is $\sigma_w = 10,000$ psi and the angle $\theta = 30°$.

SOLUTION. When the load P is applied to the ring, each wire is subjected to tension and therefore exerts on the ring a force S directed along the axis of that wire as shown in Fig. 1.6a. Actually, under tension, the wires stretch slightly, so that after the load P is applied, the angles of inclination θ will be slightly greater than 30°. However, in computing the magnitudes of the forces S, we will neglect this

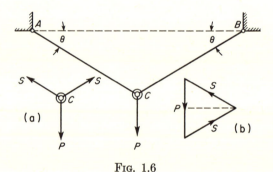

FIG. 1.6

slight change in configuration of the system due to deformation and assume in Fig. 1.6a that each force S is inclined to the horizontal by 30°. Thus, the corresponding closed triangle of forces in Fig. 1.6b is equilateral, and we conclude that $S = P = 5000$ lb. Then, from eq. (1.1), the necessary cross-sectional area of each wire is

$$A = \frac{S}{\sigma_w} = \frac{5000}{10,000} = 0.5 \text{ sq in.}$$

EXAMPLE 2. The piston of a deep-well pump is operated by a vertical prismatic steel rod of length $l = 320$ ft attached to a crank at its upper end as shown in Fig. 1.7. Determine the extreme values of tensile and compressive stress σ in the rod if the resistance on the piston during the downstroke is 200 lb and during the upstroke is

2000 lb. The cross-sectional area of the rod is $A = 0.338$ sq in. and its density is 490 lb/cu ft.

SOLUTION. The weight of the rod is

$$wlA = \frac{490}{1728} \times 320(12) \times 0.338 = 368 \text{ lb.}$$

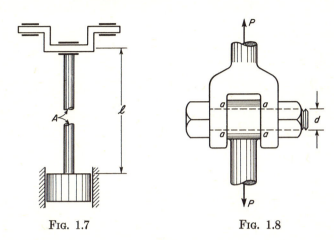

FIG. 1.7   FIG. 1.8

The maximum tensile stress will occur at the top of the rod during the upstroke. Under these conditions, the total tension is

$$S_{\max} = 368 + 2000 = 2368 \text{ lb.}$$

The corresponding maximum tensile stress is

$$\sigma_t = \frac{S}{A} = \frac{2368}{0.338} = 7000 \text{ psi.}$$

The greatest compressive stress will occur at the lower end of the rod during the downstroke. For these conditions $S = 200$ lb compression and

$$\sigma_c = \frac{S}{A} = \frac{200}{0.338} = 592 \text{ psi.}$$

EXAMPLE 3. A tension rod made up of two parts as shown in Fig. 1.8 is designed to carry a total load $P = 20,000$ lb. What is the proper diameter d for the connecting bolt if the allowable working stress in shear is $\tau_w = 10,000$ psi?

SOLUTION. Assume that the bolt fits snugly in the holes through the prongs of the fork and that the clearances are small so that bending action on the bolt will be minimized. Then the bolt is essentially in a condition of direct shear across the sections aa and eq. (1.2) may be used. This gives

$$A_s = 2A = \frac{P}{\tau_w} = \frac{20,000}{10,000} = 2 \text{ sq in.}$$

Thus the required cross-sectional area A of the bolt is 1 sq in. and the corresponding diameter $d = 1.13$ in.

PROBLEM SET 1.2

1. A short, hollow, cast-iron cylinder with a wall thickness of 1 in. is to carry a compressive load $P = 150$ kips (1 kip = 1000 lb). Compute the required outside diameter d_o if the working stress in compression is $\sigma_w = 12,000$ psi. *Ans.* $d_o = 4.98$ in.

2. Solve the preceding problem if the wall thickness is to be one-tenth of the outside diameter d_o. *Ans.* $d_o = 6.65$ in.

3. A steel wire hangs vertically under its own weight. What is the greatest length it can have if the allowable tensile stress is $\sigma_t = 30,000$ psi? The specific weight of steel is 490 lb/cu ft. *Ans.* $l = 8820$ ft.

4. Three pieces of wood having $1\frac{1}{2}$ in. $\times$ $1\frac{1}{2}$ in. square cross-sections are glued together and to the foundation as shown in Fig. A. If the horizontal force $P = 5000$ lb, what is the average shearing stress in each of the glued joints? *Ans.* $\tau_{av} = 417$ psi.

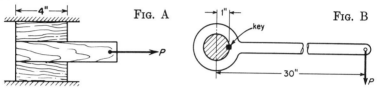

FIG. A   FIG. B

5. In Fig. B a lever is attached to a spindle by means of a square key $\frac{1}{4}$ in. $\times$ $\frac{1}{4}$ in. by 1 in. long. If the average shear stress in the key is not to exceed 10,000 psi, what is the safe value of the load P applied to the end of the lever? *Ans.* $P = 83.3$ lb.

6. If the allowable working stress for the pump rod in Fig. 1.7 is reduced to $\sigma_w = 6000$ psi, what is the required cross-sectional area A of the rod? Use all other data as given in Example 2. *Ans.* $A = 0.407$ sq in.

7. Each prong of the cast aluminum fork shown in Fig. 1.5a has a 2 in. $\times$ $\frac{1}{4}$ in. rectangular cross-section. The allowable tensile stress for the cast aluminum is $\sigma_w = 10,000$ psi and the allowable shear stress for the steel pin is $\tau_w = 6000$ psi. What is the optimum diameter d for the pin, i.e., for what diameter d of the pin will the safe load P on the assembly be a maximum? *Ans.* $d = 0.80$ in.

8. The frame shown in Fig. C is made up of 4 in. $\times$ 4 in. square wood posts, for which the allowable stress in shear parallel to the grain is $\tau_w = 100$ psi, while that in compression perpendicular to the grain is $\sigma_w = 400$ psi. Calculate the minimum safe values of the dimensions a, b, and c. The vertical post is pinned to the sill at its lower end. *Ans.* $a = 1\frac{3}{4}$ in., $b = \frac{7}{16}$ in., $c = \frac{21}{64}$ in.

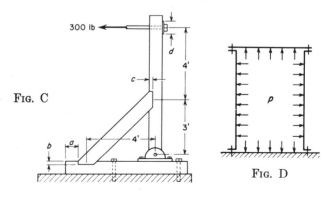

FIG. C   FIG. D

9. The 300-lb horizontal force in Fig. C is transmitted to the top end of the vertical post through a $\frac{1}{4}$-in.-diameter steel bolt as shown. If the allowable pressure between the loose-fitting washer and the back face of the post is 200 psi, what is the minimum diameter d of the washer? *Ans.* $d = 1.43$ in.

10. If the bearing pressure between the wood and the pin at the bottom end of the vertical post in Fig. C is 200 psi, what is the required diameter d of the pin? Assume the width of the bearing area in this case to be the projected diameter of the pin. *Ans.* $d = 0.825$ in.

11. The cylinder shown in Fig. D has inside diameter $D = 10$ in. and is subjected to internal gas pressure of intensity $p = 300$ psi, gauge. How many $\frac{1}{2}$-in.-diameter steel bolts will be required to fasten the top cover plate to the cylinder if the working stress for the bolts is 10,000 psi? *Ans.* 12 bolts.

1.3 Elasticity; Strain

Let us consider again the case of a prismatic bar in simple tension as shown in Fig. 1.9. As the tensile load P is gradually increased, the bar will be observed to stretch slightly so that for each value of P there will be a corresponding small elongation δ of the bar as shown. Subsequently, if the load is gradually diminished to zero, the elongation δ will either completely or partially disappear, i.e., the bar tends to reassume its initial length l. This property of a material to return partially or completely to its initial shape after unloading is called *elasticity*. If the bar completely recovers its initial length, the material is said to be *perfectly elastic;* if not, it is said to be only *partially elastic.* Experiments show that many structural materials like steel, aluminum, wood, and even concrete may be considered as perfectly elastic within limits, i.e., if not excessively loaded. For the present discussion, we confine our attention to this elastic behavior of a material.

FIG. 1.9

Referring to Fig. 1.9, it may be assumed that, under tension, all longitudinal fibers of the bar are stretched uniformly. Thus, we define the elongation per unit length of the bar by

$$\epsilon = \frac{\delta}{l}, \tag{1.3}$$

which is called the *tensile strain* or simply the *strain*. Similarly, for a bar in axial compression, the quantity ϵ will define the contraction per unit length of the bar, or the *compressive strain.* Tensile strain will be considered *positive;* compressive strain, *negative.* It should also be noted that strain is a dimensionless quantity, being a length divided by a length.

Hooke's Law. Experiments with prismatic bars of various materials in tension have shown that, within the range of elastic behavior of the material, the elongation δ is proportional to both the tensile force P and the length l of the bar. It is also observed to be inversely proportional to the cross-sectional area A of the bar. Expressed algebraically, $\delta \sim Pl/A$, or

$$\delta = \frac{1}{E} \cdot \frac{Pl}{A} = \frac{Pl}{AE}, \tag{1.4}$$

where E is a constant for any given material and is called the *modulus of elasticity* of the material in tension. The relationship expressed by eq. (1.4) is known as *Hooke's law*, after Robert Hooke who first established it by experiment in 1678.

Using the notations

$$\sigma = \frac{P}{A} \quad \text{and} \quad \epsilon = \frac{\delta}{l},$$

from eqs. (1.1) and (1.3), Hooke's law can also be written in the form

$$\epsilon = \frac{\sigma}{E} \quad \text{or} \quad \sigma = E\epsilon. \tag{1.5}$$

Expressed in words, this leads to the familiar concise statement of Hooke's law: *Stress is proportional to strain.* The modulus of elasticity E is seen to represent the factor of proportionality between stress and strain and may be directly defined by

$$E = \frac{\sigma}{\epsilon} = \frac{\text{stress}}{\text{strain}}. \tag{a}$$

Since the strain ϵ is dimensionless, it may be concluded that the modulus of elasticity E has the dimension of stress, i.e., pounds per square inch, psi.

From its definition (a), the modulus of elasticity E is seen to represent that stress σ which would produce a tensile strain $\epsilon = 1$; in other words, that tensile stress under which a bar would be stretched to twice its original length if the material could remain perfectly elastic throughout such excessive strain. As this observation would lead us to expect, the modulus of elasticity is a very large quantity for most materials. In the case of steel, for example, a value $E = 30(10)^6$ psi is usually used.[*]

If we plot tensile stress σ against strain ϵ from eq. (1.5), we obtain the straight line OA in Fig. 1.10. This plot represents the so-called *stress-strain diagram* for a material in simple tension within the range of its elastic behavior. The modulus of elasticity E is seen to represent the slope of the stress-strain line OA.

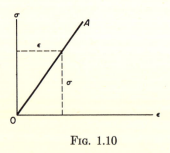

FIG. 1.10

[*]For values of E for other commonly used structural materials, see Tables I, II, III, Appendix A, p. 341.

All of the above discussion applies also to the case of axial compression of a bar. For most materials, the modulus of elasticity in compression is found to have the same value as for tension. In calculation, tensile stress and strain are to be considered as positive; compressive stress and strain, as negative.

Poisson's Ratio. Referring again to the tension member in Fig. 1.9, experiments show further that its axial elongation is always accompanied by a lateral contraction and that the ratio

$$\mu = \frac{\text{unit lateral contraction}}{\text{unit axial elongation}}$$

is constant for a given material within its range of elastic behavior. This constant μ is called *Poisson's ratio*, after the French mathematician who predicted its existence and value, using the molecular theory of structure of the material. For so-called *isotropic* materials which have the same properties in all directions, Poisson concluded that $\mu = \frac{1}{4}$. Careful measurements on structural metals in tension usually confirm this value quite closely. For structural steel, experiments indicate that approximately $\mu = 0.3$.

Knowing the modulus of elasticity E and Poisson's ratio μ for a given material, the change in dimensions and volume of a prismatic bar in tension can easily be calculated. Before deformation, the volume of the bar is $V = Al$. After deformation, the new length $l_1 = l(1 + \epsilon)$ and the new cross-sectional area $A_1 = A(1 - \mu\epsilon)^2$. Thus the new volume is $V_1 = A_1 l_1 = Al(1 + \epsilon)$ $(1 - \mu\epsilon)^2$ or, since ϵ is a small quantity compared with unity, $V_1 \approx Al$ $(1 + \epsilon - 2\mu\epsilon)$. Thus the change in volume is $\Delta V = V_1 - V = Al\epsilon$ $(1 - 2\mu)$ and the *unit volume change* becomes

$$\frac{\Delta V}{V} = \frac{Al\epsilon(1 - 2\mu)}{Al} = \epsilon(1 - 2\mu). \tag{1.6}$$

Since it is unlikely that any material would diminish in volume under tension, we may expect that μ will always be less than $\frac{1}{2}$. For such a material as rubber, μ approaches the above limit and the volume of rubber remains practically constant during extension. Other materials like concrete and cork have smaller values of μ and increase slightly in volume when subjected to tension. The value of μ for concrete can be taken as 0.1, while for cork, $\mu = 0$.

The above discussion of axial tension can be applied also to the case of axial compression of a prismatic bar. In such case, the longitudinal contraction of the bar will be accompanied by lateral expansion and for calculating this expansion, the same value for μ as in the case of tension can be used.

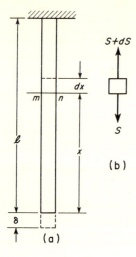

FIG. 1.11

EXAMPLE 1. Referring to Fig. 1.11, calculate the total elongation δ of a prismatic bar of length l and cross-sectional area A which hangs vertically under its own weight.

SOLUTION. Due to the uniformly distributed weight of the bar, the total tension at the cross-section mn will be

$$S_x = \gamma A x, \tag{b}$$

where γ is the weight per unit volume of the material. As a result of this tension, an element of length dx situated just above the cross-section mn will elongate by the small amount

$$d\delta = \frac{S_x\,dx}{AE}, \tag{c}$$

obtained from eq. (1.4) by replacing P by S_x and l by dx. Summing up all such elemental elongations along the full length of the bar gives

$$\delta = \int_0^l \frac{S_x\,dx}{AE}. \tag{d}$$

Substituting $S_x = \gamma A x$ from eq. (b), this becomes

$$\delta = \frac{\gamma}{E}\int_0^l x\,dx = \frac{\gamma l^2}{2E}. \tag{e}$$

It may be noted that since the total weight of the bar is $W = \gamma A l$, eq. (e) can also be expressed in the form

$$\delta = \frac{Wl}{2AE}. \tag{e'}$$

Comparing the result with eq. (1.4), we see that in this case of uniformly varying tension, the total elongation δ is just half what it would be if the tension were equal to W throughout the length of the bar.

EXAMPLE 2. A homogeneous slender prismatic bar of total length $2l$ rotates with constant angular velocity ω in a horizontal plane about a fixed axis through its midpoint as shown in Fig. 1.12. The cross-sectional area of the bar is A and its weight

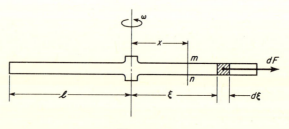

FIG. 1.12

per unit volume is γ. Find the maximum stress σ and the total elongation δ of either half of the bar due to centrifugal tension.

SOLUTION. Consider an element of the bar of length $d\xi$ at the distance ξ from the axis of rotation. The centrifugal force on this element due to its motion in a circular path of radius ξ is

$$dF = \frac{\gamma A d\xi}{g} \cdot \omega^2 \xi,$$

where $\gamma A d\xi/g$ is the mass of the element and $\omega^2 \xi$ is its radial acceleration. This force acts radially outwards along the axis of the bar as shown. The total tension S_x at any cross-section mn defined by the distance x from the axis of rotation is the sum of all these centrifugal forces between $\xi = x$ and $\xi = l$. Thus

$$S_x = \int_x^l \frac{\gamma A d\xi}{g} \omega^2 \xi = \frac{\gamma A \omega^2}{2g}(l^2 - x^2). \tag{f}$$

This tensile force has its maximum value at the mid-point of the bar ($x = 0$) where

$$S_{\max} = \frac{\gamma A \omega^2 l^2}{2g}.$$

The corresponding maximum stress is

$$\sigma_{\max} = \frac{S_{\max}}{A} = \frac{\gamma \omega^2 l^2}{2g}.$$

It will be noted that this maximum stress increases as the square of both the angular velocity ω and the length l, so that a long bar rotating at high speed will be very severely stressed.

The total elongation of either half of the bar will be

$$\delta = \int_0^l \frac{S_x \, dx}{AE}.$$

Substituting the value of S_x from eq. (f) and integrating

$$\delta = \int_0^l \frac{\gamma A \omega^2 dx}{2gAE}(l^2 - x^2) = \frac{\gamma \omega^2 l^3}{3gE}. \tag{g}$$

This may also be written in the form

$$\delta = \frac{\gamma A l \omega^2}{3g} \frac{l^2}{AE} = \frac{W}{3g}\omega^2 l \cdot \frac{l}{AE}. \tag{g'}$$

where $W = \gamma A l$ is the weight of one half of the bar. This is the same elongation that would take place in a massless string of length l having a mass particle of weight $W/3$ at its outer end.

EXAMPLE 3. A vertical load $P = 10$ kips is supported by a strut AC and a tie-rod BC arranged as shown in Fig. 1.12'. The strut has length $l = 100$ in. and both members have the same axial stiffness $AE = (10)^7$ lb. Find the horizontal and vertical components Δ_h and Δ_v of the displacement of joint C due to axial deformation of the members.

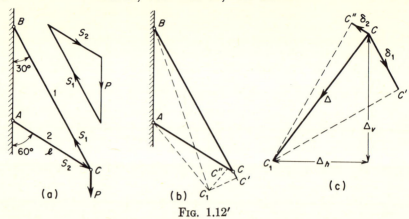

FIG. 1.12'

SOLUTION. We must first find the axial force in each member. To do this, we consider the equilibrium of joint C acted upon by forces P, S_1, and S_2 as shown in Fig. 1.12'a and construct the corresponding closed triangle of forces. From this we see that $S_1 = \sqrt{3}P$, tension, and $S_2 = P$, compression. Then from eq. (1.4), the corresponding elongations of the two members will be

$$\left. \begin{aligned} \delta_1 &= \frac{S_1 l_1}{AE} = \frac{3Pl}{AE} = +0.300 \text{ in.} \\[2mm] \delta_2 &= \frac{S_2 l_2}{AE} = -\frac{Pl}{AE} = -0.100 \text{ in.} \end{aligned} \right\} \tag{g}$$

i.e., the tie-rod lengthens and the strut shortens.

As a result of these small changes in the lengths of the members, joint C will move to a new position C_1. In principle, this new position is found as follows: In Fig. 1.12'b, we remove the pin at C and give to the members their changes of length δ_1 and δ_2 so that BC' represents the elongated tie-rod and AC'', the shortened strut. Next, we rotate each member around its fixed end so that C' moves along the arc $C'C_1$ centered at B while C'' moves along the arc $C''C_1$ centered at A. The point C_1 where these two arcs intersect determines the new position C_1 of the joint.

Since the changes of length δ_1 and δ_2 of the members are very small compared with the lengths of the members themselves, the construction shown in Fig. 1.12'b cannot be made to scale with any accuracy. However, we may note that for all practical purposes the arcs $C'C_1$ and $C''C_1$ may be considered as normals to the lines BC' and AC'' representing the axes of the members. Hence the small diagram $CC'C_1C''$ can be divorced from Fig. 1.12'b and constructed separately to a much larger scale as shown in Fig. 1.12'c, where the arcs $C'C_1$ and $C''C_1$ are now replaced by normals $C'C_1$ and $C''C_1$ to CC' and CC'', respectively.

The diagram in Fig. 1.12'c is called a *Williot diagram*. It can be constructed to any suitable scale, and the total displacement CC_1, denoted by Δ, can then be measured directly from the diagram. Or, if preferred, we can consider the geometry of the Williot diagram and calculate trigonometrically the horizontal and vertical components Δ_h and Δ_v of the resultant displacement Δ. Using the values (g) for δ_1 and δ_2, we find in this way that

$$\Delta_h = 0.473 \text{ in.}, \qquad \Delta_v = 0.620 \text{ in.}$$

PROBLEMSET 1.3

1. An aluminum bar 6 ft long has a 1-in.-square cross-section over 2 ft of its length and a 1-in.-diameter circular cross-section over the other 4 ft. How much will the bar elongate under a tensile load $P = 3500$ lb if $E = 10.5(10)^6$ psi? *Ans.* $\delta = 0.0284$ in.

2. Referring to Example 2, page 6, assume that the vertical rod of the mine pump is made of steel ($E = 30(10)^6$ psi) and that all other data are as given in the example. With these data, find the proper radius r of the crank at the upper end in order to attain an 8-in. stroke of the piston at the lower end. *Ans.* $r = 4.42$ in.

3. A prismatic steel bar having cross-sectional area $A = 0.50$ in.² is subjected to axial loading as shown in Fig. A. Neglecting localized irregularities in stress distribution near the points of application of the loads, find the net increase δ in the length of the bar. Assume $E = 30(10)^6$ psi. *Ans.* $\delta = -0.0048$ in.

4. At what distance x from the fixed end of the bar in Fig. A should the 4-kip force be applied in order that the net overall change in length of the bar will be zero? *Ans.* $x = 4.5$ ft.

5. A rigid bar AB is hinged to a vertical wall and supported horizontally by a tie-bar CD as shown in Fig. B. The tie-bar has cross-sectional area $A = 0.10$ in.² and its allowable stress in tension is $\sigma_w = 20,000$ psi. Find the safe value of the magnitude of the load P and the corresponding vertical deflection Δ_B of point B. The tie-bar has modulus of elasticity $E = 30(10)^6$ psi. *Ans.* $P = 600$ lb, $\Delta_B = 0.133$ in.

6. Calculate the vertical deflection Δ of the ring C in Fig. 1.6 using the data from Example 1, page 6. Assume that the horizontal distance AB between the supports is 30 ft. *Ans.* $\Delta = 0.138$ in.

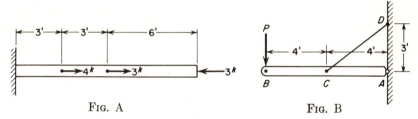

FIG. A   FIG. B

7. A rigid bar AB, 9 ft long, is suspended by two vertical rods at its ends and hangs in a horizontal position under its own weight as shown in Fig. C. The rod at A is brass; length, 6 ft, cross-sectional area, 1.76 in.², modulus of elasticity, $15(10)^6$ psi. The rod at B is steel; length, 10 ft, cross-sectional area, 0.785 in.², modulus of elasticity, $30(10)^6$ psi. At what distance x from A may a vertical load P be applied

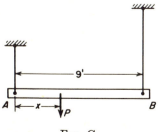

FIG. C

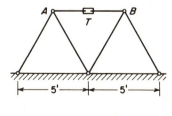

FIG. D

if the bar is to remain horizontal after the load is applied? *Ans.* $x = 3.13$ ft.

8. Each bar of the plane truss-work shown in Fig. D is 5 ft long and has cross-sectional area $A = 0.307$ in.² and modulus of elasticity $E = 30(10)^6$ psi. The turnbuckle T in the bar AB is double acting with 32 threads per inch. How many turns n of the turnbuckle will be required to produce an axial tension of 10,000 lb in the bar AB? *Ans.* $n = 5.22$ turns.

9. If the cross-sectional area of each half of the rotating bar in Fig. 1.12 varies linearly from A at the hub to $A/2$ at the tip, calculate the tensile stress σ_0 at the hub during uniform rotation with angular velocity ω. *Hint:* See Example 2 for general mode of attack. *Ans.* $\sigma_0 = \gamma \omega^2 l^2 / 3g$.

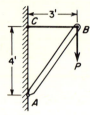

10. For the simple structure shown in Fig. E, member BC is a steel wire having diameter $d = \frac{1}{8}$ in. and member AB is a wood strut of 1-in.-square cross-section. Calculate the horizontal and vertical components of the displacement of point B due to a vertical load $P = 400$ lb acting as shown. For steel, $E_s = 30(10)^6$ psi; for wood, $E_w = 1.5(10)^6$ psi. *Ans.* $\delta_h = 0.029$ in., $\delta_v = 0.047$ in.

Fɪɢ. E

1.4 Statically Indeterminate Problems in Tension and Compression

Consider the simple structure made up of three tension members, each of cross-sectional area A and modulus of elasticity E, arranged as shown in Fig. 1.13a and subjected to a vertical load P at A. Under the action of this load, each bar will be subjected to some tension and a free-body diagram of the connecting pin A will be as shown in Fig. 1.13b, where Y denotes the tensile force in each inclined bar (equal because of symmetry) and X, the

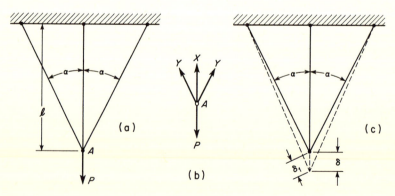

Fɪɢ. 1.13

tensile force in the vertical bar. Then from equilibrium considerations of this pin, we must have

$$X + 2Y \cos \alpha = P. \tag{a}$$

It will be seen at once that while this *equation of statics* defines a relationship between the tensile forces X and Y, it is insufficient to determine their values uniquely. For this reason, the structure is said to be *statically indeterminate*. Looked at in another way, it will be observed that any two of the three bars are sufficient to constrain the pin A in the vertical plane of the figure and therefore a third bar is said to be *redundant*. In general, any structure which contains more constraints than are necessary for geometric rigidity of the structure, i.e., one which contains redundant constraints, will prove to be statically indeterminate. In fact, if the bars of the structure in Fig. 1.13a were absolutely rigid and undeformable, as assumed in statics, there would be no way to ascertain how the load P was divided among the three bars. However, the bars are actually elastic and stretch slightly under tension. Furthermore, since they are connected together by the pin A, it is evident that the amounts they stretch must be related in some way by the geometry of the structure. This is shown in Fig. 1.13c. Denoting by δ the elongation of the vertical bar and by δ_1 the elongation of each inclined bar and keeping in mind that these elongations are extremely small compared with the dimensions of the structure, it will be seen from Fig. 1.13c that

$$\delta_1 = \delta \cos \alpha. \tag{b}$$

This equation of *geometry* or *consistent deformation* is the key to the problem. Assuming that the tensions in the bars do not exceed their elastic limits and using Hooke's law, we have

$$\delta = \frac{Xl}{AE} \quad \text{and} \quad \delta_1 = \frac{Yl_1}{AE} = \frac{Yl}{AE \cos \alpha},$$

where $l_1 = l/\cos \alpha$ is the length of each inclined bar. Substitution of these values into eq. (b) gives

$$\frac{Yl}{AE \cos \alpha} = \frac{Xl}{AE} \cos \alpha,$$

from which $$Y = X \cos^2 \alpha. \tag{c}$$

Substitution of this value of Y into eq. (a) gives

$$X + 2X \cos^3 \alpha = P,$$

from which

$$X = \frac{P}{1 + 2 \cos^3 \alpha} \quad \text{and} \quad Y = \frac{P \cos^2 \alpha}{1 + 2 \cos^3 \alpha}.$$

Then the total vertical deflection of point A becomes

$$\delta = \frac{Xl}{AE} = \frac{Pl}{AE}\left(\frac{1}{1 + 2\cos^3\alpha}\right).$$

This example illustrates one case of a statically indeterminate system involving axially loaded members. Others will be shown in the examples which follow. In all cases of systems involving redundant elements, equations of statics will be insufficient to determine all of the unknown forces and must be supplemented by equations of consistent deformation based on the geometry of the system.

EXAMPLE 1. A hollow steel cylinder of length $l = 12$ in., inside diameter $d = 6$ in., and uniform wall thickness $t = \frac{1}{8}$ in. is filled with concrete and compressed between rigid parallel plates by a load $P = 100,000$ lb as shown in Fig. 1.14a. Calculate the compressive stress in each material and the total shortening of the cylinder if $E_s = 30(10)^6$ psi for the steel and $E_c = 2(10)^6$ psi for the concrete. Assume that both materials obey Hooke's law.

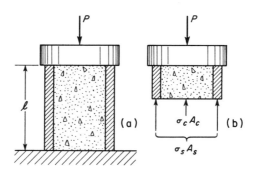

FIG. 1.14

SOLUTION. A free-body diagram for the upper portion of the cylinder is shown in Fig. 1.14b and from statics it follows that

$$\sigma_s A_s + \sigma_c A_c = P. \tag{d}$$

Also since both materials must shorten the same amount, because of the rigid end plates, and are of the same height l, their unit compressive strains must be equal, so that

$$\frac{\sigma_s}{E_s} = \frac{\sigma_c}{E_c}. \tag{e}$$

This shows that the two materials are stressed in the ratio of their moduli of elasticity. For the given values of these moduli, $E_s/E_c = 15$, so that $\sigma_s = 15\sigma_c$, and eq. (d) gives

$$\sigma_c = \frac{P}{15A_s + A_c}, \quad \sigma_s = \frac{15P}{15A_s + A_c}.$$

Using the given numerical data, $A_s = 2.41$ in.^{2} and $A_c = 28.3$ in.^{2}, so that $\sigma_c = 1560$ psi and $\sigma_s = 23,400$ psi. The total shortening of the cylinder is $\delta = \epsilon l = (\sigma_c/E_c)l = 0.000775 \times 12 = 0.00936$ in.

EXAMPLE 2. A rigid bar AB is hinged at A and supported in a horizontal position by two identical vertical steel wires as shown in Fig. 1.15a. Find the tensile forces S_1 and S_2 induced in these wires by a vertical load P applied at B as shown.

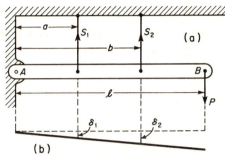

FIG. 1.15

SOLUTION. Considering the equilibrium of the bar AB, the proper weight of which is neglected, and taking moments of all forces with respect to the hinge A, we have

$$S_1 a + S_2 b = Pl. \tag{f}$$

Also, since AB remains straight and simply rotates slightly around point A, the elongations δ_1 and δ_2 of the wires must be in the ratio a/b, i.e.,

$$\frac{\delta_1}{\delta_2} = \frac{a}{b}. \tag{g}$$

Then from Hooke's law, $S_1/S_2 = a/b$ and eq. (f) gives

$$S_1 = \frac{Pal}{a^2 + b^2}, \quad S_2 = \frac{Pbl}{a^2 + b^2}.$$

EXAMPLE 3. A cold-rolled steel bolt of length $l = 12$ in. passes through a hard-drawn copper tube of the same length, as shown in Fig. 1.16a, and the nut at the left end is turned up just snug at room temperature $T = 70°$ F. Subsequently the nut is tightened up $n = \frac{1}{4}$ turn and the entire assembly is raised to a temperature $T = 140°$ F. What stresses will exist in the bolt and the tube under these conditions? The cross-sectional area of the steel bolt is $A_s = \frac{1}{2}$ sq in., its modulus of elasticity $E_s = 30(10)^6$ psi, the coefficient of thermal expansion is $\alpha_s = 6.5(10)^{-6}$ in./in./F° and the thread pitch $p = \frac{1}{8}$ in. For the copper tube, $A_c = \frac{3}{4}$ sq in., $E_c = 16(10)^6$ psi, and $\alpha_c = 9.3(10)^{-6}$ in./in./F°.

SOLUTION. Consider that portion of the assembly to the right of a section mn as a free body (Fig. 1.16b). Then from static equilibrium it is seen that the compressive force S_c in the copper tube must balance the tensile force S_s in the steel bolt, i.e.,

$$S_s = S_c. \tag{h}$$

Furthermore, since the final length of the bolt and the tube must be the same, it follows that the shortening of the tube plus the extension of the bolt must be equal to the thread displacement of the nut along the bolt. Expressed algebraically,

$$\delta_s + \delta_c = pn \tag{i}$$

wherein the total extension of the steel bolt is

$$\delta_s = \frac{S_s l}{A_s E_s} + \alpha_s l \Delta T$$

and the total shortening of the copper tube is

$$\delta_c = \frac{S_c l}{A_c E_c} - \alpha_c l \Delta T$$

ΔT being the net increase in temperature. Substituting these expressions for δ_s and δ_c into eq. (i) and noting from eq. (h) that $S_s = S_c = S$, we obtain

$$S\left(\frac{1}{A_s E_s} + \frac{1}{A_c E_c}\right) + (\alpha_s - \alpha_c)\Delta T = \frac{pn}{l}.$$

For the given numerical data this gives

$$S_s = S_c = S = 18{,}670 \text{ lb.}$$

The corresponding stresses are $\sigma_s = 37{,}340$ psi, tension and $\sigma_c = 24{,}900$ psi, compression.

We may now ask: to what temperature T_0 must we cool the assembly in order to relieve all stress? To answer this, we note that the unit strains due to cooling and stress relief will be both negative and equal. These strains, representing shortening per unit length, will be

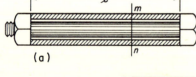

(a)

(b)

Fig. 1.16

$$\epsilon_s = \alpha_s \cdot \Delta T + \frac{\sigma_s}{E_s},$$

$$\epsilon_c = \alpha_c \cdot \Delta T - \frac{\sigma_c}{E_c}.$$

Equating these unit strains, we obtain

$$(\alpha_c - \alpha_s)\Delta T = \frac{\sigma_s}{E_s} + \frac{\sigma_c}{E_c}, \tag{j}$$

where ΔT is the required drop in temperature. Substituting $\sigma_s = 37{,}340$ psi and $\sigma_c = 24{,}900$ psi together with the given numerical data into eq. (j) gives $\Delta T = 1000°F$. Thus the assembly would have to be cooled down to $140°F - 1000°F = -860°F$ to relieve all stress. To do this, of course, would be highly impracticable.

PROBLEM SET 1.4

1. A rigid steel plate is supported by three concrete posts each having a 4 in. × 4 in. square cross-section as shown in Fig. A. By accident, the middle post is 0.02 in. shorter than the other two before load P is applied. Find the safe value of load P if the working stress for the concrete in compression is 3000 psi and the modulus of elasticity $E_c = 2(10)^6$ psi. *Ans.* $P = 117,300$ lb.

2. A concrete pedestal 8 in. high and having a 4 in. × 4 in. square cross-section is reinforced longitudinally by four steel bars as shown in Fig. B. Each reinforcing bar has circular cross-section with diameter $d = \frac{3}{4}$ in. Calculate the safe value of the compressive load P if the working stresses for steel and concrete are $\sigma_s = 20,000$ psi and $\sigma_c = 3000$ psi, respectively. Assume $E_s = 30(10)^6$ psi and $E_c = 3(10)^6$ psi. *Ans.* $P = 63,800$ lb.

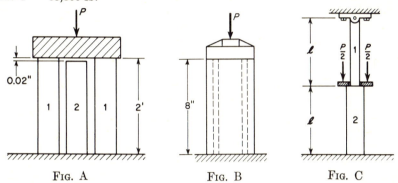

FIG. A   FIG. B   FIG. C

3. A brass rod of length $2l$ and having cross-sectional area $A_1 = 1$ sq in. over the upper half of its length and $A_2 = 1.5$ sq in. over the lower half is supported and loaded as shown in Fig. C. Determine the axial stresses in the two parts of the rod if load $P = 10,000$ lb. *Ans.* $\sigma_1 = -\sigma_2 = 4000$ psi.

4. Prestressed concrete beams are made in the following manner: Steel wires are stretched between rigid end plates to a tensile stress σ_0 as shown in Fig. D_1. Concrete is then poured around them to form the beam as shown in Fig. D_2. After the concrete sets, the external forces Q are removed and the beam is left in a prestressed condition. If the moduli of elasticity of steel and concrete are in the ratio 12:1 and their cross-sectional areas are in the ratio 1:15, what are the final residual stresses in the two materials? *Ans.* $\sigma_s = 5\sigma_0/9$; $\sigma_c = -\sigma_0/27$.

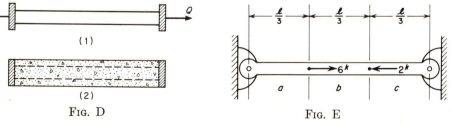

FIG. D   FIG. E

5. The steel bar in Fig. E is axially loaded at its third-points and rigidly restrained at its ends. The cross-sectional area of the bar is A. Calculate the axial force in each of the sections a, b, and c if the bar was free from stress before the external loads were applied. *Ans.* $S_a = +3333$ lb; $S_b = -2667$ lb; $S_c = -667$ lb.

6. A steel rod 2.5 ft long has a cross-sectional area of 2 sq in. over 1 ft of its length and 3 sq in. over the remaining 1.5 ft. At a temperature $T = 50°F$ the rod fits exactly between unyielding walls at its two ends. Compute the maximum compressive stress that will exist in the rod at temperature $T = 100°F$. The modulus of elasticity is $E = 30(10)^6$ psi, and the coefficient of thermal expansion is $\alpha = 0.0000065$ in./in./°F. *Ans.* $\sigma = 12,190$ psi.

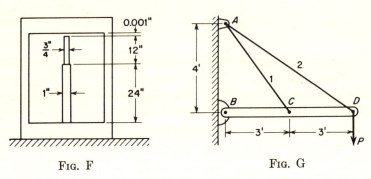

FIG. F   FIG. G

7. A copper rod with circular cross-sections rests inside a steel frame as shown in Fig. F. The two side members of the frame are each 1 sq in. in cross-section. At a temperature $T = 0°F$ there is 0.001 in. clearance between the upper end of the rod and the top of the frame as shown. Calculate the compressive force S in the copper rod when the temperature of the entire system is raised to 100°F. Neglect bending of the frame. *Ans.* $S = 2270$ lb.

8. In Fig. G, an absolutely rigid bar BD is hinged at B and supported by two guy wires attached to the vertical wall at A. The steel guy wires are identical except for length and are just taut but free from stress before the load P is applied. Find the tensile forces S_1 and S_2 produced in the guy wires by the load P. *Ans.* $S_1 = 1.07P$; $S_2 = 1.03P$.

9. A brass sleeve is slipped over a steel bolt and held in place by a nut that is turned just snug (Fig. 1.16a). Compute the temperature rise required to stress the brass to 4000 psi compression. Use the following data: $A_s = 0.785$ sq in., $A_b = 0.750$ sq in., $E_s = 30(10)^6$ psi, $E_b = 14(10)^6$ psi, $\alpha_s = 0.0000065$ in./in./°F, $\alpha_b = 0.0000100$ in./in./°F. *Ans.* $T = 118°F$.

10. A square frame with pinned joints is made of four aluminum bars each 1 ft in length and 1 sq in. in cross-section. Diagonally opposite corners of the frame are connected by steel wires each $\frac{1}{4}$ in. in diameter. At room temperature (70°F) the assembly is free from stress. What tensile forces S will exist in the diagonals when the temperature is raised to 170°F? Use the following numerical data: $E_s = 30(10)^6$ psi, $E_a = 10(10)^6$ psi, $\alpha_s = 70(10)^{-7}$ in./in./°F, $\alpha_a = 128(10)^{-7}$ in./in./°F. *Ans.* $S = 772$ lb.

11. A twisted steel clothesline wire of length l has cross-sectional area A and modulus of elasticity E in tension. This wire is stretched horizontally between two fixed points A and B, but without appreciable initial tension. A load P is then suspended from the mid-point C of the line. Find the vertical deflection of point C, assuming that this deflection is small compared with the length l of the clothesline. *Ans.*

$$\Delta = \frac{l}{2}\sqrt[3]{\frac{P}{AE}}.$$

1.5 Thin Rings

Consider the case of a thin circular ring subjected to the action of uniformly distributed radial loading as shown in Fig. 1.17. If the cross-sectional area A of the ring is constant along the circumference and the

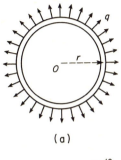

(a)

thickness t is small compared with the radius r, such loading will produce uniform circumferential stress and strain in the ring and the problem can be treated as one of simple tension or compression. If the radial loading is directed outwards, the ring will be in circumferential tension; if it is directed inwards, the ring will be in compression. The distributed load may be due either to internal or external pressure or to centrifugal force as in the case of a

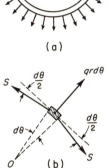

(b)

Fig. 1.17

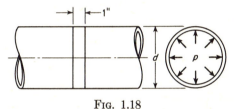

Fig. 1.18

rotating ring. In any case, its *intensity* q may be defined in pounds per inch of circumference of the mean center line of radius r.

To examine the internal forces induced by such loading, consider the equilibrium of an element of length $ds = rd\theta$ as shown in Fig. 1.17b, where S denotes the circumferential tension or *hoop tension* and $qrd\theta$ is the element of external radial load. Equating to zero the algebraic sum of projections of these forces in the radial direction and noting that $\sin (d\theta/2) \approx d\theta/2$ gives

$$qr \cdot d\theta - 2S\frac{d\theta}{2} = 0,$$

from which
$$S = qr. \tag{1.7}$$

If the distributed loading q is directed radially inwards, eq. (1.7) will give the uniform circumferential compression in the ring.

Since the thickness t of the ring is assumed small compared with its mean radius r, the hoop tension S can be taken as uniformly distributed[*] over the cross-section A and the unit stress becomes

$$\sigma = \frac{S}{A} = \frac{qr}{A}. \tag{1.8}$$

[*]For thick rings, see Timoshenko, *Strength of Materials*, Vol. 2, p. 205.

Likewise, the circumferential strain, uniform around the circumference, becomes

$$\epsilon_c = \frac{\sigma}{E} = \frac{qr}{AE}. \tag{1.9}$$

It may be noted that, because the circumference and diameter of the ring are in the constant ratio π, the unit diametral extension will be the same as the circumferential strain, i.e., $\epsilon_d = \epsilon_c$. This notion of diametral strain ϵ_d will be useful in discussing problems of *shrink fits* as will be shown in some of the examples to follow.

The uniform radial loading shown in Fig. 1.17a can be produced by rotating the ring about its geometric axis normal to the plane of the ring with some constant angular velocity ω usually specified in radians per second (sec^{-1}). Under such conditions, each element of the ring of mass dm moves with constant speed in a circular path of radius r and has a normal acceleration $a_n = \omega^2 r$, directed towards the center of the ring. The corresponding centrifugal force is $\omega^2 r \, dm$ directed radially outward. For an element of length $ds = r d\theta$ and weight w per unit length of circumference the mass $dm = wr d\theta / g$. Such centrifugal forces on all elements of the ring are seen to represent a uniform distribution of radial loading of intensity

$$q = \frac{w}{g}\omega^2 r. \tag{a}$$

Substituting this into eq. (1.7), the corresponding hoop tension S becomes

$$S = qr = \frac{w}{g}\omega^2 r^2. \tag{b}$$

Likewise from eq. (1.8), the hoop stress is

$$\sigma = \frac{S}{A} = \frac{w}{Ag}\omega^2 r^2 = \frac{\gamma}{g}\omega^2 r^2, \tag{c}$$

where $\gamma = w/A$ is the weight per unit volume of the material from which the ring is made. It will be noted that this hoop stress due to centrifugal forces increases as the square of both the angular velocity ω and the radius r of the ring. Thus a ring of large radius that rotates at high speed can be subjected to very high stresses.

EXAMPLE 1. A long steel water pipe having a mean diameter $d = 24$ in. and wall thickness $t = \frac{1}{8}$ in. carries an internal pressure $p = 150$ psi, Fig. 1.18. Calculate the magnitude of the hoop stress in the pipe wall and the increase in its diameter.

SOLUTION. Consider a section of the pipe having a length of 1 in. as a ring, the cross-sectional area of which will be $A = 1$ in. $\times \frac{1}{8}$ in. $= 0.125$ in.^{2}. Then the internal pressure p represents the intensity of load per unit length of circumference of this ring and eq. (1.8) gives

$$\sigma = \frac{pr}{A} = \frac{150 \times 12}{0.125} = 14{,}400 \text{ psi.}$$

From eq. (1.9), the circumferential strain becomes

$$\epsilon = \frac{\sigma}{E} = \frac{14,400}{30(10)^6} = 4.8\,(10)^{-4}.$$

Since this also represents the diametral strain, the increase in diameter is

$$\Delta d = 4.8(10)^{-4} \times 24 = 0.0115 \text{ in.}$$

EXAMPLE 2. A steel sleeve of small thickness t is to be shrunk onto a solid shaft of diameter d as shown in Fig. 1.19. Neglecting deformation of the shaft, find the proper initial inside diameter of the sleeve if the hoop stress induced in the sleeve by the shrink fit is not to exceed a prescribed working stress σ_w.

PROBLEM SET 1.5

1. At what rpm will the thin cast-iron rim of a fly-wheel having mean diameter $d = 4$ ft reach its ultimate strength ($\sigma_u = 20,000$ psi) in hoop tension? Neglect any restraining effect of the spokes of the wheel. *Ans.* 2170 rpm.

2. A steel sleeve of small thickness t is to be shrunk onto a solid shaft of diameter $d = 4$ in. Neglecting deformation of the shaft, find the proper initial inside diameter of the sleeve if the allowable hoop stress in the sleeve is $\sigma_w = 30,000$ psi. *Ans.* 3.996 in.

3. When both parts of the ring in Fig. A are at a temperature of 420°F, the outside brass hoop just fits snugly over the inside steel hoop without any stress in either hoop. Both hoops are 1 in. wide in the direction normal to the plane of the figure. Subsequently, the system cools down to a temperature of 70°F. Calculate the radial pressure q set up between the two hoops and the hoop stress σ_b in the brass due to the shrink fit. To simplify the calculations without much loss of accuracy, assume that each hoop has a mean radius $r = 10$ in. The following data are given: $E_s = 30(10)^6$ psi, $E_b = 13(10)^6$ psi, $\alpha_s = 6.5(10)^{-6}$ in./in./°F, $\alpha_b = 10.4(10)^{-6}$ in./in./°F. *Ans.* $q = 730$ lb/in., $\sigma_b = 14,600$ psi.

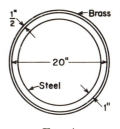

FIG. A

4. What angular velocity ω of the shrunk-fit ring in Fig. A will be required to relieve the shrink-fit pressure existing under the conditions given in the preceding problem? What will be the hoop stress in the brass ring at this angular velocity? *Ans.* $\omega = 605$ rad/sec; $\sigma_b = 29,400$ psi.

5. The steel strap of a brake band is 2 in. wide and $\frac{1}{32}$ in. thick. The brake lining is $\frac{1}{4}$ in. thick and is riveted to the strap by pairs of $\frac{3}{16}$-in.-diameter rivets. Thus the effective width of the strap is reduced to $1\frac{5}{8}$ in. because of the rivet holes. The brake drum is 10 in. in diameter. The coefficient of friction between brake lining and drum is assumed to be 0.4. How much braking torque can be developed without exceeding a tensile stress of 18,000 psi in the steel strap? *Ans.* $T = 862$ ft-lb.

6. Assume that the ring in Fig. A is the cross-section of a cast-iron pipe reinforced by an outer steel pipe; all dimensions are as shown in the figure. There is no prestressing, i.e., the two pipes fit snugly without initial radial pressure between them. What is the allowable internal pressure p for the reinforced pipe if the working stress in tension for cast-iron is $\sigma_c = 6000$ psi and that for steel is $\sigma_s = 20,000$ psi? *Ans.* $p = 1234$ psi.

2

Tension, Compression, and Shear: II

2.1 Variation of Stress with Aspect of Cross-Section

In the case of axial tension of a prismatic bar, Fig. 2.1a, the stress on a normal cross-section mn is uniform and has the magnitude $\sigma = P/A$ as discussed in Art. 1.2. Let us consider now the state of stress on an oblique cross-section pq cutting the bar at an angle ϕ with the normal cross-section

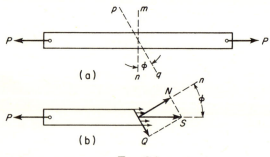

FIG. 2.1

mn. First, we isolate that portion of the bar to the left of the oblique section pq as a free body and represent the action of the removed portion on this free body by the stress resultant S as shown in Fig. 2.1b. From the equilibrium condition, this internal force S must be equal, opposite, and collinear with the external force P, as shown. Resolving the force S into components N and Q, normal and tangential, respectively, to the plane pq, we find

$$N = P \cos \phi; \quad Q = P \sin \phi. \tag{a}$$

Then since the area of the oblique section pq is

$$A' = A/\cos \phi,$$

the corresponding stresses are

$$\left. \begin{aligned} \sigma_n &= \frac{N}{A'} = \frac{P}{A} \cos^2 \phi, \\[2mm] \tau &= \frac{Q}{A'} = \frac{1}{2}\frac{P}{A} \sin 2\phi. \end{aligned} \right\} \tag{2.1}$$

26

These are called, respectively, the *normal stress* and the *shear stress* on the oblique section *pq*, the aspect of which is defined by ϕ. We see that when $\phi = 0$ and the section *pq* coincides with the normal section *mn*, eqs. (2.1) give

$$(\sigma_n)_{\max} = \frac{P}{A} \tag{b}$$

and $\tau = 0$, as they should. However, as ϕ is increased, the normal stress σ_n diminishes, until when $\phi = \pi/2$, $\sigma_n = 0$. Thus there is seen to be no normal lateral stress between the longitudinal fibers of a prismatic bar in tension. On the other hand, with increase in the angle ϕ, the shear stress τ increases to a maximum value

$$\tau_{\max} = \frac{1}{2}\frac{P}{A} \tag{c}$$

when $\phi = \pi/4$ and then diminishes to $\tau = 0$ when $\phi = \pi/2$.

These observations lead us to consider more carefully the question of the strength of a bar in simple tension. If the bar is made of a material that is much weaker in shear than it is in cohesion, it may happen that failure will take place due to relative slipping between two parts of the bar along a 45°-plane where the shear stress is a maximum, rather than due to direct rupture across a normal section where the normal stress is a maximum. For example, a short wood post loaded in axial compression, as shown in Fig. 2.2a, may

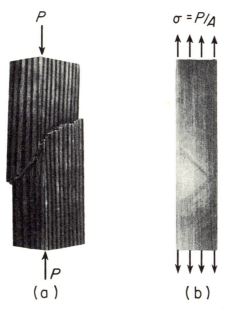

Fig. 2.2

actually fail by shearing along a jagged plane inclined roughly by 45° to the axis of the post. In such case, we may still specify the value of P/A at which this failure occurs as the ultimate strength of the wood in compression, even though the failure is not a true compression failure of the material.

Similarly, during a tensile test of a flat bar of low-carbon steel with polished surfaces, it is possible to observe a very interesting phenomenon. At a certain value of the tensile stress $\sigma = P/A$, visible slip bands approximately inclined by 45° to the axis of the bar will appear on the flat sides of the specimen as shown in Fig. 2.2b. These lines, called *Lueders' lines*, indicate that the material is failing in shear, even though the bar is being loaded in simple tension. This relative sliding along 45°-planes causes the specimen to elongate axially, and after unloading it will not return to its original length. Such apparent stretching of the bar due to this slip phenomenon is called *plastic yielding*. Again, the axial tensile stress $\sigma_{\text{y.p.}} = P/A$ at which this occurs may be designated as the *yield stress* in tension, even though the failure is not a true tension failure of the material. These matters will be discussed further in the next article.

Formulas (2.1), derived for the case of axial tension, can be used also for axial compression, simply by changing the sign of P/A. We then obtain negative values for both the normal stress σ_n and the shear stress τ. The complete state of stress on a thin element between two parallel oblique sections for axial tension and axial compression are compared in Fig. 2.3. The directions of these stresses associated with axial tension (Fig. 2.3a) will be considered as *positive;* those associated with axial compression (Fig. 2.3b), as *negative*. Thus σ_n is positive when it is a tensile stress and negative when it is a compressive stress. By reference to Fig. 2.3, the rule for sign of

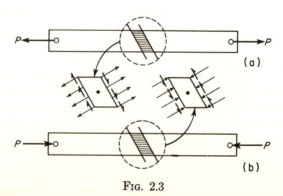

Fig. 2.3

shear stress τ will be as follows: The shear stress τ on any face of the element will be considered *positive* when it has a *clockwise* moment with respect to a center *inside* the element (Fig. 2.3a). If the moment is *counter-*

clockwise with respect to a center *inside* the element, the shear stress is *negative*. Stated in a different way, the shear stress on any surface of a body will be considered to be of positive sign if it points in a direction corresponding to clockwise rotation about a center inside the body, otherwise

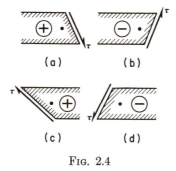

(a)   (b)

(c)   (d)

Fig. 2.4

of negative sign. Several examples of both positive and negative shear stress are shown in Fig. 2.4. These sign conventions, while arbitrary, must nonetheless be carefully observed to avoid confusion.

Returning to the case of a bar in axial tension, let us consider now the stresses on an oblique section $p'q'$ at right angles to the section pq, as shown in Fig. 2.5. To obtain the stresses σ'_n and τ' on this section, we need

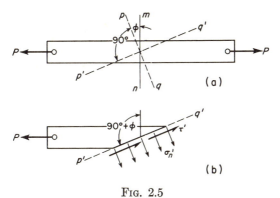

Fig. 2.5

only to replace ϕ by $90° + \phi$ in eqs. (2.1). Then remembering that sin $(90° + \phi) = \cos \phi$, while $\cos(90° + \phi) = - \sin \phi$, this gives

$$\sigma_n' = \frac{P}{A} \cos^2(90° + \phi) = \frac{P}{A} \sin^2 \phi,$$

$$\tau' = \frac{1}{2}\frac{P}{A} \sin(180° + 2\phi) = -\frac{1}{2}\frac{P}{A} \sin 2\phi. \quad \left.\right\} \quad (2.1')$$

These stresses on the plane $p'q'$ act as shown in Fig. 2.5b.

The complete set of stresses given by eqs. (2.1) and (2.1′) are called *complementary stresses* because they occur on mutually perpendicular planes. Comparing the two sets of formulas, we observe that

$$\left.\begin{array}{c} \sigma_n + \sigma'_n = P/A, \\ \tau' = -\tau. \end{array}\right\} \tag{d}$$

Thus the sum of normal stresses σ_n and σ'_n on any two mutually perpendicular sections of a bar in axial tension is constant and equal to P/A, the normal stress on the normal section *mn*. Also, *complementary shear stresses are always equal in magnitude but opposite in sign.*

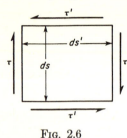

FIG. 2.6

The equality of complementary shear stresses such as τ and τ' on the faces of a rectangular element (Fig. 2.6) also can be established from the equilibrium conditions of the element itself, as follows: Let dz denote the thickness of the element normal to the plane of the paper and ds, ds', the lengths of its edges. Then the areas on which τ and τ' act will be, respectively, $dsdz$ and $ds'dz$. Multiplying the shear stresses by the areas on which they act, we obtain two counteracting couples, the moments of which must balance each other. Thus

$$\tau(dsdz) \times ds' = \tau'(ds'dz) \times ds,$$

from which $\tau = \tau'$, where τ' has already been represented as negative in Fig. 2.6.

EXAMPLE 1. A short steel bar having a 1-in. × 1-in. square cross-section is subjected to compressive forces $P = 25,000$ lb axially applied as shown in Fig. 2.7.

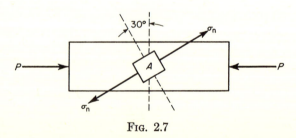

FIG. 2.7

Compute the complete set of complementary stresses on the sides of the rectangular element A oriented as shown.

SOLUTION. Taking $P/A = -25,000$ psi and $\phi = 30°$ in eqs. (2.1)

$$\sigma_n = -25,000 \times (0.866)^2 = -18,750 \text{ psi},$$
$$\tau = -12,500 \times 0.866 = -10,820 \text{ psi}.$$

Similarly, from eqs. (2.1′)

$$\sigma'_n = -25{,}000 \times (0.50)^2 = -6250 \text{ psi},$$

$$\tau' = -(-12{,}500) \times 0.866 = +10{,}820 \text{ psi}.$$

PROBLEM SET 2.1

1. A mild steel tensile-test specimen having a circular cross-section of diameter $d = 0.505$ in. shows an elongation reading of 0.00200 in. over a gauge length of 2 in. Calculate the maximum shear stress in the material, assuming that $E = 30(10)^6$ psi. *Ans.* $\tau_{max} = 15{,}000$ psi.

2. A brass wire of diameter $d = \frac{1}{16}$ in. and length $l = 24$ in. is tightly stretched between fixed points A and B at its ends so that it is under a tension of 31 lb. If the temperature of the wire subsequently drops 50°F, what is the maximum shear stress in the material? The coefficient of thermal expansion for brass is $\alpha_b = 10.4(10)^{-6}$ in./in./°F and the modulus of elasticity is $E_b = 14(10)^6$ psi. *Ans.* $\tau_{max} = 8640$ psi.

3. A prismatic bar carrying an axial tensile stress σ_x is cut by an oblique section pq as shown in Fig. A. If the normal and shear stresses, respectively, on this section are $\sigma_n = 12{,}000$ psi and $\tau = 4000$ psi, find the value of σ_x and the angle ϕ defining the aspect of the section pq. *Ans.* $\sigma_x = 13{,}333$ psi; $\phi = 18° 26'$.

4. Referring to the case of axial tension of a prismatic bar as shown in Fig. 2.5, we are given the following numerical data: $A = 0.785$ in.^{2}, $P = 10{,}000$ lb, $\phi = 20°$. Calculate the stresses σ_n, σ'_n, τ, τ', for sections pq and $p'q'$. *Ans.* $\sigma_n = 11{,}250$ psi, $\sigma'_n = 1490$ psi, $\tau = -\tau' = 4095$ psi.

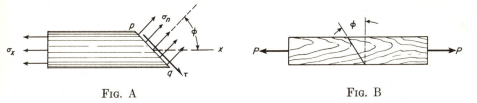

FIG. A   FIG. B

5. A steel rod of circular cross-section is to carry a tensile load $P = 30$ kips. The allowable working stress in shear for the rod is $\tau_w = 8000$ psi. Find the required diameter d for the rod. *Ans.* $d = 1.55$ in.

6. A prismatic bar is subjected to axial tension. Find the aspect angle ϕ which defines an oblique section on which the normal and shearing stresses are equal. *Ans.* $\phi = 45°$.

7. Referring to Fig. A, assume that the angle $\phi = 30°$ and that $\sigma_n = 10{,}000$ psi. In such case, what is the shear stress τ? *Ans.* $\tau = 5780$ psi.

8. The normal stresses on the edges of the element A in Fig. 2.7 are given as follows: $\sigma_n = -14{,}500$ psi, $\sigma'_n = -9500$ psi. What is the aspect angle ϕ defining the orientation of the element and what is the axial stress P/A to which the bar is subjected? *Ans.* $\phi = 39°$, $P/A = -24{,}000$ psi.

9. A concrete test cylinder having length $l = 12$ in. and diameter $d = 6$ in. is subjected to axial compressive forces P in a testing machine. If the maximum shear stress in the concrete is not to exceed 2000 psi, what is the safe value for the axial load P? *Ans.* $P = 113{,}200$ lb.

10. A wood stick having a 2-in. by 2-in. square cross-section has a glued joint at its mid-section as shown in Fig. B. If the allowable working stresses for the glue in tension and shear are, respectively, $\sigma_w = 1000$ psi and $\tau_w = 600$ psi, what is the optimum angle $\phi < 45°$ for the joint? What is the corresponding safe tensile load P_w for the stick? *Ans.* $\phi = 31°$; $P_w = 5440$ lb.

2.2   Stress-Strain Diagrams; Working Stress

In the preceding discussions of tension or compression of a prismatic bar, it was always assumed that the material was perfectly elastic and obeyed Hooke's law. This assumption is justifiable for many materials so long as they are not too highly stressed. However, when loaded beyond a certain stress, called the *proportional limit*, most materials begin to exhibit some nonlinearity between stress and strain, i.e., they no longer obey Hooke's law. With this nonlinear behavior, the problem of calculating deformations and distributions of internal forces in structures and machines becomes much more complicated than for a linear stress-strain behavior.

In investigating the mechanical properties of materials beyond the proportional limit, the relation between stress and corresponding strain is usually represented graphically by a *stress-strain diagram* obtained experimentally from a *standard tensile test.** A typical stress-strain diagram for structural steel is shown in Fig. 2.8a, where the axial strains are plotted as abscissa and the corresponding stresses are given by ordinates of the curve $OABCDE$. From O to A the stress is proportional to the strain. Beyond A, the deviation from Hooke's law becomes marked; hence the stress at A represents the *proportional limit*. This is usually found to lie somewhere between 30,000 psi and 36,000 psi for a low-carbon steel. Upon loading of the specimen beyond the proportional limit, the elongation increases more rapidly and the diagram becomes nonlinear. At B, elongation of the specimen begins to take place without any appreciable increase in load and the material is said to become *plastic*. This phenomenon, called *yielding*, continues until the test bar may stretch plastically as much as ten to fifteen times the elastic stretch up to the proportional limit. The stress at which this yielding begins is called the *yield point* of the steel. This behavior results from the slip associated with the appearance of Lueder lines as discussed in the preceding article. This indicates that the phenomenon of yielding is really an indirect manifestation of failure of the material in shear along 45°-planes of maximum shear stress. At point C, the material begins to *strain harden*, recovers some of its elastic property, and with further elongation the stress-strain curve climbs to point D, representing the maximum tensile stress or the *ultimate strength*. Beyond

*For more details of the standard tensile test and various aspects of the stress-strain diagram, see Art. 11.1, p. 294.

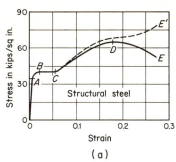

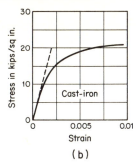

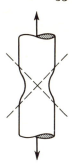

(a)   (b)

Fig. 2.8   Fig. 2.9

point D, further stretching of the bar is accompanied by a decrease in the load and fracture takes place suddenly at point E. The fact that the nominal stress at E is less than the ultimate strength at D is somewhat misleading. As failure develops between D and E, the test bar gradually necks down in a short region somewhere along its length as shown in Fig. 2.9. This phenomenon again is the result of shear slip along 45°-planes which causes a pronounced decrease in the cross-sectional area at the narrowest part of the neck. If for each value of the strain between C and E, the tensile load P is divided by the reduced cross-sectional area, it will be found that the *true stress-strain curve* will follow the dotted line CE'. However, it is established practice to calculate the ultimate strength on the basis of the original or nominal cross-sectional area A of the specimen without regard to the reduction of area.

Structural steel is the only material that exhibits a pronounced yield point. Other materials show a more or less gradual transition from the linear range to the nonlinear range, and even the proportional limit is rather indefinite. As an example of the tensile behavior of a brittle material, a stress-strain diagram for cast-iron is shown in Fig. 2.8b. This material has a very low proportional limit and shows no yield point.

Stress-strain diagrams for axial compression of various materials can also be obtained and such characteristic stresses as the proportional limit, yield point, and ultimate strength can be found. In the case of steel these characteristics for compression are found to have the same values as for tension.

Working Stress. A tensile test diagram gives valuable information on the mechanical properties of a material. Knowing the limit of proportionality, the yield point, and the ultimate strength of the material, it is possible to establish for each particular engineering problem the magnitude of the stress which may be considered a *safe stress*. This stress is usually called the *working stress*.

In choosing the magnitude of the working stress for steel it must be taken into consideration that at stresses below the proportional limit the material may be considered as perfectly elastic, and beyond this limit a part of the strain usually remains after unloading the bar, i.e., *permanent set* occurs. In order to have the structure in an elastic condition and to remove the possibility of a permanent set, it is usual practice to keep the working stress well below the proportional limit. In the experimental determination of this limit, sensitive measuring instruments (extensometers) are necessary and the position of the limit depends to some extent upon the accuracy with which the measurements are made. In order to eliminate this difficulty, one usually takes the *yield point* or the *ultimate strength* of the material as a basis for determining the magnitude of the working stress. Denoting by σ_w, $\sigma_{y.p.}$, σ_{ult} respectively the working stress, the yield point, and the ultimate strength of the material, the magnitude of the working stress will be determined by one of the two following equations:

$$\sigma_w = \frac{\sigma_{y.p.}}{n}, \text{ or } \sigma_w = \frac{\sigma_{ult}}{n_1}. \tag{2.2}$$

Here n and n_1 are usually called *factors of safety*, which determine the magnitude of the working stress. In the case of structural steel, it is logical to take the yield point as the basis for calculating the working stress, because here a considerable permanent set may occur which is not permissible in engineering structures. In such a case a factor of safety $n = 2$ will give a conservative value for the working stress provided that only constant loads are acting upon the structure. In the cases of suddenly applied loads or variable loads (and these occur very often in machine parts), larger factors of safety may be necessary. For brittle materials such as cast iron, concrete, and various kinds of stone and for such material as wood, the ultimate strength is usually taken as a basis for determining the working stresses.

The magnitude of the factor of safety depends very much upon the accuracy with which the external forces acting upon a structure are known, upon the accuracy with which the stresses in the members of a structure may be calculated, and also upon the homogeneity of the materials used.

The common practice of speaking of working stresses and factors of safety based on some characteristic stress such as the yield point of steel or the ultimate strength of cast-iron is somewhat dangerous and mis-leading. If $P_1, P_2, \ldots, P_k$ are a set of external loads for a structure, what we really mean by a factor of safety n is that if all these loads are in-creased to $nP_1, nP_2, \ldots, nP_k$, the structure will be just on the verge of failure, where "failure" of course must be clearly defined. It may mean, in the case of a steel structure, that collapse due to yielding of some members can occur, or, in the case of a concrete structure, that some member is on the verge of fracture.

In the case of a statically determinate system, the stresses in its members will all increase in proportion to the applied loads even though the proportional limit is exceeded. Thus, in such systems, the use of a working stress as defined by one of eqs. (2.2) gives the desired objective as stated above. However, in the case of a statically indeterminate system like those discussed in Art. 1.4, the stresses in the members depend on their deformations. Thus, the methods of analysis used in Art. 1.4 and based on Hooke's law will not be applicable beyond the proportional limit. In such cases, the use of working stresses as defined by eqs. (2.2) will be more or less meaningless.

2.3 Limit Design

We shall here discuss briefly a method of analysis of statically indeterminate structures which will enable us to predict the loading under which the structure will collapse due to simultaneous yielding of some or all of its members. Such an analysis, of course, is automatically limited to steel structures where there is a pronounced yield point of the material. The discussion will also be confined to structures made up of simple tension or compression members under uniform stress. When we have the magnitudes of such collapse loads and specify the working loads as $1/n$ times these values, we have realized a true factor of safety n against complete failure of the structure. This philosophy of *limit design* or *plastic analysis* is gaining wide favor among structural engineers because its use results generally in more efficient and economical designs.*

To begin with, it is customary to idealize the portion $OABC$ of the stress-strain diagram for steel as shown in Fig. 2.8a to that in Fig. 2.10, where it is assumed that proportionality holds up to the yield stress and that thereafter the material yields indefinitely. Such an idealized material is said to

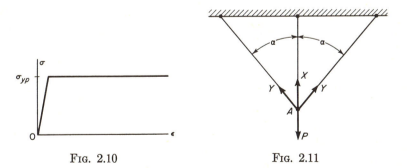

FIG. 2.10   FIG. 2.11

*For those especially interested in this subject the following references may be consulted: J. A. Van den Brock, *Limit Design*, John Wiley & Sons, N. Y., 1935. American Institute of Steel Construction, *Plastic Design in Steel*, 1959.

be either perfectly elastic or perfectly plastic depending on whether the stress is below or at the yield point.   It will be further assumed here that the yield point has the same value in both tension and compression.   For ordinary structural steel, this value may be taken as 40,000 psi.

To illustrate now the idea of limit design, let us consider the statically indeterminate system shown in Fig. 2.11 and already discussed in Art. 1.4, assuming now that all three members are made of steel and have the same cross-sectional area A.   As the external load P is gradually increased, the axial forces X and Y in the members increase and are in the constant ratio shown by eq. (c), p. 17.   This continues until at a certain value P_1 of the external load, the force in the vertical bar reaches the magnitude $X = \sigma_{y.p.}A$, i.e., the stress in this bar reaches the yield point.   At this time, the vertical bar becomes plastic and will stretch thereafter without further increase in tension, but the inclined bars having the tensions $Y = X \cos^2 \alpha$ are still elastic and capable of carrying further load.   Thus as the load P is increased beyond the value P_1, the tensile force X in the vertical bar remains constant $(\sigma_{y.p.}A)$ and the tensile forces Y in the inclined bars continue to increase.   However during this phase of the increase in external load the relationship (c) on p. 17 is no longer valid.   We have only to satisfy the equilibrium condition

$$X + 2Y \cos \alpha = P. \tag{a}$$

Finally, the inclined bars also reach the yield point and we have $X = Y = \sigma_{y.p.}A$.   Substituting these values into eq. (a), we obtain

$$P_L = \sigma_{y.p.}A\,(1 + 2 \cos \alpha). \tag{b}$$

This represents the so-called *collapse load* or *limit load* of the structure.*
A working load $P = P_L/n$ will now have a true factor of safety n against complete collapse of the system.   We see from this discussion that a plastic analysis of a statically indeterminate system is actually simpler than an elastic analysis as given in Art. 1.4.

As another example of plastic analysis of a structure, consider the case of a square frame with pin joints loaded as shown in Fig. 2.12a.   This structure is seen to be statically indeterminate, having one redundant member, but we make no attempt here to calculate the axial forces induced in the bars during elastic behavior.   As the external load P is gradually increased, the axial forces in the bars all increase elastically until finally a load P_1 is reached at which some bar begins to yield and thereafter will take no further load.   With further increase in P beyond the value P_1, the truss behaves as

---

*Throughout the above discussion it is assumed that the deformed configuration of the structure is not essentially different from the undeformed configuration, i.e., small change in the angle α due to deformation of the bars is neglected.

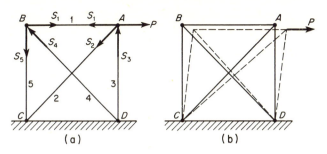

Fig. 2.12

a statically determinate system. During this phase of loading, the axial forces in the bars that are still elastic continue to increase until finally a second bar reaches the yield stress. When this happens, the truss becomes non-rigid and can collapse more or less freely due to steady yielding of the two bars that have reached the plastic condition. Thus the limit load has been reached.

To discover which two bars will first reach the yield stress, we consider the equilibrium conditions of joints A and B and conclude that at all times we must have

$$S_1 = S_5 = \frac{S_4}{\sqrt{2}} \quad \text{and} \quad S_3 = \frac{S_2}{\sqrt{2}}. \tag{c}$$

Thus if each bar has the same cross-sectional area A, it is clear that the diagonals 2 and 4 will be the first two bars to reach the yield stress. When this happens, the system can collapse as shown in Fig. 2.12b and from equilibrium considerations of joint A, we may write

$$P = S_1 + \frac{S_2}{\sqrt{2}}. \tag{d}$$

Expressing S_1 in terms of S_4 from eqs. (c), this becomes

$$P = \frac{S_4 + S_2}{\sqrt{2}}.$$

Finally, setting $S_4 = S_2 = \sigma_{\text{y.p.}}A$, we obtain for the collapse load

$$P_L = \frac{2\sigma_{\text{y.p.}}A}{\sqrt{2}} = \sqrt{2}\,\sigma_{\text{y.p.}}A. \tag{e}$$

The above result is based on the assumption that each bar has the same cross-sectional area A. Thus even in the limit condition, the bars 1, 3, 5 are not working to full capacity, their unit stresses being only $\sigma_{\text{y.s.}}/\sqrt{2}$; see eqs. (c). Clearly, we may now reduce the cross-sectional areas A_1, A_3, A_5 to $A/\sqrt{2}$ without reducing the value of the limit load P_L given by eq. (e).

Therefore, taking $A_2 = A_4 = A$ and $A_1 = A_3 = A_5 = A/\sqrt{2}$, we obtain the optimum limit design of the truss where all bars will reach the yield condition simultaneously when the applied load $P = \sqrt{2}\sigma_{\text{y.p.}}A$. Using a factor of safety n, the corresponding safe working load will be $P_w = \sqrt{2}\sigma_{\text{y.p.}}A/n$.

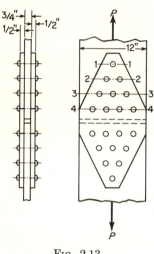

FIG. 2.13

EXAMPLE 1. A tension member consists of two pieces of flat steel plate 12 in. wide by $\frac{3}{4}$ in. thick connected together by a riveted joint as shown in Fig. 2.13. The rivets are $\frac{7}{8}$ in. in diameter and the material has a yield stress in shear, $\tau_{\text{y.p.}} = 20,000$ psi. Calculate the safe tensile load P_w for the member based on a factor of safety $n = 2$ against complete failure due to shear yielding of the rivets.

SOLUTION. This system is highly statically indeterminate, and during elastic behavior of the material a calculation of the true distribution of load among the various rows of rivets represents an impossible problem. However, as the load P on the member is gradually increased, one row of rivets after another reaches the yield condition in shear and thereafter transmits a constant amount of load until finally all rivets will have reached the yield condition. At such time, the total load that the splice can transmit is

$$P_L = 10 \times 2(\tau_{\text{y.p.}}A),$$

where $A = \pi(0.875)^2/4 = 0.601$ sq in. is the cross-sectional area of one rivet. Thus the limit load for the spliced member is $P_L = 10 \times 2(20,000 \times 0.601) = 240,400$ lb, and the safe working load $(n = 2)$ becomes $P_w = 120,200$ lb. This example shows the advantage of plastic analysis in riveted joints.

PROBLEM SET 2.3

1. A vertical load P is supported by five steel wires symmetrically arranged as shown in Fig. A and each having cross-sectional area $A = 0.10$ sq in. Calculate the limit load for the system if $\sigma_{\text{y.p.}} = 40,000$ psi and $\alpha = 45°$, $\beta = 30°$. *Ans.* $P_L = 16,600$ lb.

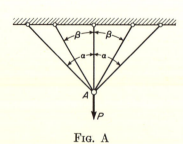

FIG. A

FIG. B

2. A horizontal rigid bar AB is hinged to a vertical wall at A and supported by two steel guy wires BD and CD and a strut EC as shown in Fig. B. Each guy wire has a cross-sectional area of 0.100 in.^{2} and a yield stress $\sigma_{y.p.} = 40,000$ psi. The strut has a buckling strength of $Q = 3000$ lb in compression. Calculate the limit load P_L for the system. *Ans.* $P_L = 4570$ lb.

3. A 6-ft length of railroad rail weighing 150 lb per linear foot is suspended horizontally from a girder by three steel wires each $\frac{1}{8}$ in. in diameter and spaced 3 ft apart; the wires are vertical. What is the maximum distance x from the mid-point of the rail that a vertical load $P = 300$ lb can be hung without causing failure of the supporting wires due to yielding? The yield stress for the wires is $\sigma_{y.p.} = 40,000$ psi and they cannot carry compression. *Ans.* $x = 2.76$ ft.

4. Referring to Fig. 1.15, we are given the following numerical data: $a = 2$ ft, $b = 5$ ft, $l = 8$ ft. Each of the supporting steel wires has diameter $d = \frac{1}{8}$ in. and a yield stress of 36,000 psi. The weight of the rigid bar AB is $W = 400$ lb. Calculate the limit value of the load P at B. *Ans.* $P_L = 187$ lb.

5. A steel rod $\frac{1}{2}$ in. in diameter and 20 in. long stretched between fixed points A and B at its ends is under a tensile stress $\sigma_{y.p.} = 36,000$ psi. Subsequently, an external force P, directed along the axis of the rod, is applied at an intermediate point C. What is the limit value of this force P based on yielding? *Ans.* $P_L = 14,100$ lb.

6. Referring to Fig. 1.16, assume that both the sleeve and the bolt are made of steel ($\sigma_{y.p.} = 36,000$ psi). The cross-sectional area of the sleeve is $A_s = 0.75$ sq in. and that of the bolt is $A_b = 0.50$ sq in. The nut at the end of the bolt is first turned up until the yield stress in the bolt is reached. Subsequently, external tensile forces P are applied at the two ends of the bolt. What is the limit value of these tensile forces based on yielding? *Ans.* $P_L = 18,000$ lb.

2.4 Strain Energy in Tension and Compression

Let us consider again the case of a prismatic bar in simple tension, Fig 2.14a. Assuming elastic behavior of the material, the load-deflection diagram will be a straight line OA as shown in Fig. 2.14b, and for any value of the tensile load P' the corresponding elongation of the bar is denoted by δ'. Now if an increment of load dP' is added to P', the elongation δ' will increase by the amount $d\delta'$ and the load P' does positive work, $P'd\delta'$. This work, represented by the area of the shaded strip in Fig. 2.14b, is stored in the bar in the form of potential energy or *strain energy*, as it is more commonly called. Subsequently, if the increment of load dP' is removed, the lower end of the bar moves up through the distance $d\delta'$ and the stored energy $P'd\delta'$ is transformed back into the work of raising the external load P' through the distance $d\delta'$. Thus, we may regard the elastic bar as a *spring* in which energy can be stored or released accordingly as the load P' is increased or decreased. This property of an elastic bar to absorb and release energy with changes in loading is very important in dealing with time-varying or *dynamic loads* in structures and machines.

The total energy stored in the bar of Fig. 2.14a under a tensile load P is equal to the sum of all such elemental strips as $P'd\delta'$ between O and B and is represented by the area OAB in Fig. 2.14b. Denoting this strain energy by U, we have

$$U = \frac{P\delta}{2}. \tag{2.3}$$

Since expression (2.3) is valid only if Hooke's law applies, we also have, between P and δ, the relationship:

$$\delta = \frac{Pl}{AE}.$$

Using this, expression (2.3) can be written in either of the following two forms:

$$U = \frac{P^2l}{2AE} \quad \text{or} \quad U = \frac{AE\delta^2}{2l}. \tag{2.4}$$

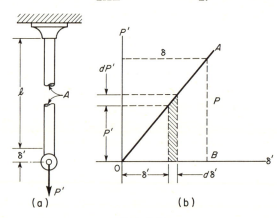

(a)   (b)

FIG. 2.14

In the first form, the strain energy is represented as a function of the tension P in the bar; in the second form, it is represented as a function of the elongation δ of the bar. Both forms will be found useful in further discussions.

In some applications, the strain energy of tension per unit volume is of importance. In the case of uniform tension of a prismatic bar as in Fig. 2.14a, this may be obtained simply by dividing expressions (2.4) by the volume Al of the bar. Thus, using the notation $u = U/Al$,

$$u = \frac{\sigma^2}{2E} \quad \text{or} \quad u = \frac{E\epsilon^2}{2}, \tag{2.5}$$

where $\sigma = P/A$ is the tensile stress and $\epsilon = \delta/l$ is the tensile strain in the bar.

The greatest amount of strain energy per unit volume that a material can absorb without exceeding its proportional limit can be found by substituting this proportional limit for σ in the first of eqs. (2.5). This quantity is then called the *modulus of resilience* of the material in tension. Taking the case of steel, for example, with a proportional limit $\sigma_{\text{p·l}} = 30{,}000$ psi, we obtain for the modulus of resilience $u_R = 15$ in.-lb per in³.

The foregoing discussion of strain energy of a bar in tension can be used also in the case of axial compression. In such case both σ and ϵ are negative, but since they appear to the square in the expressions for strain energy, this has no effect on the final result. In short, strain energy is always a positive scalar quantity.

As an example of an application of eqs. (2.4) to a problem of dynamic loading, consider the simple arrangement for producing tension by impact, Fig. 2.15. A weight W, starting from rest, falls through a height h and strikes a flange at the lower end B of a prismatic bar of length l, the upper end A of the bar being fixed. It is desired to find the maximum elongation and stress that will be induced in the bar by such an impact. In discussing the problem for elastic deformation, we assume that the mass of the bar is negligible compared with that of the the weight W, and that there is no loss of energy due to impact between the weight and the flange at B.

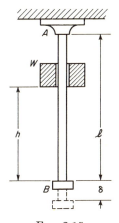

FIG. 2.15

After striking the flange, the weight W continues to move downwards, stretching the bar as it goes. However, due to the resisting force offered by the stretched bar, it decelerates rapidly and soon comes to rest having stretched the bar an amount δ. At this instant the tension in the bar is a maximum and the strain energy stored is given by the second of eqs. (2.4). Neglecting minor losses, this stored strain energy in the bar must be equal to the work done by W in falling through the total vertical distance $h + \delta$. Thus, we write

$$W(h + \delta) = \frac{AE}{2l}\delta^2. \qquad \text{(a)}$$

Introducing the notation $\delta_{\text{st}} = Wl/AE$ for the static elongation of the bar under the action of the dead weight W, and solving equation (a), we obtain

$$\delta = \delta_{\text{st}} + \sqrt{\delta_{\text{st}}^2 + 2\delta_{\text{st}}h} = \delta_{\text{st}} + \sqrt{\delta_{\text{st}}^2 + \frac{1}{g}\delta_{\text{st}}v^2}, \qquad \text{(b)}$$

where $v = \sqrt{2gh}$ is the velocity of the falling body at the moment of striking the flange at B.

If the height h is large in comparison with δ_{st}, eq. (b) reduces to

$$\delta \approx \sqrt{\frac{1}{g}\delta_{\text{st}}v^2}.$$

The corresponding tensile stress * in the bar is then

$$\sigma = \frac{\delta E}{l} = \frac{E}{l}\sqrt{\frac{1}{g}\delta_{st}v^2} = \sqrt{\frac{2E}{Al}\cdot\frac{Wv^2}{2g}}.$$ (c)

The expression under the radical is directly proportional to the kinetic energy of the falling body and to the modulus of elasticity of the material of the bar, and inversely proportional to the volume Al of the bar. Hence the stress can be diminished not only by an increase in the cross-sectional area but also by an increase in the length of the bar or by a decrease in the modulus E. This is quite different from static tension of a bar, where the stress is independent of the length l and the modulus E.

Consider now another extreme case, in which h is equal to zero, that is, the body W is suddenly put on the flange B without an initial velocity. Although in this case there is no kinetic energy at the beginning of extension of the bar, the problem is quite different from that of a static loading of the bar. In the case of a static tension, we assume a gradual application of the load and consequently there is always equilibrium between the acting load and the resisting forces of elasticity in the bar. The question of the kinetic energy of the load does not enter into the problem at all under such conditions. In the case of a sudden application of the load, the elongation of the bar and the stress in the bar are zero at the beginning, and the suddenly applied load begins to fall under the action of its own weight. During this motion the resisting force of the bar gradually increases until it just equals W when the vertical displacement of the weight is δ_{st}. But at this moment the load has a certain kinetic energy acquired during the displacement δ_{st}; hence it continues to move downward until brought to rest by the resisting force in the bar. The maximum elongation for this condition is obtained from eq. (b) by setting $v = 0$. Then

$$\delta = 2\delta_{st},$$ (d)

that is, a suddenly applied load, due to dynamic conditions, produces a deflection which is twice as great as that obtained when the load is applied gradually.

The above discussion of impact is based on the assumption that the stress in the bar remains within the proportional limit. Beyond this limit the problem becomes more involved because the elongation of the bar is no longer proportional to the tensile force. Assuming that the tensile test diagram does not depend upon the speed of straining the bar, elongation beyond the elastic limit during impact can be determined from an ordinary tensile test diagram such as is shown in Fig. 2.16. For any assumed maxi-

*It is here assumed that the general scheme of stress distribution under impact is the same as that under static load.

mum elongation δ, the corresponding area $OADF$ gives the work necessary to produce such an elongation; this must equal the work $W(h + \delta)$ produced by the weight W. When $W(h + \delta)$ is equal to or larger than the

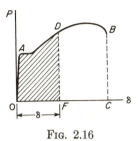

FIG. 2.16

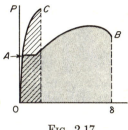

FIG. 2.17

total area $OABC$ of the tensile test diagram, the falling body will fracture the bar.

From this it follows that any change in the form of the bar which results in diminishing the total area $OABC$ of the diagram diminishes also the resisting power of the bar to impact. In the grooved specimens shown in Fig. 2.18, for instance, the plastic flow of the metal will be concentrated at the groove and the total elongation and the work necessary to produce fracture will be much smaller than in the case of the prismatic bar shown in the same figure. Such grooved specimens are very weak in impact; a slight shock may produce fracture although the material itself is ductile. Members having rivet holes or any sharp variation in cross-section are similarly weak against impact.

The resistance of a bar to impact also depends upon the ductility of the material. This may be seen from the tensile test diagrams of Fig. 2.17. The curve OAC represents the load-deformation curve for a material of high strength but low ductility, while the curve OAB is for a material of lower strength but high ductility. The more ductile material is capable of absorbing much more energy before rupture than the less ductile material, as is evident by comparing the shaded areas under OAB and OAC. For this reason, a ductile material has a greater resistance to fracture under shock loading.

EXAMPLE 1. Three tension members having the dimensions shown in Fig. 2.18 each carry the same tensile load P. Compute the amounts of strain energy stored in the three cases, assuming that the stress is uniformly distributed over each cross-section.

SOLUTION. From the first of eqs. (2.4), the strain energy in the prismatic bar, case (a), is

$$U_0 = \frac{P^2 l}{2AE},$$

where $A = \pi d^2/4$. For case (b), the strain energy is

$$U_1 = \frac{P^2\left(\frac{3}{4}l\right)}{2(4A)E} + \frac{P^2\left(\frac{l}{4}\right)}{2AE} = \frac{7}{16}\frac{P^2l}{2AE} = \frac{7}{16}U_0.$$

For case (c), the strain energy is

$$U_2 = \frac{P^2(0.9l)}{2(9A)E} + \frac{P^2(0.1l)}{2AE} = 0.2\frac{P^2l}{2AE} = 0.2\,U_0.$$

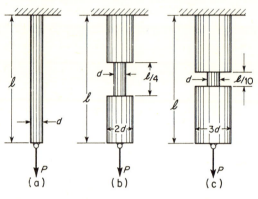

FIG. 2.18

Comparison of these three expressions shows that the strain energy becomes smaller and smaller as the cross-sectional area of the bar is increased over more and more of its length. Thus, as already noted above, a grooved bar is very ineffective in absorbing energy under dynamic loading. It has been found, for example, that the life of engine-head stud bolts as shown in Fig. 2.19 can be increased by turning

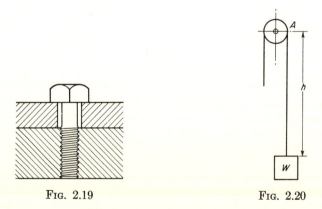

FIG. 2.19   FIG. 2.20

down the shank of the bolt to the root diameter of the threaded portion, thus increasing the amount of strain energy that the bolt can absorb under a given tensile load.

EXAMPLE 2. A weight $W = 10,000$ lb attached to the end of a steel wire rope (Fig. 2.20) moves downward with a constant velocity $v = 3$ ft per second. What stresses are produced in the rope when its upper end is suddenly stopped? The free length of the rope at the moment of impact is $l = 60$ ft, its net cross-sectional area is $A = 2.5$ sq in., and $E = 15 \times 10^6$ psi.

SOLUTION. Neglecting the mass of the rope, the total energy in the system at any instant is the sum of the elastic strain energy in the rope, the kinetic energy of the falling weight, and its potential energy relative to some datum plane (here assumed to be a horizontal plane through the lowest position of the weight). Let δ_{st} represent the static elongation of the rope caused by W, and δ, the total elongation of the rope when the weight is at its lowest position. If it is assumed that there are no losses in the over-all energy of the system, the energy just prior to stopping (when at an elevation $\delta - \delta_{st}$ above the datum plane) may be equated to the energy in the lowest position. Hence

$$\frac{AE\delta_{st}^2}{2l} + \frac{W}{2g}v^2 + W(\delta - \delta_{st}) = \frac{AE\delta^2}{2l}.$$

Introducing W obtained from the known relation $\delta_{st} = Wl/AE$ into the above equation, one gets

$$\delta = \delta_{st} + \sqrt{\frac{Wv^2l}{AEg}}.$$

Upon sudden stopping of point A, the tensile stress in the rope increases in the ratio

$$\frac{\delta}{\delta_{st}} = 1 + \frac{v}{\delta_{st}}\sqrt{\frac{Wl}{AEg}} = 1 + \frac{v}{\sqrt{g\delta_{st}}}.$$

Thus using the given numerical data, we have

$$\sigma = \frac{W}{A}\left(1 + \frac{v}{\sqrt{g\delta_{st}}}\right) = 4000(1 + 4.17) = 20,700 \text{ psi.}$$

We see that the dynamic stress in this case is more than five times the corresponding static stress W/A.

PROBLEM SET 2.4

1. A prismatic steel bar, 10 in. long, is subjected to axial compressive forces $P = 4000$ lb. Compute the amount of strain energy: (a) if the cross-sectional area $A = 4$ sq in., (b) if $A = 2$ sq in.; $E = 30(10)^6$ psi. *Ans.* (a) $U = \frac{2}{3}$ in.-lb.; (b) $U = \frac{4}{3}$ in.-lb.

2. A prismatic steel rod of length l and cross-sectional area A hangs vertically under its own weight. How much strain energy is stored in the bar if its weight per unit volume is γ? *Ans.* $U = \gamma^2Al^3/6E$.

3. Compute the strain energy per unit volume and per pound which can be stored in the following materials in tension without exceeding their proportional limits: (1) structural steel; $\gamma = 0.284$ lb/in.^{3}, $E = 30(10)^6$ psi, $\sigma_{p.l.} = 30,000$ psi. (2) tool steel; $\gamma = 0.284$ lb/in.^{3}, $E = 30(10)^6$ psi, $\sigma_{p.l.} = 120,000$ psi. (3) rubber; $\gamma = 0.0336$ lb/in.^{3}, $E = 300$ psi, $\sigma_{p.l.} = 300$ psi.

4. For the system shown in Fig. 2.15, determine the height h through which the weight W must fall in order to produce a maximum stress $\sigma = 20{,}000$ psi in the rod. The following numerical data are given: $W = 20$ lb, $l = 5$ ft, $A = 0.50$ in.^{2}, $E = 10(10)^6$ psi. *Ans.* $h = 30$ in.

5. A small weight W is attached to one end of an elastic string and the other end of the string is tied to a fixed support. If the weight W is allowed to fall freely through the full length of the string, what maximum tensile stress σ will be produced in the string? The string has cross-sectional area A, length l, modulus of elasticity E, and can be considered weightless. *Ans.*

$$\sigma = \frac{W}{A}\left(1 + \sqrt{1 + \frac{2EA}{W}}\right).$$

6. A toy consists of a small wooden paddle with a small, soft rubber ball connected to its face by a piece of thin rubber band. It somewhat provides a way of playing ping-pong by oneself. The ball weighs 1 oz and the unstretched length of the rubber band is 12 in. If, after bouncing off the face of the paddle, the ball stretches the rubber band to four times its natural length before returning, what was the velocity of the ball when it left the face of the paddle? Ignore gravity. The natural cross-sectional area of the band is $A = 0.0025$ in.^{2}, the modulus of elasticity $E = 300$ psi, and Poisson's ratio $\mu = 0.5$. Assume that the load-deflection curve for the band in tension is linear up to $\delta = 3l$. What is the *true stress* in the rubber band at the instant of greatest extension? *Ans.* $v = 59$ ft/sec; $\sigma_t = 3600$ psi.

7. When the vertical load P in Fig. 1.13a is gradually applied at point A, deflection is proportional to load and the work done externally is $P\delta/2$, where δ is the deflection of point A. Using the fact that this external work must be stored in the wires in the form of strain energy, calculate the deflection δ. Assume that each wire has the same cross-sectional area A and modulus of elasticity E in tension. *Ans.*

$$\frac{Pl}{AE}\left(\frac{1}{1 + 2\cos^3\alpha}\right).$$

2.5   Stress Concentration in Tension or Compression Members

So far in the discussion of simple tension and compression, it has been assumed that the bar has a prismatical form. Then for centrally applied forces, the stress at some distance from the ends is uniformly distributed over the cross-section. Abrupt changes in cross-section give rise to great irregularities in stress distribution. These irregularities are of particular importance in the design of machine parts subjected to variable external forces and to reversal of stresses. Irregularity of stress distribution at such places means that at certain points the stress is far above the average value and, under the action of reversal of stresses, progressive cracks are likely to start gradually from such points. The majority of fractures of machine parts in service can be attributed to such progressive cracks.

Stress concentration is a matter which is frequently overlooked by designers. That the neglect of this factor in much of our engineering design has not led to more frequent disaster is due to the employment of a large factor of safety in stress analysis and to the beneficial effect of local yielding upon stress distribution. This does not justify, however, our disregarding it in those cases where it does enter. In this article, the importance of this factor will be illustrated by treating briefly stress concentration as it occurs in tension or compression members. This abbreviated treatment* will be sufficient, it is hoped, to acquaint the student with the occurrence and effects of stress concentration and to help him recognize those cases where it must be given consideration.

A simple example of stress concentration occurs in a rectangular plate with a small *circular hole* at the center when subjected to tension (Fig. 2.21a).

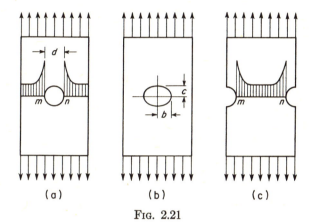

Fig. 2.21

Investigation shows† that in this case there is a high stress concentration at points m and n at the edge of the hole. The distribution of stresses over the cross-section through the center of the hole is shown by the shaded area. If the size of the hole is small compared with the width of the plate, the magnitude of the normal stress at any point of this cross-section, a distance r from the center of the hole, is given by the equation‡

$$\sigma = \frac{\sigma_0}{2}\left(2 + \frac{d^2}{4r^2} + \frac{3}{16}\frac{d^4}{r^4}\right) \tag{a}$$

*For a rather complete discussion of this subject, together with references to current literature, see Timoshenko's *Strength of Materials*, 3d ed., Part II, Chap. VIII, and the article by M. M. Frocht, "Photoelastic Studies in Stress Concentration," *Mechanical Engineering*, August 1936. Also see papers in the *Proc. Soc. Exp. Stress Analysis* by the following: Durelli and Murray, Vol. I, No. I, p. 19; Berkey, Vol. I, No. II, p. 56; Hetenyi, Vol. I, No. I, p. 147. For numerous excellent illustrations of failures originating at points of stress concentration, see Battelle Memorial Institute, *Prevention of Fatigue of Metals*, John Wiley and Sons, New York, 1941.

†See Timoshenko, *Strength of Materials*, 3d ed., Part II, Art. 56, p. 301.

‡For derivation, see Timoshenko's *Theory of Elasticity*, McGraw-Hill Book Co., 1934, p. 75.

in which σ_0 is the uniform stress at the ends of the plate. It can be seen that the stress concentration is highly localized in this case. At m and n, where $\sigma = 3\sigma_0$, the stresses decrease rapidly with increase in the distance from these overstressed points; at a distance from the edge of the hole equal to the radius of the hole ($r = d$), $\sigma = 1\frac{7}{32}\,\sigma_0$ only. Due to the bending action around the hole of Fig. 2.21a, a compressive stress of magnitude σ_0 is set up at the top and the bottom of the hole.

In the case of a small elliptical hole (Fig. 2.21b), the maximum stress is likewise at the ends of the horizontal axis of the hole and is given by the equation

$$\sigma_{\max} = \sigma_0\left(1 + 2\frac{b}{c}\right). \tag{b}$$

This stress increases with the ratio b/c so that a very narrow hole perpendicular to the direction of tension produces a very high stress concentration. This explains why cracks perpendicular to the direction of forces tend to spread. This spreading can be stopped by drilling holes at the ends of the crack to eliminate the small radii at the ends of the crack which produce high stress concentration.

Small *semicircular grooves* in a plate subjected to tension (Fig. 2.21c) also produce high stress concentration. Experiments show that at points

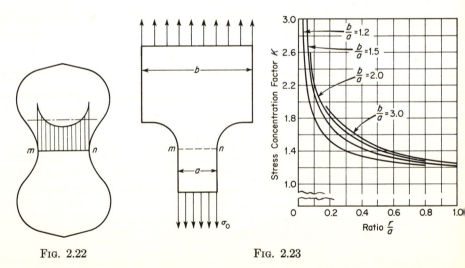

Fig. 2.22   Fig. 2.23

m and n the stresses are about three times the average stress applied at the ends of the plate. The standard tensile test specimen for cement (Fig. 2.22) is another example of a tension member with sharp variation in cross-section. Experiments show that the maximum stress occurs at points m

and n and that this stress is about $1\frac{3}{4}$ times the average stress over the cross-section mn.

In the case of a plate having two portions of different widths, the maximum stress occurs at points m and n in Fig. 2.23. This stress may be represented by the equation:

$$\sigma_{max} = K\sigma_0$$

in which σ_0 is the uniform tensile stress applied at the end of the narrow portion and K is a numerical factor which depends on the radius r of fillets and on the ratio b/a. This factor is usually called the *factor of stress concentration*. Values of this factor are given by the curves in Fig. 2.23.* It will be seen that the stress concentration factor increases with reduction of the radius of the fillet. A more satisfactory stress distribution at the weak section can be obtained by using a larger radius of fillet. This fact is now well recognized in machine design and in shop practice. It is the custom to specify a minimum radius of fillet to prevent the development of progressive cracks at reentrant corners.

All the conclusions reached above concerning stress distribution assume that the maximum stresses are within the proportional limit of the material. Beyond the proportional limit, stress distribution depends on the *ductility* of the material. A *ductile material* can be subjected to considerable stretching beyond the yield point without great increase in stress. Due to this fact, the stress distribution beyond the yield point becomes more and more uniform as the material stretches. This explains why, with ductile materials, holes and notches do not lower the *ultimate strength* when the notched piece is tested statically. However, in the case of a *brittle material*, such as glass, the high stress concentration remains up to the point of breaking. This causes a substantial weakening effect, as demonstrated by the decrease in ultimate strength of any notched bar of brittle material.

The above discussion shows, therefore, that the use of notches and reentrant corners in design is a matter of judgment. In the case of ductile structural steel, high stress concentration is not dangerous provided there is no alternating stress. In the case of brittle material, points of stress concentration may have a great weakening effect and such places should be eliminated or the stress concentration reduced by using generous fillets. In members subjected to reversal of stress, the effect of stress concentration must always be considered, as progressive cracks are likely to start at such points even if the material is ductile.

*These curves are taken from article by M. M. Frocht, *loc. cit.*, p. 47.

3

Biaxial Tension and Compression

3.1 Stresses in Thin-Walled Pressure Vessels

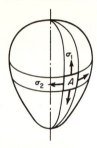

FIG. 3.1

In preceding chapters we have discussed simple tension or compression of a material in one direction. Very often we may encounter the case of a plate or sheet of material subjected to tension or compression in two perpendicular directions at the same time. To see how such *biaxial stress* may arise, let us consider, in Fig. 3.1, a closed thin-walled container having the form of a surface of revolution and subjected to internal pressure of intensity p. Such containers as this are often encountered in stress analysis problems and are called *pressure vessels*. If the wall thickness t of such a vessel is small compared with its principal radii of curvature, the wall will have very little bending resistance and acts primarily as a *membrane* in which the stresses are tangential to the middle surface of the wall and uniformly distributed across its thickness. Such stresses are called *membrane stresses* and are easily calculated from equations of statics. Isolating an element A of the wall cut out by two parallel circles and two meridians, we conclude from symmetry conditions that only normal stresses σ_1 and σ_2 act on its edges and we obtain the case of biaxial stress.

To calculate these stresses, we refer to Fig. 3.2 and introduce the following notations:

σ_1 = tensile stress in meridional direction or *meridional stress*.
σ_2 = tensile stress in circumferential direction or *hoop stress*.
t = uniform thickness of shell wall.
r_1 = radius of curvature of meridian at A.
r_2 = radius of curvature perpendicular to meridian at A.
$d\theta_1$ = angle subtended by meridian arc of element.
$d\theta_2$ = angle subtended by arc normal to meridian at A.
$ds_1 = r_1 d\theta_1$ = dimension of element in meridional direction.
$ds_2 = r_2 d\theta_2$ = dimension of element in circumferential direction.

Then the stress resultants acting on the edges of the element are $\sigma_1 ds_2 t$ and $\sigma_2 ds_1 t$ as shown in Fig. 3.2b. The two stress resultants in the meridional

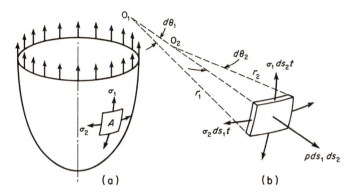

Fig. 3.2

direction have a resultant in the direction of the normal to the element equal to

$$\sigma_1 ds_2 t\, d\theta_1 = \frac{\sigma_1 ds_1 ds_2 t}{r_1}. \tag{a}$$

In the same manner, the stress resultants in the circumferential direction have a normal resultant equal to

$$\sigma_2 ds_1 t\, d\theta_2 = \frac{\sigma_2 ds_1 ds_2 t}{r_2}. \tag{b}$$

The sum of these normal forces is in equilibrium with the normal pressure force on the inside surface of the element; thus

$$\frac{\sigma_1 ds_1 ds_2 t}{r_1} + \frac{\sigma_2 ds_1 ds_2 t}{r_2} = p ds_1 ds_2, \tag{c}$$

from which

$$\frac{\sigma_1}{r_1} + \frac{\sigma_2}{r_2} = \frac{p}{t}. \tag{3.1}$$

Applications of this formula for thin-walled pressure vessels of various shapes will now be illustrated by several examples.

EXAMPLE 1. Calculate the membrane stresses σ_1 and σ_2 for a thin-walled spherical vessel of radius r and wall thickness t if it is subjected to uniform internal pressure of intensity p.

SOLUTION. In this case $\sigma_1 = \sigma_2 = \sigma$ and $r_1 = r_2 = r$. Then eq. (3.1) at once reduces to

$$\sigma = \frac{pr}{2t}. \tag{d}$$

EXAMPLE 2. Calculate the membrane stresses σ_1 and σ_2 for a cylindrical tank of radius r and wall thickness t if it carries a uniform internal pressure p (see Fig. 3.3).

SOLUTION. In this case $r_1 = \infty$ and $r_2 = r$ so that eq. (3.1) reduces at once to

$$\sigma_2 = \frac{pr}{t}. \tag{e}$$

To find the principal stress σ_1 in the longitudinal direction, we cut the cylinder in two by a section normal to its axis of revolution and consider the equilibrium of that portion to one side of this section as shown in Fig. 3.3b. In this case the re-

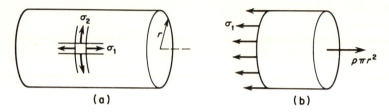

(a)   (b)

FIG. 3.3

sultant thrust on the end of the tank is $p\pi r^2$ and this force must be balanced by the uniform longitudinal stress σ_1 around the circumference of the cylinder. Thus

$$\sigma_1 \cdot 2\pi r t = \pi r^2 p,$$

from which

$$\sigma_1 = \frac{pr}{2t}. \tag{f}$$

Comparing eqs. (e) and (f), we see that in the case of a circular cylindrical tank subjected to uniform internal pressure, the longitudinal stress σ_1 is just half as large as the hoop stress σ_2.

EXAMPLE 3. An open conical tank uniformly suspended around its upper rim is filled with water to a depth h as shown in Fig. 3.4. Calculate the membrane stresses at the level mn and find the value of y for which each of these stresses will be a maximum.

SOLUTION. Since $r_1 = \infty$ in the case of a conical vessel, eq. (3.1) gives

$$\sigma_2 = \frac{pr_2}{t}, \tag{g}$$

where r_2 is the radius of circumferential curvature at the level mn as shown. Substituting, in eq. (g)

$$p = w(h - y), \qquad r_2 = \frac{y \tan \alpha}{\cos \alpha},$$

where w denotes the weight per unit volume of water, we obtain

$$\sigma_2 = \frac{w(h - y)}{t} \frac{y \tan \alpha}{\cos \alpha} \tag{h}$$

for the hoop stress at the level mn.

To find the value of y for which this stress is a maximum, we set the derivative of expression (h) equal to zero and obtain

$$\frac{d\sigma_2}{dy} = \frac{w \tan \alpha}{t \cos \alpha}(h - 2y) = 0,$$

from which $y = h/2$. Then with this value of y, eq. (h) gives

$$(\sigma_2)_{max} = \frac{wh^2 \tan \alpha}{4t \cos \alpha}. \qquad \text{(h')}$$

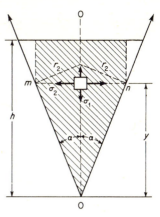

FIG. 3.4

The stress σ_1 in the meridional direction at the level mn is found from the condition that the total weight of the shaded volume of water in Fig. 3.4 is supported by the vertical component of the meridional tension on the circumference mn. Thus

$$\sigma_1 \cdot 2\pi y \tan \alpha \cdot t \cdot \cos \alpha = w(h - y)\pi y^2 \tan^2 \alpha + w\pi y^2 \tan^2 \alpha \frac{y}{3}.$$

This gives

$$\sigma_1 = \frac{w \tan \alpha}{2t \cos \alpha}\left(hy - \frac{2}{3}y^2\right). \qquad \text{(i)}$$

Again setting $d\sigma_1/dy = 0$, we obtain

$$\frac{d\sigma_1}{dy} = \frac{w \tan \alpha}{2t \cos \alpha}\left(h - \frac{4}{3}y\right) = 0,$$

from which $y = 3h/4$. With this value of y, eq. (i) becomes

$$(\sigma_1)_{max} = \frac{3wh^2 \tan \alpha}{16t \cos \alpha}. \qquad \text{(i')}$$

PROBLEM SET 3.1

1. Calculate the safe internal gas pressure p for a spherical pressure vessel made of thin magnesium plate 0.10 in. thick if the mean diameter of the sphere is $D = 250$ in. and the allowable stress in tension is 12,500 psi. *Ans.* $p = 20$ psi.

2. A virtually weightless, thin spherical shell of mean radius R and wall thickness t is held half submerged in water by a fixed ceiling as shown in Fig. A. Calculate the principal stresses σ_1 and σ_2 at the water line, i.e., at the equator of the sphere. The specific weight of water is w. *Ans.* $\sigma_1 = -wR^2/3t$; $\sigma_2 = +wR^2/3t$.

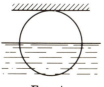

FIG. A

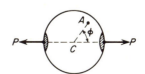

FIG. B

3. A thin spherical shell of mean radius r and wall thickness t is subjected to tensile forces P acting along a diameter of the sphere as shown in Fig. B. There is no internal pressure, i.e., $p = 0$. Find the principal membrane stresses σ_1 and σ_2 at a point A on the shell defined by the angle ϕ as shown. *Ans.* $\sigma_1 = -\sigma_2 = P/2\pi rt$ $\sin^2 \phi$.

4. A thin-walled cone (wall thickness t) is supported on a horizontal base as shown in Fig. C and subjected to internal gas pressure p. Neglecting the weight of the cone itself, find the principal membrane stresses σ_1 and σ_2 at the level h below the apex. The apex angle of the cone is 2α as shown. *Ans.*

$$\sigma_1 = \frac{ph \tan \alpha}{2t \cos \alpha}; \qquad \sigma_2 = \frac{ph \tan \alpha}{t \cos \alpha}.$$

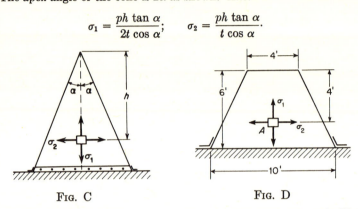

Fig. C   Fig. D

5. A truncated conical tank having the dimensions shown in Fig. D is filled with water ($w = 62.4$ lb/ft³). Calculate the membrane stresses σ_1 and σ_2 for an element A of the wall situated as shown in the figure if $t = 0.012$ in. *Ans.* $\sigma_1 = 1620$ psi; $\sigma_2 = 7760$ psi.

6. If the cone in Fig. C has no internal pressure but instead there is a concentrated vertical force P acting downwards at the apex, find the principal membrane stresses σ_1 and σ_2 at the level h below the apex. *Ans.* $\sigma_1 = -P/2\pi ht \sin \alpha$; $\sigma_2 = 0$.

7. Two vertical steel standpipes each 100 ft high have diameters D_1 and D_2 that are in the ratio $D_1/D_2 = 2$ and are filled with water. What is the ratio t_1/t_2 of their wall thicknesses if they are to have equal maximum hoop stresses σ_2? *Ans.* $t_1/t_2 = D_1/D_2 = 2$.

8. An inverted hemispherical shell of mean radius $R = 5$ ft and wall thickness $t = 0.01$ in. is bolted to a horizontal slab and filled with water (specific weight $w = 62.4$ lb/ft³) to a depth $h = 4$ ft as shown in Fig. E. There is a small air hole at the top of the shell to equalize inside and outside air pressure. Calculate the principal membrane stresses σ_1 and σ_2 for an element A of the shell one foot above the base. *Ans.* $\sigma_1 = 980$ psi; $\sigma_2 = 6820$ psi.

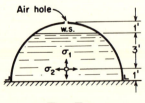

Fig. E

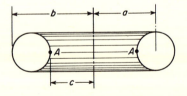

Fig. F

9. A rubber torus inflated to a pressure $p = 10$ psi has the dimensions shown in Fig. F. Calculate the membrane stresses on an element at A if $a = 12$ in., $b = 15$ in., $c = 9$ in., $t = 0.1$ in. *Ans.* $\sigma_1 = 350$ psi; $\sigma_2 = 150$ psi.

3.2 Further Analysis of Biaxial Stress

In the preceding article, it has been shown how an element of material may be subjected to tensile or compressive stresses in two perpendicular directions simultaneously. Considering again such an element (Fig.3.5a), we shall now study this state of biaxial stress in more detail. Here, we denote the two given stresses by σ_x in the x direction and σ_y in the y-direction and assume that tensile stresses are positive.

We may now inquire what stresses exist on a plane whose normal n makes the angle ϕ with the x-axis, as shown. To answer this question, we isolate, as a free body, a triangular portion abc of the element as shown in Fig.3.5b. Let dA_x and dA_y denote the areas of the faces ab and bc, respectively, and dA_n, the area of the face ac. Then the stress resultants on the faces ab and

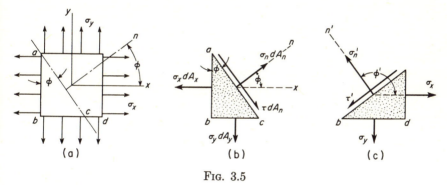

Fig. 3.5

bc, respectively, will be $\sigma_x dA_x$ and $\sigma_y dA_y$, as shown. To balance these forces, we shall need both a normal stress resultant $\sigma_n dA_n$ and a shear stress resultant τdA_n on the inclined face ac of the element as shown. Noting that $dA_x = dA_n \cos \phi$ and $dA_y = dA_n \sin \phi$ and projecting all forces on the n-axis, we must have, for equilibrium of the element,

$$\sigma_n dA_n = \sigma_x(dA_n \cos \phi) \cos \phi + \sigma_y(dA_n \sin \phi) \sin \phi,$$

from which

$$\sigma_n = \sigma_x \cos^2 \phi + \sigma_y \sin^2 \phi = \tfrac{1}{2}(\sigma_x + \sigma_y) + \tfrac{1}{2}(\sigma_x - \sigma_y) \cos 2\phi. \qquad (3.2a)$$

In the same way, projecting all forces on the element on the direction ac, we obtain

$$\tau dA_n = \sigma_x(dA_n \cos \phi) \sin \phi - \sigma_y(dA_n \sin \phi) \cos \phi.$$

from which

$$\tau = (\sigma_x - \sigma_y) \sin \phi \cos \phi = \tfrac{1}{2}(\sigma_x - \sigma_y) \sin 2\phi. \qquad (3.2b)$$

Thus expressions (3.2a) and (3.2b) give the magnitudes of normal stress σ_n and shear stress τ on any plane, the orientation of which is defined by ϕ.

If we take an angle $\phi' = 90° + \phi$ and remember that $\sin (90° + \phi) = \cos \phi$ while $\cos (90° + \phi) = -\sin \phi$, eqs. (3.2a) and (3.2b) become

$$\sigma_n' = \sigma_x \sin^2 \phi + \sigma_y \cos^2 \phi = \tfrac{1}{2}(\sigma_x + \sigma_y) - \tfrac{1}{2}(\sigma_x - \sigma_y) \cos 2\phi, \qquad (3.3a)$$

$$\tau' = -(\sigma_x - \sigma_y) \cos \phi \sin \phi = -\tfrac{1}{2}(\sigma_x - \sigma_y) \sin 2\phi. \qquad (3.3b)$$

These expressions represent the normal and shear stresses on a plane perpendicular to the plane ac (see Fig. 3.5c). As discussed in Art. 2.1, the stresses given by eqs. (3.2) and by eqs. (3.3) are called *complementary stresses*. In this more general case of biaxial stress, the sum of complementary normal stresses σ_n and σ'_n is again constant and equal to $\sigma_x + \sigma_y$, the sum of the two given stresses. Likewise the complementary shear stresses τ and τ' are seen to be equal in magnitude but of opposite sign. We continue to use, for shear stress τ, the sign convention discussed in Art. 2.1: that is, a shear stress on any face of an element is *positive* when its sense of rotation about a center *inside* the element is clockwise. Thus the negative shear stress τ' has been shown in its proper direction in Fig. 3.5c.

Further examination of eq. (3.2a) will show that as the angle ϕ varies from zero to $\pi/2$, the normal stress σ_n varies between the values σ_x, when $\phi = 0$, and σ_y, when $\phi = \pi/2$. Thus the stresses σ_x and σ_y represent maximum and minimum values of the normal stress. For this reason, they are called *principal stresses*.

Similarly, it will be seen from eq. (3.2b) that the absolute value of the shear stress τ is a maximum when $\phi = \pi/4$ and that the magnitude of this maximum shear stress is

$$\tau_{\max} = \tfrac{1}{2} (\sigma_x - \sigma_y), \qquad (3.4)$$

i.e., half the difference between the two principal stresses. If the two principal stresses are equal ($\sigma_x = \sigma_y$), there will be no shear stress on any plane such as ac in Fig. 3.5b.

With proper changes in sign, eqs. (3.2) and (3.3) can be used also if one or both of the principal stresses σ_x and σ_y should be compression, i.e., negative. For example, if one principal stress σ_x is tension and the other σ_y is compression of the same magnitude ($\sigma_y = -\sigma_x$), we see from eq. (3.4) that the maximum shear stress becomes $\tau_{\max} = \sigma_x$.

So far, we have considered only the state of stress for an element subjected to biaxial tension or compression. We shall now discuss briefly the strain or deformation of such an element. When there are tensile stresses σ_x and σ_y in both principal directions x and y of the element (Fig. 3.5a), the strain in either of these directions will depend not only upon the stress in that direction but also upon the stress in the orthogonal direction because of the *Poisson ratio* effect discussed in Art. 1.3. In the x direction, within the elastic limit, the plate will have positive strain $\epsilon_x = \sigma_x/E$ and at the same time it will have negative strain $\epsilon_z = -\mu\sigma_y/E$ due to lateral contraction associated with the tensile stress σ_y. The same reasoning holds for total strain in the y direction. Thus in the case of biaxial tension of a thin plate, the total or net strains will be

$$\left.\begin{array}{l} \epsilon_x = \dfrac{\sigma_x}{E} - \dfrac{\mu\sigma_y}{E}, \\[2mm] \epsilon_y = \dfrac{\sigma_y}{E} - \dfrac{\mu\sigma_x}{E}. \end{array}\right\} \tag{3.5a}$$

In the z-direction normal to the plane of the plate there will be lateral contraction due to the Poisson ratio effect associated with each of the principal stresses σ_x and σ_y so that the net strain in this direction becomes

$$\epsilon_z = -\frac{\mu}{E}(\sigma_x + \sigma_y). \tag{3.5b}$$

Very often in experimental stress analysis the principal strains ϵ_x and ϵ_y will be measured directly by strain gages. Then the corresponding stresses σ_x and σ_y can be calculated from eqs. (3.5a), which, for this purpose, may be put in the form

$$\left.\begin{array}{l} \sigma_x = \dfrac{(\epsilon_x + \mu\epsilon_y)E}{1 - \mu^2}, \\[2mm] \sigma_y = \dfrac{(\epsilon_y + \mu\epsilon_x)E}{1 - \mu^2}. \end{array}\right\} \tag{3.6}$$

If the plate carries compression in either principal direction, eqs. (3.5) or (3.6) can still be used by taking the corresponding σ or ϵ with negative sign.

EXAMPLE 1. A thin circular steel plate of radius r and thickness t is subjected to radial stress σ uniformly distributed around its circumference as shown in Fig. 3.6. Determine the state of stress on any element such as A and also the unit volume change of the entire plate.

SOLUTION. Since both the plate and the external loading are symmetrical around the center O, it follows that the deformed plate will remain perfectly circular in form. This means that both the radial and circumferential strains must be uniform and equal throughout the plate. Then it follows from eqs. (3.5a) and reference to the elements A and B in Fig. 3.6 that $\sigma_x = \sigma_y = \sigma$. Thus we have the case of equal principal stresses.

Before deformation, the volume of the plate is $V_0 = \pi r^2 t$. After deformation, the new radius is $r(1 + \epsilon_x)$ and the new thickness is $t(1 + \epsilon_z)$. Thus the new volume is $V =$

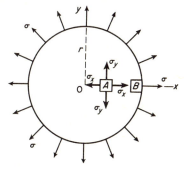

FIG. 3.6

$\pi r^2 t(1 + \epsilon_x)^2(1 + \epsilon_z)$. Expanding this and keeping only terms of first order in the small quantity ϵ, it reduces to $V \approx \pi r^2 t(1 + \epsilon_z + 2\epsilon_x)$, and the total volume change becomes $\Delta V = V - V_0 = \pi r^2 t(\epsilon_z + 2\epsilon_x)$. The corresponding unit volume change is $\Delta V/V_0 = \epsilon_z + 2\epsilon_x$. Now for $\sigma_x = \sigma_y = \sigma$, eqs. (3.5a) give

$$\epsilon_x = \epsilon_y = \frac{\sigma}{E}(1 - \mu),$$

and eq. (3.5b) gives

$$\epsilon_z = -2\mu\frac{\sigma}{E}.$$

With these values for strains, the unit volume change becomes

$$\frac{\Delta V}{V_0} = \frac{2\sigma}{E}(1 - 2\mu). \tag{a}$$

Taking $E = 30(10)^6$ psi, $\sigma = 30(10)^3$ psi, and $\mu = \frac{1}{4}$, this gives

$$\frac{\Delta V}{V_0} = \frac{\sigma}{E} = \frac{30(10)^3}{30(10)^6} = 0.001.$$

PROBLEM SET 3.2

1. Determine the stresses σ_n, σ'_n, τ, and τ' for the element in Fig. 3.5 if $\sigma_x = 8000$ psi, $\sigma_y = -4000$ psi, and $\phi = 30°$. *Ans.* $\sigma_n = 5000$ psi; $\sigma'_n = -1000$ psi; $\tau = -\tau' = 5200$ psi.

2. Determine the stresses σ_n, σ'_n, τ, and τ' for the element in Fig. 3.5 if $\sigma_x = -5000$ psi, $\sigma_y = -8000$ psi, and $\phi = 22° 30'$. *Ans.* $\sigma_n = -5440$ psi; $\sigma'_n = -7560$ psi; $\tau = -\tau' = 1060$ psi.

3. If the principal stresses $\sigma_x = 5000$ psi and $\sigma_y = 3000$ psi for the element in Fig. 3.5, find the value of the angle ϕ for which the shear stress will be a maximum and evaluate this shear stress. *Ans.* $\tau_{max} = 1000$ psi.

4. A rectangular block having dimensions a, b, c, in the x, y, z-directions, respectively, is subjected to biaxial stresses σ_x and σ_y of such magnitudes that the unit strains in the x, y, z-directions are ϵ_x, ϵ_y, and ϵ_z. Show that if these strains are small, the unit volume change of the block will be $\Delta V/V = \epsilon_x + \epsilon_y + \epsilon_z$.

5. A square plate 1 in. thick is subjected to tensile stress $\sigma_x = 20,000$ psi in one direction and compressive stress $\sigma_y = -20,000$ psi in the other direction. Find the change in volume of the plate. *Ans.* $\Delta V = 0$.

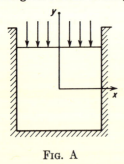

6. A hard rubber block completely confined in the x-direction but free to expand in both the y- and z-directions is subjected to compressive stress $\sigma_y = -300$ psi in the y-direction as shown in Fig. A. Calculate the stress σ_x in the x-direction. What is the change in volume of the block if $E = 300$ psi and $\mu = \frac{1}{2}$? *Ans.* $\sigma_x = -150$ psi; $\Delta V = 0$.

7. A thin circular brass membrane is supported around its rim by a rigid circular brass ring, and at room temperature (70° F) the membrane is free from stress. What principal stresses σ_x and σ_y will be induced in the membrane if its temperature falls to zero degrees? Assume

Fig. A

that $E = 14(10)^6$ psi, $\mu = 0.3$ and $\alpha = 10.4(10)^{-6}$ in./in./F°. *Ans.* $\sigma_x = \sigma_y = 14{,}550$ psi.

8. The thin-walled cylindrical tank shown in Fig. 3.3, p. 52, has radius $r = 20$ in., wall thickness $t = \frac{1}{8}$ in., and is subjected to internal gas pressure $p = 110$ psi. Calculate the magnitude of shear stress τ along a 45°-helix. *Ans.* $\tau = 4400$ psi.

9. Derive a formula for the maximum shear stress $\tau_{\max}$ in the plane of the wall of the conical tank filled with water as shown in Fig. 3.4, p. 53. *Ans.*

$$\tau_{\max} = \frac{3wh^2}{64t}\frac{\tan \alpha}{\cos \alpha}.$$

3.3 Mohr's Circle for Biaxial Stress

Consider again the case of a thin plate subjected to biaxial tension as shown in 3.7a. Then as discussed in Art. 3.2, the normal and shear stresses on any plane whose normal n makes an angle ϕ with the x-axis will be given by eqs. (3.2) repeated below for convenience of reference.

$$\left. \begin{array}{l} \sigma_n = \dfrac{1}{2}(\sigma_x + \sigma_y) + \dfrac{1}{2}(\sigma_x - \sigma_y)\cos 2\phi, \\[2mm] \tau = \dfrac{1}{2}(\sigma_x - \sigma_y)\sin 2\phi. \end{array} \right\} \tag{a}$$

It can easily be shown that these are the equations of a circle in a σ-τ plane, with the angle ϕ as a parameter. Introducing the notations

$$\frac{1}{2}(\sigma_x + \sigma_y) = \sigma_{av}, \quad \frac{1}{2}(\sigma_x - \sigma_y) = \tau_{\max}, \tag{b}$$

we can write eqs. (a) in the more compact form

$$\left. \begin{array}{l} \sigma_n = \sigma_{av} + \tau_{\max}\cos 2\phi, \\[2mm] \tau = \tau_{\max}\sin 2\phi. \end{array} \right\} \tag{c}$$

Then to eliminate the parameter ϕ, we note from the second of eqs. (c) that

$$\cos 2\phi = \sqrt{1 - \sin^2 2\phi} = \sqrt{1 - \frac{\tau^2}{\tau_{\max}{}^2}}.$$

Substitution of this in the first of eqs. (c) gives

$$\sigma_n - \sigma_{av} = \tau_{\max}\sqrt{1 - \tau^2/\tau_{\max}{}^2} \tag{d}$$

or

$$(\sigma_n - \sigma_{av})^2 + \tau^2 = \tau_{\max}{}^2.$$

This is the standard form for the equation of a circle having a radius $\tau_{\max} = \frac{1}{2}(\sigma_x - \sigma_y)$ and centered on the σ-axis at $\sigma_{av} = \frac{1}{2}(\sigma_x + \sigma_y)$ as shown in Fig. 3.7b. This is called *Mohr's circle* for biaxial stress and it is very useful as a graphical means of solving eqs. (a).

Consider, for example, any point D on the circle and denote the angle

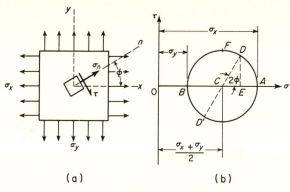

(a)   (b)

FIG. 3.7

ACD by 2ϕ as shown. Then from geometry it can be seen that the coordinates of point D are

$$OE = OC + CD \cos 2\phi = \tfrac{1}{2}(\sigma_x + \sigma_y) + \tfrac{1}{2}(\sigma_x - \sigma_y) \cos 2\phi,$$

$$DE = CD \sin 2\phi = \tfrac{1}{2}(\sigma_x - \sigma_y) \sin 2\phi.$$

Thus the coordinates of point D in Fig. 3.7b represent the values of σ_n and τ for the plane whose aspect is defined by ϕ in Fig. 3.7a.

For each different aspect of plane in Fig. 3.7a as defined by ϕ, there is a corresponding point on the circle, the coordinates of which represent the normal and shear stress on that plane. For example, when $\phi = 0$, the normal n coincides with the x-axis and point D on the circle coincides with point A, giving $\sigma_n = \sigma_x$ and $\tau = 0$. When $\phi = 45°$, $2\phi = 90°$ and point D falls at F giving $\sigma_n = \tfrac{1}{2}(\sigma_x + \sigma_y)$ and $\tau = \tau_{\max} = \tfrac{1}{2}(\sigma_x - \sigma_y)$. When $\phi = 90°$, the normal n coincides with the y-axis and the corresponding position of point D is at B on the circle indicating $\sigma_n = \sigma_y$ and $\tau = 0$. It will be noted also that the coordinates of point D' diametrically opposite point D on the circle give the complementary stresses $\sigma_n{'}$ and τ' as defined by eqs. (3.3). Thus all possible information about the stresses on various planes can be found from Mohr's circle.

If either principal stress is compression, it must be taken with negative sign so that in general the center C of Mohr's circle may lie to either side of the origin in Fig. 3.7b, but always on the σ-axis. Several particular cases are illustrated in the following examples.

EXAMPLE 1. Construct Mohr's circle for the case of biaxial stress of a thin plate where $\sigma_y = -\sigma_x$ as shown in Fig. 3.8a.

SOLUTION. In this case $\sigma_{av} = \tfrac{1}{2}(\sigma_x + \sigma_y) = \tfrac{1}{2}(\sigma_x - \sigma_x) = 0$ and the center C of the circle falls at the origin, while its radius $\tau_{\max} = \tfrac{1}{2}(\sigma_x - \sigma_y) = \tfrac{1}{2}(2\sigma_x) = \sigma_x$. Thus Mohr's circle will be as shown in Fig. 3.8b.

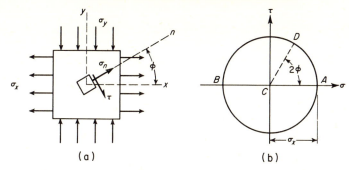

FIG. 3.8

EXAMPLE 2. Construct Mohr's circle for the case of simple tension where $\sigma_y = 0$, as shown in Fig. 3.9a.

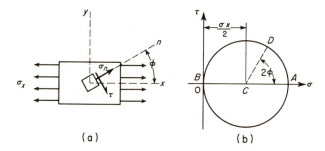

FIG. 3.9

SOLUTION. In this case, $\sigma_{av} = \frac{1}{2}(\sigma_x + \sigma_y) = \frac{1}{2}\sigma_x$ and the circle is tangent to the τ-axis as shown in Fig. 3.9b.

EXAMPLE 3. Construct Mohr's circle for the case of biaxial stress shown in Fig. 3.10a where σ_x is tension and σ_y is compression. Assume $|\sigma_y| = 2\,|\sigma_x|$.

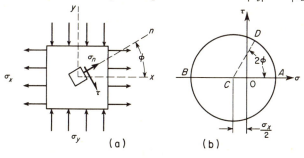

FIG. 3.10

SOLUTION. In this case $\sigma_{av} = \frac{1}{2}(\sigma_x + \sigma_y) = -\frac{1}{2}\sigma_x$ and $\tau_{max} = \frac{1}{2}(\sigma_x - \sigma_y) = \frac{3}{2}\sigma_x$. Thus Mohr's circle will be as shown in Fig. 3.10b.

PROBLEMSET 3.3

1. Construct Mohr's circle for the particular case of biaxial tension where $\sigma_x = \sigma_y$. What is the maximum shear stress τ_{max} in such case? *Ans.* $\tau_{max} = 0$.

2. Construct Mohr's circle for the case of biaxial stress shown in Fig. 3.7 if $\sigma_x = 8000$ psi and $\sigma_y = -4000$ psi. From this circle, find the stresses σ_n, σ'_n, τ, and τ' for the inner element where $\phi = 30°$. *Ans.* $\sigma_n = 5000$ psi; $\sigma'_n = -1000$ psi; $\tau = -\tau' = 5200$ psi.

3. Referring to the conical tank in Fig. 3.4, construct Mohr's circle for the state of stress on an element situated at the elevation $y = h/3$ above the apex 0. The following numerical data are given: $h = 9$ ft, $\alpha = 22° 30'$, $t = 0.010$ in. From this circle, find the maximum shear stress in the plane of the wall at this point. *Ans.* $\tau_{max} = 87.5$ psi.

4. Referring to Fig. 3.10, assume $\sigma_x = 5000$ psi and $\sigma_y = -10,000$ psi. For these principal stresses, construct Mohr's circle and find the value of ϕ defining the plane on which σ_n vanishes. What is the magnitude of the shear stress on this plane? *Ans.* $\phi = 35° 16'$; $\tau = 7070$ psi.

5. Referring to Fig. 3.7, assume that $\sigma_x = 12,000$ psi and $\sigma_y = 6000$ psi. For these principal stresses, construct Mohr's circle and find the angle ϕ defining the plane on which the resultant stress makes the minimum angle with the plane on which it acts, or, stated another way, the plane for which the ratio of normal stress to shearing stress is a minimum. *Ans.* $\phi = 54° 44'$.

3.4 Pure Shear

Let us consider now the particular case of biaxial stress where $\sigma_x = -\sigma_y$ as shown in Fig. 3.11a. In such case, Mohr's circle will be as shown in Fig. 3.11b. From points F and F_1 on the circle, it is seen that the maximum shear stress on 45°-planes is $\tau_{max} = \pm\sigma_x$ and also that the normal stresses σ_n and σ'_n on these planes vanish. This means that the square element $acbd$ oriented at 45° to the directions of principal stress is subjected to shear stresses only on its edges and is said to be in a state of *pure shear*. This particular state of stress is of special interest and it will now be examined in further detail.

Consider first, the deformation of the element $acbd$ shown to a larger

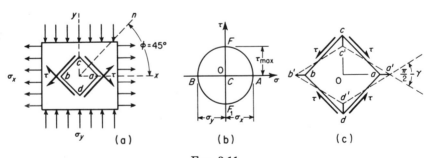

(a)   (b)   (c)

FIG. 3.11

scale in Fig. 3.11c. Since there are no normal stresses on its edges, these edges will remain unchanged in length during deformation of the element. The horizontal diagonal ab will simply elongate and the vertical diagonal will contract so that, after deformation, the element will have the form of a rhombus as shown by dotted lines.   Thus, the original right angles at a and b become $\pi/2 - \gamma$ and those at c and d become $\pi/2 + \gamma$, wher γ defines the amount of angular deformation of the element. This quantity is a pure number and is called the *shearing strain*, analogous to the tensile or compressive strain ϵ as discussed in Art. 1.3. To visualize this stress condition more readily, we rotate the element $acbd$ by 45° and place the edges $b'd'$ and bd in juxtaposition as shown in Fig. 3.12. Then the angle cbc' is seen to represent the previously defined γ and its value is

$$\gamma = \frac{cc'}{bc} = \frac{\delta}{l},$$

where δ is the lateral displacement of the upper face of the element relative to the lower face and l is the distance between these faces. Thus the analogy between shearing strain γ and extensional strain ϵ is complete, except that in the case of shear the displacement δ takes place at right angles to the length l so that we see the ratio δ/l as an angle.

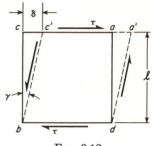

Fig. 3.12

Within the elastic limit of a material, it is reasonable to assume that the shearing strain γ is proportional to the shearing stress τ that produces it. In fact, experiments with materials in pure shear show this to be the case, so that

$$\gamma = \frac{\tau}{G}, \tag{3.7}$$

where the constant of proportionality, denoted by G, is called the *modulus of elasticity in shear* or simply the *shear modulus*. Like the tension or compression modulus E, it has the dimension of stress, lb/in.[2].

In the case of pure shear, there will be no change in volume of an element during deformation. To show this, we must return to eqs. (3.5) defining the strains ϵ_x, ϵ_y, ϵ_z, in the case of biaxial stress. Then for $\sigma_x = -\sigma_y = \tau$, these expressions become

$$\epsilon_x = \frac{\tau}{E}(1 + \mu), \quad \epsilon_y = -\frac{\tau}{E}(1 + \mu), \quad \epsilon_z = 0. \tag{3.8}$$

Then the unit volume change $\Delta V/V = \epsilon_x + \epsilon_y + \epsilon_z = 0$.

It is seen from Fig. 3.11a that the linear strains ϵ_x and ϵ_y in the directions of the diagonals ab and cd of the element $acbd$ must be geometrically related to the shearing strain γ. Out of this, comes an important relationship between the shear modulus G and the tension modulus E, which will now be established. Using the first two of eqs. (3.8) for linear strains and referring to Fig. 3.11c, we see that

$$oa' = oa(1 + \epsilon_x) = oa \left[1 + \frac{\tau}{E} (1 + \mu) \right],$$

$$oc' = oc(1 + \epsilon_y) = oc \left[1 - \frac{\tau}{E} (1 + \mu) \right]. \qquad \text{(a)}$$

Also from the geometry of the right triangle $c'oa'$, we have

$$\tan oa'c' = \tan \left(\frac{\pi}{4} - \frac{\gamma}{2} \right) = \frac{oc'}{oa'} = \frac{1 - \frac{\tau}{E} (1 + \mu)}{1 + \frac{\tau}{E} (1 + \mu)}, \qquad \text{(b)}$$

since $oa = oc$ in eqs. (a). Finally, from trigonometry,

$$\tan \left(\frac{\pi}{4} - \frac{\gamma}{2} \right) = \frac{\tan \frac{\pi}{4} - \tan \frac{\gamma}{2}}{1 + \tan \frac{\pi}{4} \tan \frac{\gamma}{2}} \approx \frac{1 - \gamma/2}{1 + \gamma/2}, \qquad \text{(c)}$$

since $\tan \pi/4 = 1$ and $\tan \gamma/2 \approx \gamma/2$, the angle γ being very small. Comparing expressions (b) and (c), we conclude that

$$\frac{\gamma}{2} = \frac{\tau}{E} (1 + \mu). \qquad \text{(d)}$$

Substituting for γ its value from eq. (3.7), this becomes

$$\frac{\tau}{2G} = \frac{\tau}{E} (1 + \mu),$$

from which
$$G = \frac{E}{2(1 + \mu)}. \qquad \text{(3.9)}$$

This theoretical relationship between the shear modulus G and the tension modulus E is in good agreement with that found by experiment. Taking, for mild steel, $E = 30 \ (10)^6$ psi and $\mu = 0.3$, we find $G = 11.5 \ (10)^6$ psi. If Poisson's theoretical value of the ratio $\mu = 0.25$ is used, eq. (3.9) gives $G = 12 \ (10)^6$ psi. It will be shown in the next chapter that the modulus of elasticity in shear can be determined experimentally by twisting a circular shaft. Such experiments with structural steel specimens show the value of G to lie within the above limits.

1. In Fig. 3.11a, $\sigma_x = -\sigma_y = 30,000$ psi. Calculate the shearing strain γ for the element *acbd* if $E = 30(10)^6$ psi and $\mu = 0.25$. *Ans.* $\gamma = 0.0025$.

2. Calculate the modulus of elasticity in shear for rubber ($E = 300$ psi, $\mu = \frac{1}{2}$) and for concrete ($E = 2(10)^6$ psi, $\mu = 0.1$). *Ans.* $G_r = 100$ psi; $G_c = 910(10)^3$ psi.

3. A short length of thin-walled steel tube with closed ends has mean diameter $D = 2$ in., wall thickness $t = 0.025$ in., and carries internal gas pressure of intensity $p = 500$ psi. What externally applied axial compressive loads P must be applied at the ends to put the wall of the tube in a condition of pure shear? *Ans.* $P = 4710$ lb.

3.5 Riveted and Welded Joints in Pressure Vessels

In the actual fabrication of various kinds of pressure vessels as discussed in Art. 3.1, it is usually necessary to have one or more joints or seams as shown in Fig. 3.13. Such joints may be either *riveted* or *welded* depending upon the material and the service conditions to which the vessel will be subjected. The general objective is to obtain a joint that will be tight and as strong as possible.

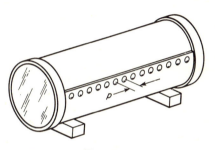

FIG. 3.13

Riveted joints for boilers and other pressure vessels are of two kinds: *lap joints* and *butt joints* as illustrated in Fig. 3.14a and 3.14b. The rivets in such joints are driven red hot so that, after cooling, they squeeze the plates tightly together. Thus when internal pressure is applied to the shell, relative motion between the plates is prevented by friction as well as by the shearing strength of the rivets. In fact, only after the friction is overcome do the rivets begin to work in shear. We see that the behavior of a riveted joint under load is extremely complex. To simplify the problem, it is usual practice to neglect the friction completely and to assume that the rivets carry the load in shear with the shearing stress

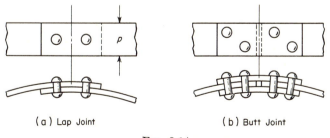

(a) Lap Joint   (b) Butt Joint

FIG. 3.14

uniformly distributed over the cross-section of each rivet as discussed on p. 5.

Besides shearing of the rivets, there are other ways by which a riveted boiler joint like those in Fig. 3.14 can fail after friction has been overcome. These possible modes of failure are illustrated in Fig. 3.15. As already dis-

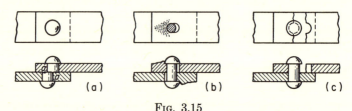

FIG. 3.15

cussed, the rivet may fail in shear across the section *aa*, Fig. 3.15a, although it may undergo considerable bending before this happens. The resistance of the rivet to shear failure can be increased by increasing its diameter. A second possible mode of failure is shown in Fig. 3.15b, where crushing of the plate in compression just behind the rivet allows the joint to open up. Resistance to this type of failure can also be increased by using a larger diameter for the rivet or by increasing the thickness of the plate. If the rivets are too closely spaced, the net section of the plate between rivet holes will be so much reduced that failure of the joint may take place due to tearing of the plate in tension between rivets as shown in Fig. 3.15c. Resistance to such failure can be increased by increasing the spacing or *pitch* p of the rivets along the seam (see Fig. 3.13).

The problem of design of a riveted boiler joint consists of adjusting the plate thickness, the rivet diameter, and the rivet pitch so that the joint is equally strong against each mode of failure. For such design of steel pressure vessel joints, the ASME Boiler Code recommends the following ultimate stresses to be used with a suitable factor of safety, say $n = 5$.

$$\text{TENSION:} \quad \sigma_t = 55,000 \text{ psi,}$$
$$\text{SHEAR:} \quad \tau = 44,000 \text{ psi,}$$
$$\text{CRUSHING:} \quad \sigma_c = 95,000 \text{ psi.}$$

Obviously, the strength of a riveted boiler joint can never be as great as that of the shell itself. The ratio of the strength of the joint to that of the shell proper is called the *efficiency* of the joint.

EXAMPLE 1. The cylindrical container shown in Fig. 3.13 is made of steel plate of thickness $t = \frac{1}{4}$ in. The rivets are $\frac{11}{16}$ in. in diameter and the pitch $p = 1\frac{5}{8}$ in. Calculate the ultimate strength of the joint and its efficiency, using A.S.M.E. Boiler Code specifications.

SOLUTION. Referring to Fig. 3.14a, let us consider a length of joint equal to the rivet pitch $p = 1.625$ in. Then for this length of undisturbed shell the strength is $55,000 \times 0.25 \times 1.625 = 22,300$ lb. The cross-sectional area of one rivet is 0.371 in.^{2} and the corresponding shearing strength is $44,000 \times 0.371 = 16,300$ lb. The projected area behind one rivet is $0.25 \times \frac{11}{16} = 0.172$ in.^{2}, and the corresponding crushing strength is $95,000 \times 0.172 = 16,300$ lb. The net cross-sectional area of plate between rivet holes is $0.25(1.625 - 0.688) = 0.234$ in.^{2} and the corresponding tearing strength of the joint is $55,000 \times 0.234 = 12,900$ lb. This is the smallest value and therefore represents the strength of the joint. The corresponding efficiency is $12,900 \div 22,300 = 0.58$ or 58 per cent.

The efficiency of the joint can be improved by increasing the rivet pitch so that the tearing strength comes up to the strengths in shear and crushing. The equation for determining the optimum pitch is

$$(p - 0.688)0.25 \times 55,000 = 16,300,$$

from which $p = 1.875$ in. The strength of this length of undisturbed shell is $55,000 \times 0.25 \times 1.875 = 25,800$ lb. Hence the new efficiency becomes $16,300 \div 25,800 = 0.63$ or 63 per cent.

Welded Joints. With the present day advances in welding techniques, welded joints for pressure vessels and structural connections are rapidly replacing riveted joints. Typical *lap welds* and *butt welds* for the longitudinal seam of a pressure vessel are shown in Fig. 3.16. In the case of a lap weld,

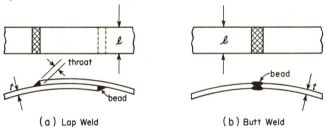

(a) Lap Weld   (b) Butt Weld

FIG. 3.16

Fig. 3.16a, the thinnest section through the weld bead is called the *throat* of the weld. The length l of the bead is called the *length* of the weld. The product of the throat dimension and the length of the bead gives the cross-sectional area of the weld to be used in calculating its strength. In the case of a butt weld, Fig. 3.16b, the thickness t of the plate would be taken as the throat dimension.

The calculation of the strength of a welded joint is largely empirical because of the difficulties involved in making any rational analysis owing to the presence of stress concentrations. It is usually assumed that the strength of a weld is the cross-sectional area of the throat multiplied by an arbitrary working stress. The Code for Fusion Welding specifies a working stress of 16,000 psi in tension or compression for shielded arc welds. If the weld is subjected to shear, the working stress is 13,600 psi.

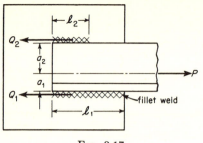

FIG. 3.17

EXAMPLE 2. A 6 in. $\times$ 6 in. $\times$ $\frac{1}{2}$ in. angle-section tension member is welded to a gusset plate by two side fillet welds which act in direct shear as shown in Fig. 3.17. The total tensile force in the member is $P = 103,500$ lb and each weld bead has a throat dimension $t = 0.353$ in. If the working stress for the weld metal in shear is $\tau_w = 13,600$ psi, what lengths l_1 and l_2 should the two weld beads have?

SOLUTION. The line of action of the tensile force P acts through the centroid of the angle section defined by the distances $a_1 = 1.68$ in. and $a_2 = 4.32$ in. In order to balance the load P, the resultant of the two weld resistances must have the same line of action as P. Then if these two forces are denoted by Q_1 and Q_2, it follows that the following conditions of equilibrium must be fulfilled:

$$Q_1 + Q_2 = P, \qquad Q_1 a_1 = Q_2 a_2,$$

from which, with the given numerical data,

$$Q_1 = 74,500 \text{ lb}, \quad Q_2 = 29,000 \text{ lb}.$$

Since the shearing resistance of each bead per inch of length is

$$q = 0.353 \times 13,600 = 4,800 \text{ lb/in.},$$

the corresponding lengths l_1 and l_2 should be

$$l_1 = \frac{74,500}{4800} = 15.5 \text{ in.}, \quad l_2 = \frac{29,000}{4800} = 6.05 \text{ in.}$$

PROBLEM SET 3.5

1. A 4 in. $\times$ 1 in. steel strap is spliced as shown in Fig. A. The rivets are $\frac{7}{8}$ in. in diameter. The allowable working stresses are $\tau_w = 8000$ psi in shear and $\sigma_w = 16,000$ psi in tension. Calculate the safe load P for the spliced strap and the efficiency of the joint. *Ans.* $P = 28,850$ lb; eff. $= 45\%$.

2. What rivet diameter d should be used for the splice shown in Fig. A in order to attain maximum efficiency of the joint? All other data remain the same as given in Problem 1. What is this maximum efficiency? *Ans.* $d = 0.944$ in.; max. eff. $= 52.8\%$.

3. Referring to the splice in Fig. A, calculate the limit load P_L if $\tau_{y.p.} = 20,000$ psi in shear, $\sigma_{y.p.} = 40,000$ psi in tension, and the rivets are 1 in. in diameter. What is the efficiency of the joint in this case? *Ans.* $P_L = 80,000$ lb; eff. $= 50\%$.

4. A tank like that shown in Fig. 3.13 is made of $\frac{1}{2}$-in. steel plate and has a riveted lap joint with two rows of 1-in.-diameter rivets. The pitch in the first row of rivets is 5 in. and that in the second row is $2\frac{1}{2}$ in. Using working stresses $\sigma_w = 16,000$ psi in tension and $\tau_w = 12,000$ psi in shear, find the safe internal pressure p if the tank is 5 ft in diameter. What is the efficiency of the joint? *Ans.* $p = 160$ psi; eff. $= 60\%$.

5. A detail sketch of the lower chord joint of a truss is shown in Fig. B. The diagonal members are $2\frac{1}{2}$ in. $\times$ $1\frac{1}{2}$ in. $\times$ $\frac{1}{4}$ in. angle sections (see Table B-6) and the gusset plate is $\frac{1}{4}$ in. thick. If each diagonal is stressed to 15,000 psi, how many $\frac{7}{8}$-in.-diameter rivets will be required to fasten it to the gusset plate? Assume that the working stress in shear for the rivets is $\tau_w = 8000$ psi. In such case, how many rivets will be required to fasten the gusset plate to the lower chord member? *Ans.* 3 and 5 rivets, respectively.

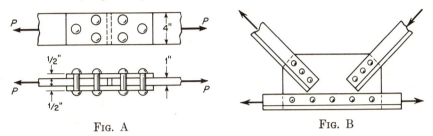

FIG. A   FIG. B

6. A steel plate 4 in. wide by $\frac{1}{2}$ in. thick is connected to a gusset plate by three fillet welds as shown in Fig. C. Calculate the required bead length l of the two side welds if $P = 40,000$ lb and the working stresses for the weld metal are $\sigma_w = 16,000$ psi in tension and $\tau_w = 13,600$ psi in shear. *Ans.* $l = 1.81$ in.

7. Solve the preceding problem if only the two side welds are used. All other data remain the same. *Ans.* $l = 4.17$ in.

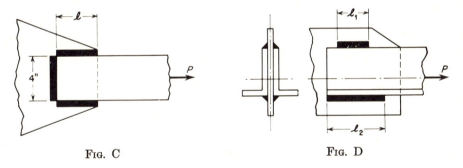

FIG. C   FIG. D

8. One chord member of a truss consists of two 3 in. $\times$ 2 in. $\times$ $\frac{3}{8}$ in. angle sections (see Table B-6) welded to a $\frac{1}{2}$-in. gusset plate as shown in Fig. D. The member carries a total tension $P = 70,000$ lb. Calculate the proper lengths l_1 and l_2 of the welds if the working stress for the weld metal in shear is $\tau_w = 13,600$ psi. *Ans.* $l_1 = 3.36$ in.; $l_2 = 6.34$ in.

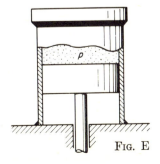

FIG. E

9. The cylinder of a steam engine has outside diameter $D_1 = 8.25$ in. and inside diameter $D_2 = 8.00$ in. The head has diameter $D_o = 8.50$ in. The head is attached to the cylinder by a circumferential bead weld as shown. What is the safe maximum pressure p in the cylinder if the working stress for the weld metal in tension is $\sigma_w = 16,000$ psi and the same for the cylinder wall itself. *Ans.* $p = 493$ psi.

4

Torsion

4.1 Torsion of a Circular Shaft

Consider a circular shaft built in at the upper end and twisted by a couple applied to the lower end (Fig. 4.1a). It can be shown by measurements at the surface that circular sections of the shaft remain circular during twist, and that their diameters and the distances between them do not change provided the angle of twist is small.

A disc-like element of the shaft, such as that adjacent to the section mn and shown as a free body in Fig. 4.1b, will be in the following state of strain. There will be a rotation of its bottom cross-section with reference to its top through an angle $d\phi$. A thin element $abcd$ of the surface of the disc whose sides were vertical before strain takes the form shown in Fig. 4.1b. The

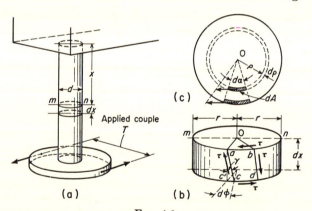

Fɪɢ. 4.1

lengths of the sides remain essentially the same and only the angles at the corners change. Thus we may conclude that the element is in a state of *pure shear* (see Art. 3.4) and the magnitude of the shearing strain, measured by the angle cac', is given very closely by

$$\gamma = \frac{c'c}{ac'}.$$

70

Since $c'c$ is the small arc of radius r subtended by the angle $d\phi$, $c'c = r\,d\phi$. Thus

$$\gamma = \frac{c'c}{ac'} = \frac{r\,d\phi}{dx}. \tag{a}$$

For a shaft twisted by a torque at the end, the angle of twist ϕ is proportional to the distance x of the cross-section from the fixed end and hence $d\phi/dx$ is a constant. This constant represents *the angle of twist per unit length of the shaft* and will be called θ. Then, from (a),

$$\gamma = r\theta. \tag{b}$$

The shearing stresses which act on the sides of the element and produce the above shearing strain have the directions shown. The magnitude of each, from eq. (3.7), is

$$\tau = G\gamma = Gr\theta. \tag{4.1}$$

So much for the state of stress on an element at the surface of the shaft.

As for the state of stress within the shaft, the assumption will now be made that not only the circular boundaries of the cross-sections of the shaft remain undistorted but also that the cross-sections themselves remain plane and rotate as if absolutely rigid; that is, every diameter of the cross-section remains straight and rotates through the same angle. Tests of circular shafts show that the theory developed on this assumption is in very good agreement with experimental results. Such being the case, the discussion for the element $abcd$ at the surface of the shaft (Fig. 4.1b) will hold also for a similar element within the shaft, whose radius ρ replaces r (Fig. 4.1c). The thickness of the element in the radial direction is considered as very small. Such elements are then also in pure shear, and the shearing stress on their side is

$$\tau = G\rho\theta. \tag{4.2}$$

This states that the shearing stress varies directly as the distance ρ from the axis of the shaft. Fig. 4.2 pictures this stress distribution in the plane of

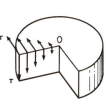

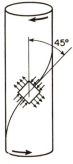

FIG. 4.2   FIG. 4.3

the cross-section and also the complementary shearing stresses in an axial plane (see p. 30). The maximum stress occurs in the outer surface of the shaft, where $\rho = r$.

For a ductile material, plastic flow begins first in this outer surface. For a material which is weaker in shear longitudinally than transversely — for instance, a wooden shaft with the fibers parallel to the axis — the first cracks will be produced by shearing stresses acting in the axial sections and they will appear on the surface of the shaft in the longitudinal direction. In the case of a material which is weaker in tension than in shear — for instance, a circular shaft of cast iron or a cylindrical piece of chalk —a crack along a helix inclined at 45° to the axis of the shaft often occurs (Fig. 4.3). The explanation is simple. The state of pure shear is equivalent to one of tension in one direction and equal compression in the perpendicular direction (see Art. 3.4). A rectangular element cut from the outer layer of a twisted shaft with sides at 45° to the axis will be subjected to such stresses, as shown in Fig. 4.3. The tensile stresses shown produce the helical crack mentioned.

The relationship between the applied torque T and the stresses which it produces will now be found. From the equilibrium of that portion of the shaft between the bottom and the section mn, it can be concluded that the shearing stresses distributed over the cross-section mn are statically equivalent to a couple equal and opposite to the external torque T. For each element of area dA (Fig. 4.1c). the shearing force is τdA. The moment of this force about the axis of the shaft is $(\tau dA)\rho = G\theta\rho^2 dA$, from eq. (4.2). The total resisting torque T about the axis of the shaft is the summation, taken over the entire cross-sectional area, of these moments of the individual elements; that is,

$$T = \int_A G\theta\rho^2\, dA = G\theta \int_A \rho^2\, dA = G\theta J, \qquad (c)$$

where

$$J = \int_A \rho^2\, dA$$

is defined as the polar moment of inertia of the circular cross-section.* For a circle of diameter d, $J = \pi d^4/32$; so that

$$T = G\theta \frac{\pi d^4}{32}$$

and

$$\theta = \frac{T}{GJ} = \frac{T}{G}\frac{32}{\pi d^4}. \qquad (4.3)$$

Thus θ, the angle of twist per unit of length of the shaft, varies directly as the applied torque and inversely as the modulus of shear G and the fourth

*See Appendix B, p. 346.

power of the diameter. If the shaft is of length l, the total angle of twist will be

$$\phi = \theta l = \frac{Tl}{GJ}. \tag{4.4}$$

This equation is useful in the experimental verification of the theory, and is checked by numerous experiments which justify the assumptions made in its derivation. It should be noted that experiments in torsion are commonly used for determining the modulus of materials in shear. If the angle of twist produced in a given shaft by a given torque be measured, the magnitude of G can be easily calculated from eq. (4.4).

Substituting θ from eq. (4.3) in eq. (4.1), we obtain an equation for calculating the maximum shearing stress in twist of a *solid* circular shaft

$$\tau_{max} = \frac{Tr}{J} = \frac{16T}{\pi d^3}. \tag{4.5}$$

We see that this stress is proportional to the applied torque T and inversely proportional to the cube of the diameter of the shaft.

If we substitute θ from eq. (4.3) in eq. (4.2), we obtain an expression for the shearing stress τ at *any* point in a *solid* circular shaft;

$$\tau = \frac{T\rho}{J}. \tag{4.6}$$

In practical applications, the diameter of the shaft must sometimes be calculated from the horsepower which it is required to transmit. Given the horsepower hp, the speed n in rpm, and the torque T in inch pounds, a formula connecting these quantities is derived as follows: Since the work done by the applied torque T per revolution of the shaft is $T \cdot 2\pi$, the work done per minute at n rpm will be $T \cdot 2\pi n$. Then with 1 hp = 33,000 ft-lb per minute, we have

$$\text{hp} = \frac{2\pi nT}{12 \times 33,000}$$

or

$$T = \frac{12 \times 33,000 \times \text{hp}}{2\pi n} = \frac{63,000\ (\text{hp})}{n} \text{ in.-lb.} \tag{4.7}$$

When the horsepower and rpm are given, the corresponding torque T can be computed from this formula and then used in any of the foregoing equations to compute the shear stress or angle of twist in the shaft.

In this way it may be shown that the required diameter d of a solid circular shaft, to transmit hp horsepower at n rpm with a maximum allowable working stress τ_w in shear, will be

$$d = 68.5 \sqrt[3]{\frac{\text{hp}}{n\tau_w}}. \tag{d}$$

Hollow Shaft. From the preceding discussion of torsion of a solid shaft of circular cross-section, it is seen that only the material at the outer surface of the shaft can be stressed to the limit assigned as an allowable working stress. All of the material within the shaft will work at a lower stress and is not being used to full capacity. Thus in those cases where weight reduction is important, it is advantageous to use hollow shafts. In discussing the torsion of a hollow shaft, the same assumptions will be made as in the case of a solid shaft. The general expression for shearing stress, eq. (4.2), will apply. However, in calculating the internal resisting torque T in this case, we sum the moments of the elemental forces τdA only over the region from $\rho_1 = \frac{1}{2}d_i$ to $\rho_0 = \frac{1}{2}d_0$ as shown in Fig. 4.4. In this way, we again obtain, for the internal resisting torque,

$$T = G\theta \int_A \rho^2 \, dA = G\theta J,$$

where now

$$J = \frac{\pi}{32} (d_0{}^4 - d_i{}^4) \qquad (e)$$

is the polar moment of inertia of the hollow cross-section. Thus, with the proper value of J, the basic equations (4.2), (4.4), and (4.6) apply also to a circular shaft of hollow cross-section.

Taking, for example, the case where $d_i = \frac{1}{2}d_0$, the angle of twist ϕ, eq. (4.4), and the maximum shear stress τ_{max}, eq. (4.6), will be found to be about 6 per cent larger than in the case of a solid shaft having the same outside diameter d_0. But the reduction in weight will be 25 per cent.

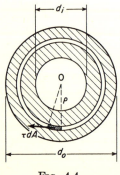

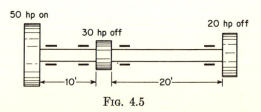

FIG. 4.4   FIG. 4.5

EXAMPLE 1. The solid line shaft shown in Fig. 4.5 is made of steel, has diameter $d = 1.5$ in., and runs at 525 rpm. It is supported in bearings so placed that bending of the shaft will be negligible. A driving belt feeds 50 hp to the left hand pulley while 30 hp and 20 hp, respectively, are taken off by belts overrunning the middle and right hand pulleys. Compute the maximum shear stress τ induced in the shaft and the total angle of twist ϕ. Assume $G = 12(10)^6$ psi.

SOLUTION. The greater torque is in the left-hand portion of the shaft. Its value from eq. (4.7) is

$$T = 63,000 \times \frac{50}{525} = 6000 \text{ in.-lb.}$$

The corresponding maximum shear stress from eq. (4.5) is

$$\tau = \frac{16 \times 6000}{\pi \times (1.5)^3} = 9060 \text{ psi.}$$

Similarly, for the right-hand portion of the shaft which transmits 20 hp, $T = 2400$ in.-lb and $\tau = 3620$ psi.

The total angle of twist is the sum of the angles of twist ϕ_1 and ϕ_2 in the two portions of the shaft. Using eq. (4.4), this becomes

$$\phi = \phi_1 + \phi_2 = \frac{6000 \times 120}{12(10)^6 \times 0.497} + \frac{2400 \times 240}{12(10)^6 \times 0.497} = 0.217 \text{ rad} = 12°27'.$$

The value of J used in the above calculation is

$$J = \frac{\pi d^4}{32} = \frac{\pi}{32}(1.5)^4 = 0.497 \text{ in.}^4$$

EXAMPLE 2. A stepped solid circular shaft is built-in at its ends and subjected to an externally applied torque T_0 at the shoulder as shown in Fig. 4.6. Determine the angle of rotation ϕ_0 of the shoulder section where T_0 is applied.

SOLUTION. This is a statically indeterminate system because the shaft is built-in at both ends. All that we can find from statics is that the sum of the two reactive torques T_A and T_B at the built-in ends of the shaft must be equal to the applied torque T_0. Thus

$$T_A + T_B = T_0. \tag{f}$$

From consideration of consistent deformation, we see that the angle of twist in each portion of the shaft must be the same, i.e.,

$$\phi_a = \phi_b = \phi_0. \tag{g}$$

Using eq. (4.4) for angle of twist, expression (g) becomes

$$\frac{T_A a}{G J_A} = \frac{T_B b}{G J_B} = \phi_0. \tag{h}$$

This defines the ratio between T_A and T_B as

$$\frac{T_A}{T_B} = \frac{J_A b}{J_B a}. \tag{i}$$

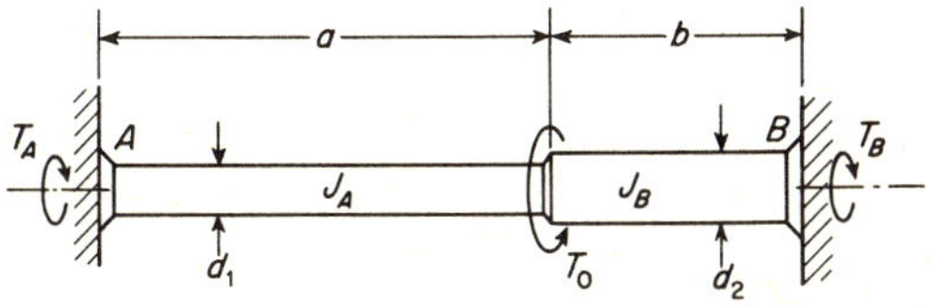

FIG. 4.6

From this expression, we see that in the particular case where $J_A = J_B$, i.e., for a shaft of uniform cross-section, the reactive torques T_A and T_B are simply in the inverse ratio of the lengths a and b. From eqs. (f) and (i), we obtain, for the more general case,

$$T_A = \frac{T_0}{1 + \dfrac{J_B a}{J_A b}}, \quad T_B = \frac{T_0}{1 + \dfrac{J_A b}{J_B a}}. \tag{j}$$

Using either of these values in eqs. (h), we have for the angle of rotation ϕ_0 of the junction

$$\phi_0 = \frac{T_0 ab}{(J_A b + J_B a)G}. \tag{k}$$

PROBLEMSET 4.1

1. A steel shaft $\frac{1}{4}$ in. in diameter turns at 10,000 rpm. What is the maximum power that such a shaft may develop if the assigned working stress in shear is $\tau_w = 5000$ psi? *Ans.* 2.44 hp.

2. Determine the proper diameter d for a solid steel shaft to transmit 200 hp at 105 rpm if the working stress in shear is 6000 psi. *Ans.* $d = 4.67$ in.

3. Determine the proper diameter d for a solid steel shaft to transmit 300 hp at 3600 rpm if the working stress in shear is 6000 psi. *Ans.* $d = 1.65$ in.

4. A hollow steel shaft is to have outside diameter d and inside diameter $d/2$. Calculate the proper value of d for the shaft if it is to transmit 200 hp at 105 rpm with a working stress in shear of 6000 psi. *Ans.* $d = 4.78$ in.

5. A propeller shaft of diameter $d = 6$ in. is spliced as shown in Fig. A. The two pins are each 1 in. in diameter and the working stress in shear is 6000 psi. What is the maximum safe horsepower that the shaft can transmit at 120 rpm? What is the efficiency of the spliced joint? *Ans.* 54 hp; eff. = 11.1%.

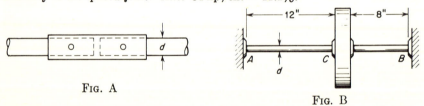

FIG. A

FIG. B

6. A solid steel shaft of diameter $d = \frac{1}{2}$ in. is built in at its ends A and B and carries a disk at C as shown in Fig. B. If the working stress in shear for the shaft is 10,000 psi, what is the maximum safe angle of rotation that can be given to the disk which is rigidly attached to the shaft? *Ans.* 1° 30′.

7. For the stepped shaft in Fig. 4.6, the following numerical data are given: $a = 8$ in., $b = 16$ in., $d_1 = 1$ in., $d_2 = 1\frac{1}{4}$ in. The allowable working stress in shear for the shaft is $\tau_w = 8000$ psi. Calculate the safe value of the torque T_0. *Ans.* $T_o = 3480$ in.-lb.

8. A hollow steel tube (outside diameter, d_o and inside diameter d_i) is to be used as a torque dynamometer. It is desired to attain an angle of twist of 1° per foot of length per 100 in.-lb of torque without exceeding an allowable shear stress $\tau_w = 6000$ psi. What are the required values of d_o and d_i? *Ans.* $d_o = 0.688$ in.; $d_i = 0.638$ in.

9. A slightly tapered shaft with circular cross-section has diameter d_o at one end and diameter $d_o/2$ at the other; the length of the shaft is l. Determine the total angle of twist ϕ between the two ends of the shaft if it is subjected to uniform torque T. *Ans.* $\phi = 14Tl/3GJ_o$, where $J_o = \pi d_o^4/32$.

10. For the system in Fig. C, power is transmitted from the gear A to the gear D. If the pitch diameters of the gears B and C are in the ratio 1:2, what is the proper ratio of shaft diameters $d_1:d_2$ for both shafts to have the same maximum shear stress τ? *Ans.* $d_1:d_2 = 0.794$.

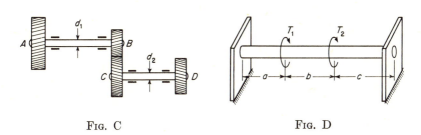

FIG. C   FIG. D

11. A prismatic shaft of diameter d has built-in ends and is subjected to the action of externally applied twisting moments T_1 and T_2 as shown in Fig. D. Find the internal torques T_a, T_b, T_c, in the three portions a, b, c, of the shaft. The following numerical data are given: $a = 30$ in., $b = 50$ in., $c = 40$ in., $T_1 = 12{,}000$ in.-lb, and $T_2 = 24{,}000$ in.-lb. *Ans.* $T_a = 17{,}000$ in.-lb; $T_b = 5000$ in.-lb; $T_c = 19{,}000$ in.-lb.

12. A solid brass rod of diameter $d = 2.5$ in. has a steel sleeve with wall thickness $t = \frac{1}{4}$ in. solidly fused onto it for reinforcement. What is the safe torque for the compound shaft if $G_s = 12(10)^6$ psi, $G_b = 5(10)^6$ psi, $(\tau_w)_s = 12(10)^3$ psi, $(\tau_w)_b = 7.5(10)^3$ psi. What is the ratio of this torque to that which the brass rod alone could safely carry? *Ans.* $T = 45{,}800$ in.-lb; ratio $= 1.99$.

4.2 Close-Coiled Helical Spring

An interesting application of the theory of torsion arises in the case of a close-coiled helical spring.* Assume that such a spring, wound from a wire of solid circular cross-section on a circular core, is subjected to the action of axial forces P (Fig. 4.7a), and that any one coil lies nearly in a plane perpendicular to the axis of the helix. Considering the equilibrium of the upper portion of the spring bounded by an axial section such as mn (Fig. 4.7b), it can be concluded from equations of statics that the stress resultant on the cross-section mn of the coil reduces to a shearing force P through the center of the cross-section and a couple acting in a counterclockwise direction in the plane of the cross-section of magnitude PR, where R is the radius of the cylindrical surface containing the center line of the spring. The

*For a complete treatise on springs, see *Mechanical Springs* by A. M. Wahl, Penton Publishing Co., Cleveland, 1944.

couple PR twists the coil and causes a maximum shearing stress given by
eq. (4.5), which becomes here

$$\tau' = \frac{16PR}{\pi d^3},$$

(a)

where d is the diameter of the cross-section mn of the wire. Upon this stress
due to twist, that due to the shearing force P is superposed. For a rough

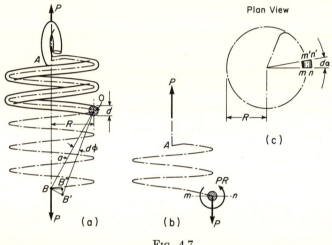

Fig. 4.7

approximation, this shearing force is assumed to be uniformly distributed
over the cross-section; the corresponding shearing stress will be

$$\tau'' = \frac{4P}{\pi d^2}.$$

(b)

At the point m the directions of τ' and τ'' coincide so that the maximum
shearing stress occurs here and has the magnitude

$$\tau_{\max} = \tau' + \tau'' = \frac{16PR}{\pi d^3}\left(1 + \frac{d}{4R}\right).$$

(4.8)

It can be seen that the second term in the parentheses, which represents the
effect of the shearing force, increases with the ratio d/R. It becomes of
practical importance in heavy helical springs, such as are used on railway
cars. Due to this term, points such as m on the inner side of a coil are in a
less favorable condition than points such as n, for at point n the shearing
stresses τ' and τ'' act in opposite directions. Experience with heavy springs
shows that cracks usually start on the inner side of the coil.

There is another reason for expecting higher stresses at the inner side of the coil. In calculating the stresses due to twist, we use eq. (a), which was derived for cylindrical bars. In reality each element of the spring will be in the condition shown in Fig. 4.8. It is seen that if the cross-section bf rotates with respect to ac, due to twist, the displacement of the point b with respect to a will be the same as that

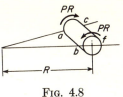

FIG. 4.8

of the point f with respect to c. Due to the fact that the distance ab is smaller than the distance cf, the *shearing strain* at the inner side ab will be larger than that at the outer side cf, and therefore the shearing stresses produced by the couple PR will be larger at b than at f. Taking this into consideration, together with the effect of the shearing force, we replace eq. (4.8) by the following equation* for calculating the maximum shearing stress:

$$\tau_{max} = \frac{16PR}{\pi d^3}\left(\frac{4m-1}{4m-4} + \frac{.615}{m}\right), \tag{4.9}$$

in which $m = 2R/d$. It can be seen that the *correction factor* in the parentheses increases with a decrease of m; for instance, in case $m = 4$ this factor is about 1.40 and for $m = 10$ it is equal to 1.14.

In calculating the deflection of the spring, usually only the effect of the twist of the coils is taken into consideration. For the angle of twist of one element between the two adjacent cross-sections mn and $m'n'$ (Fig. 4.7c), using eq. (4.4), in which $Rd\alpha$ replaces l, we obtain

$$d\phi = \frac{P \cdot R \cdot R\,d\alpha}{JG}.$$

Due to this twist, the lower portion of the spring rotates with respect to the point O (Fig. 4.7a), and the point of application B of the force P describes the small arc $BB' = a\,d\phi$. This movement of B is easily pictured by imagining all of the spring as rigid except the element between mn and $m'n'$. The vertical component of this displacement is

$$d\delta = B'B'' = BB'\frac{R}{a} = R\,d\phi = \frac{PR^3\,d\alpha}{JG}. \tag{c}$$

The complete deflection of the spring is obtained by summation of the deflections $B'B''$ due to each element $mnm'n'$, over the length of the spring. Then

$$\delta = \int_0^{2\pi n} \frac{PR^3}{JG}\,d\alpha = \frac{64nPR^3}{d^4G}, \tag{4.10}$$

*For the derivation of this formula, see A. M. Wahl, "Stresses in Heavy Closely Coiled Helical Springs, "*Trans. A.S.M.E.*, 1929, Vol. 51, Paper No. APM-51-17.

in which n denotes the number of coils. The net horizontal displacement in one complete turn is zero.

The ratio P/δ for a given spring is called the *spring constant*, denoted by k. Thus, from eq. (4.10), we have

$$k = \frac{P}{\delta} = \frac{Gd^4}{64R^3n}. \tag{4.11}$$

Two springs are said to be of the same stiffness if their spring constants are equal. The spring constant k can be varied by changing the material, the wire diameter, the core radius, or the number of coils.

EXAMPLE 1. Two close-coiled helical springs wound from the same wire but with different core radii are assembled as shown in Fig. 4.9 and compressed between rigid plates at their ends. Calculate the maximum shear stress induced in each spring if the wire diameter $d = \frac{1}{2}$ in. and $P = 100$ lb. The core radii are as shown in the figure.

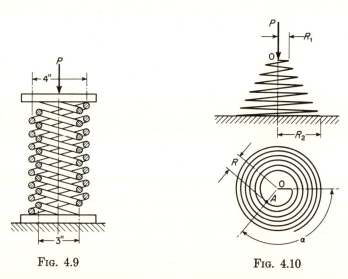

FIG. 4.9   FIG. 4.10

SOLUTION. Since the two springs must have the same over-all shortening δ, it follows from eq. (4.10) that the load P is divided between them in the inverse ratio of the cubes of their core radii. Thus denoting by P_1 the load carried by the outside spring and by P_2 that carried by the inside spring, we have

$$\frac{P_1}{P_2} = \frac{R_2^3}{R_1^3} = \frac{27}{64}.$$

Also

$$P_1 + P_2 = P = 100 \text{ lb.}$$

From these two equations, we obtain $P_1 = 29.7$ lb and $P_2 = 70.3$ lb. Substituting these values, together with the other given data, into eq. (4.9), the corresponding maximum shear stresses in the two springs are found to be $\tau_1 = 2860$ psi and $\tau_2 = 5380$ psi.

EXAMPLE 2. A conical spring as shown in Fig. 4.10 is subjected to a compressive load P. Each coil is assumed to lie essentially in a horizontal plane and the shape of the spiral in plan view is defined by the equation

$$R = R_1 + \frac{(R_2 - R_1)\alpha}{2\pi n},$$ (d)

where R is the radius at any point A on the spiral and α is the angle measured as shown. It is required to develop a formula for the spring constant k.

SOLUTION. The conditions of an element of the spring at A of length $Rd\alpha$ will be the same as for the element in Fig. 4.7c. Hence, using eq. (c),

$$\delta = \int_0^{2\pi n} \frac{PR^3}{JG} d\alpha.$$ (e)

Substituting now the value of R from eq. (d), we obtain

$$\delta = \frac{32P}{\pi d^4 G} \int_0^{2\pi n} \left[R_1 + \frac{(R_2 - R_1)\alpha}{2\pi n} \right]^3 d\alpha = \frac{16Pn}{Gd^4}(R_1{}^2 + R_2{}^2)(R_1 + R_2),$$

where J has been replaced by $\pi d^4/32$. Thus the spring constant is

$$k = \frac{P}{\delta} = \frac{Gd^4}{16n(R_1{}^2 + R_2{}^2)(R_1 + R_2)}.$$ (f)

PROBLEM SET 4.2

1. Two close-coiled helical springs are made from the same small-diameter wire, one wound on a 1-in.-diameter core and the other on a $\frac{1}{2}$-in.-diameter core. If each spring has n coils, find the ratio of their spring constants. Ans. $k_1/k_2 = \frac{1}{8}$.

2. For a close-coiled helical spring, the following numerical data are given: $R = 1$ in., $d = \frac{1}{8}$ in., $n = 30$ coils, and $G = 12(10)^6$ psi. If the allowable working stress in shear is 20,000 psi, what is the safe load P for the spring? What is the corresponding elongation δ? Ans. $P = 7.44$ lb; $\delta = 4.80$ in.

3. Assuming that the two springs shown in Fig. 4.9 are made of steel and that each one has 8 coils, calculate the spring constant k for the assembly. Ans. $k = 617$ lb/in.

4. A variable-radius coil spring is similar to that shown in Fig. 4.10 except that its plan view represents a spiral defined in polar coordinates by the equation $R = R_1 e^{a\alpha}$. Numerical data are given as follows: $R_1 = 1$ in., $a = 0.01$, $d = 0.10$ in., $G = 12(10)^6$ psi, and $n = 10$ coils. Calculate the spring constant k. Ans. $k = 0.634$ lb/in.

5. The base of an electric motor is mounted on four heavy-duty coil springs. For each spring, $R = 3$ in., $d = 1$ in., $G = 12(10)^6$ psi, and $n = 6$ coils. If the motor weighs 1000 lb, what is its maximum allowable amplitude of vertical vibration without danger of exceeding a maximum shear stress of 7500 psi in the springs? Ans. $\delta = 0.123$ in.

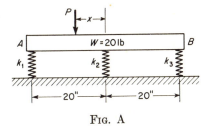

FIG. A

6. Using formula (4.9), find the safe load P for the conical spring in Fig. 4.10 if the working stress in shear is $\tau_w = 45,000$ psi. The spring is wound from a 1-in.-diameter steel rod with $R_1 = 2$ in. and $R_2 = 8$ in. *Ans.* $P_w = 1010$ lb.

7. A rigid bar AB weighing 20 lb and carrying a load $P = 80$ lb is supported by three springs having spring constants $k_1 = 100$ lb/in., $k_2 = 60$ lb/in., $k_3 = 40$ lb/in. (Fig. A). If the unloaded springs were all of the same length, find the distance x such that AB will be horizontal. *Ans.* $x = 7.5$ in.

4.3   Strain Energy in Shear and Torsion

Consider, in Fig. 4.11, an element of elastic material in a state of pure shear. The strain energy stored in such an element may be calculated in the same way as for the case of simple tension (see Art. 2.4). During deformation of the element, the top face cd moves horizontally through the distance γdy relative to the bottom face ab as the shear stress is gradually

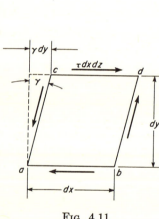

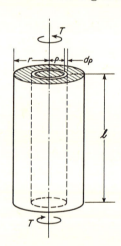

FIG. 4.11   FIG. 4.12

increasing from zero to the final value τ. Then the work done by the shear force $\tau dx\, dz$ on the top face is $\frac{1}{2}\tau dx\, dz \cdot \gamma dy$. Since the shear forces on the sides ac and bd do no work, this represents the total strain energy stored in the element. Dividing by the volume $dx\, dy\, dz$ of the element, we obtain

$$u = \frac{\gamma\tau}{2} \tag{a}$$

for the strain energy per unit volume.

Noting that, within the elastic limit of the material, $\gamma = \tau/G$, expression (a) may be written in either of the following two forms:

$$u = \frac{\tau^2}{2G} \quad \text{or} \quad u = \frac{G\gamma^2}{2}. \tag{4.12}$$

The first of these equations expresses the strain energy as a function of the shear stress τ; the second, as a function of the shear strain γ. The limit of elastic strain energy per unit volume will be obtained by setting τ in the first equation equal to the elastic limit in shear.

Having expressions (4.12) for strain energy of shear per unit volume, the total strain energy in a solid circular shaft of radius r and length l subjected to twisting moments T at its ends (Fig. 4.12) can easily be found. Denoting by $\tau_{\max}$ the maximum shear stress at the surface of the shaft, the shear stress at any intermediate radius ρ will be $\tau_{\max}(\rho/r)$. Then, from the first of eqs. (4.12), the strain energy per unit volume at the radius ρ will be

$$u = \frac{\tau_{\max}^2 \, \rho^2}{2G \, r^2}. \tag{b}$$

The energy in the elemental tube of length l, radius ρ, and thickness $d\rho$ will be

$$dU = u \, dV = \frac{\tau_{\max}^2 \, \rho^2}{2G \, r^2} \cdot l \cdot 2\pi\rho d\rho. \tag{c}$$

Summation of expression (c), from $\rho = 0$ to $\rho = r$, gives for the total strain energy in the twisted shaft,

$$U = \int_o^r \frac{\tau_{\max}^2 \rho^2}{2G \, r^2} \cdot l \cdot 2\pi\rho d\rho = \frac{1}{2}(\pi r^2 l)\frac{\tau_{\max}^2}{2G}. \tag{4.13}$$

This is seen to be just half the value that would be obtained if all the material were stressed to the maximum value $\tau_{\max}$. Noting that $\tau_{\max} = Tr/J$, where $J = \pi r^4/2$, expression (4.13) can be written also in the form

$$U = \frac{T^2 l}{2GJ}. \tag{4.13'}$$

This strain energy U for a shaft in torsion may be obtained in another way by using expression (4.4) for the angle of twist in the shaft. This shows that the relation between torque T and angle of twist ϕ is linear within the elastic limit of the material as shown by the torque-twist diagram OAB in Fig. 4.13. For any small increment $d\phi$ of the angle of twist, the work done by the acting torque is represented by the area of the shaded strip in this diagram. Thus as the torque is gradually increased from zero to any final value T, the total work, equal to the

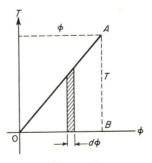

Fig. 4.13

energy stored, is represented by the area OAB in Fig. 4.13, i.e.,

$$U = \frac{T\phi}{2}, \tag{d}$$

where $\phi = Tl/GJ$, from eq. (4.4). With this relationship between T and ϕ, expression (d) may be written in either of the following two forms:

$$U = \frac{T^2 l}{2GJ} \quad \text{or} \quad U = \frac{GJ\phi^2}{2l}. \tag{4.14}$$

The first of these equations expresses the strain energy in terms of the torque T; the second, in terms of the angle of twist ϕ. Expressions (4.14) may be used either for a solid circular shaft with $J = \pi d^4/32$ or for a hollow circular shaft with $J = \pi(d_0{}^4 - d_i{}^4)/32$.

If the internal torque varies along the length of a shaft as in the case of a conical spring (Fig. 4.10), we consider one elemental disc of length dx and under twisting moment T_x. Then the angle of twist in this elemental disc is

$$d\phi = \frac{d\phi}{dx}dx, \tag{e}$$

where

$$\frac{d\phi}{dx} = \frac{T_x}{GJ} \tag{f}$$

is the angle of twist per unit thickness of the elemental disc. Substitution of $d\phi$ for ϕ and dx for l in the second of eqs. (4.14), gives, for the strain energy in the disc,

$$dU = \frac{GJ}{2}\left(\frac{d\phi}{dx}\right)^2 dx.$$

Then a summation over the full length l of the shaft gives, for the total strain energy

$$U = \frac{GJ}{2}\int_o^l \left(\frac{d\phi}{dx}\right)^2 dx. \tag{4.15}$$

EXAMPLE 1. Determine the deflection δ for the close-coiled helical spring in Fig. 4.7a, by using the expression for strain energy of torsion.

SOLUTION. The twisting moment on each and every element of the coil like that shown in Fig. 4.7c is PR. Hence, by the first of eqs. (4.14), the total strain energy in the spring of length $2\pi Rn$ is

$$U = \frac{(PR)^2 \cdot 2\pi Rn}{2GJ}.$$

Equating this to the work $P\delta/2$ of the load P during deflection, we obtain

$$\delta = \frac{2\pi n P R^3}{GJ} = \frac{64nPR^3}{Gd^4},$$

which agrees with expression (4.10) on p. 79.

EXAMPLE 2. Verify expression (e) on p. 81, for the deflection δ of the conical spring loaded as shown in Fig. 4.10, by using eq. (4.15) for strain energy of twist.

SOLUTION. At any point A on the coil (Fig. 4.10), the torque $T_\alpha = PR$, where the radius R varies with α according to eq. (d) on p. 81. Then from expression (f) above, the angle of twist per unit length of coil at A is

$$\frac{d\phi}{dx} = \frac{PR}{GJ}.$$

Substituting this into eq. (4.15), taking $dx = R\,d\alpha$ and changing the limits of integration accordingly, we obtain

$$U = \frac{GJ}{2}\int_0^{2\pi n}\left(\frac{PR}{GJ}\right)^2 R\,d\alpha.$$

Equating this to the work $P\delta/2$ of the load P, we obtain

$$\delta = \int_0^{2\pi n}\frac{PR^3}{GJ}d\alpha,$$

which agrees with eq. (e), p. 81.

EXAMPLE 3. A solid steel shaft with a flywheel at one end rotates at constant speed $n = 120$ rpm, Fig. 4.14. If the bearing A suddenly freezes, what maximum shear stress τ_{max} will be produced in the shaft due to dynamic effects? Assume $l = 5$ ft, $d = 2$ in., the weight of the flywheel $W = 100$ lb, and its radius of gyration $i = 10$ in.

SOLUTION. The maximum shear stress in the shaft will occur when the total kinetic energy of the flywheel has been absorbed by the shaft. This energy is

$$\frac{Wi^2\omega^2}{2g} = \frac{100 \times 10^2 \times (4\pi)^2}{2 \times 386} = 2050 \text{ in.-lb.}$$

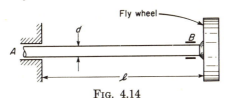

FIG. 4.14

Setting this equal to the strain energy U as given by eq. (4.13) and solving for τ_{max}, we find

$$\tau_{max} = \sqrt{\frac{4GU}{\pi r^2 l}} = \sqrt{\frac{4 \times 11.5(10)^6 \times 2050}{\pi \times 1 \times 60}} = 22,400 \text{ psi.}$$

PROBLEM SET 4.3

1. Two shafts of solid circular cross-section are identical except for their diameters d_1 and d_2. Under the same torque T, what will be the ratio of the amounts of strain energy stored in each shaft? Ans. $U_1/U_2 = (d_2/d_1)^4$.

2. A solid circular shaft and a thin-walled circular tube made of the same material and having the same weight are stressed in torsion to the same maximum shear stress τ. What is the ratio of the amounts of strain energy stored in the two shafts? Ans. $U_1/U_2 = \frac{1}{2}$.

3. For hard drawn aluminum and for cold rolled steel, the proportional limits in shear are, respectively, 10,000 psi and 36,000 psi. What are the amounts of strain energy per pound that each material can absorb in pure shear without exceeding the proportional limit stress? See Table A.1 for the densities of the materials. *Ans.* $u_a = 129$ in.-lb/lb; $u_s = 191$ in.-lb/lb.

4. Two steel shafts each of length l and outside diameter d are subjected to uniform torsion. The first shaft is solid while the second one is hollow with inside diameter $d/2$. What is the ratio of the strain energies that they can absorb without exceeding a maximum allowable shear stress τ_w? *Ans.* $U_1/U_2 = \frac{16}{15}$.

5. Referring to Fig. 4.9 and using the numerical data given there in Example 1, calculate the total strain energy in the two springs when $P = 150$ lb. Assume that each spring has 8 coils. *Ans.* 18.23 in.-lb.

6. If the shaft in Fig. 4.14 is hollow with $d_o = 2$ in. and $d_i = 1$ in., what maximum shear stress will be produced by the sudden freezing of the bearing A? *Ans.* $\tau_{max} = 23,100$ psi.

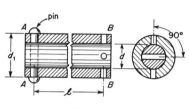

FIG. A

7. A flywheel of weight $W = 38.6$ lb and radius of gyration $i = 10$ in. is mounted at the middle of a solid steel shaft of diameter $d = 2$ in. and length l. The shaft rotates in bearings A and B at its ends with an angular speed $n = 120$ rpm. If both bearings suddenly freeze so that the ends of the shaft become locked, the shaft will have to absorb the kinetic energy of the flywheel. Calculate the shortest length l of the shaft for which this can be done without exceeding a maximum shear stress $\tau_w = 12,000$ psi in the shaft. *Ans.* $l_{min} = 7.00$ ft.

8. A solid steel shaft of diameter $d = \frac{1}{4}$ in. fits loosely inside a hollow steel shaft of inside diameter $d = \frac{1}{4}$ in. and outside diameter $d_1 = \frac{3}{8}$ in. as shown in Fig. A. A pin AA prevents relative rotation between the ends of the shafts at the left. Pinholes at the right are initially at right angles to each other as shown. The two shafts are now twisted in opposite directions until the pinholes at B line up and a pin BB is then inserted. How much strain energy will be locked in the system if $l = 100$ in.? *Ans.* $U = 45.6$ in.-lb.

4.4 Torsion of Thin-Walled Tubes

In the case of a hollow shaft of circular cross-section for which the inside diameter is very nearly equal to the outside diameter, we speak of the shaft as a *thin-walled tube*. For such a tube in torsion (Fig. 4.15), the polar moment of inertia of the cross-section cannot be calculated with good accuracy from the formula

$$J = \frac{\pi}{32}(d_o{}^4 - d_i{}^4);$$

and it is preferable to use the approximate expression

$$J = \int_A \rho^2 dA \approx r^2 \int_A dA = 2\pi r^3 t, \tag{a}$$

where r is the radius of the mean center line and t is the wall thickness. Then assuming for such a thin-walled tube that the shear stress τ is uniform across the wall and equal to the value at the mean radius r as shown in Fig. 4.15b, we obtain from eq. (4.6)

$$\tau = \frac{T}{2\pi r^2 t}. \qquad (b)$$

Likewise, from eq. (4.4), the angle of twist of the tube becomes

$$\phi = \frac{Tl}{GJ} = \frac{Tl}{2\pi r^3 t G}. \qquad (c)$$

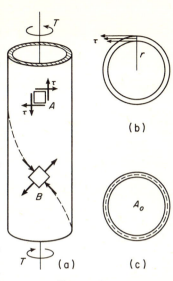

FIG. 4.15

In Art. 3.4, it was shown that the state of pure shear such as exists on the element A in Fig. 4.15a is equivalent to biaxial tension and compression on an element oriented at 45° to the axis of the tube, like the element B in Fig. 4.15a. From this, we see that a long narrow strip of the wall coinciding with the 45°-helix shown in Fig. 4.15a is subjected to axial compression, and if the wall of the tube is very thin, such a helical strip may buckle. This phenomenon can be demonstrated by rolling a sheet of paper into a tube and then subjecting it to torsion. Analysis of this problem* shows that for a long steel tube under torsion, we should have the ratio $t/r > \frac{1}{60}$ to avoid the danger of buckling at normal working stresses.

Introducing, in eqs. (b) and (c), the notations

$$A_0 = \pi r^2 = \text{area enclosed by mean center line,}$$

$$s = 2\pi r = \text{length of mean center line,}$$

we obtain

$$\tau = \frac{T}{2A_0 t}, \qquad (d)$$

and

$$\phi = \frac{\tau s l}{2A_0 G}. \qquad (e)$$

The above formulas can be used for calculating the shear stress and angle of twist in a thin-walled tube of arbitrary cross-section as shown in Fig.

*See L. H. Donnell, *Stability of Thin-Walled Tubes under Torsion*, Nat Adv, Comm. Aeronautics, Tech. Rept. 479, 1933.

4.16. When such a thin tube of noncircular cross-section is twisted, the cross-sections rotate slightly, one with respect to another, but they do not remain plane. After twist, each cross-section becomes slightly warped, and

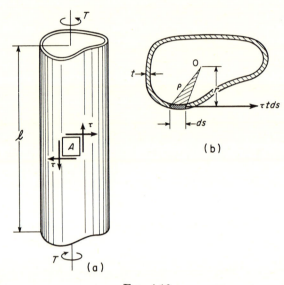

Fig. 4.16

if no restraint against this warping exists at the ends of the tube, it takes place in such a way that the shear strain γ of every element of the wall such as A has the same magnitude regardless of the radial distance ρ of that element from the centroid of the cross-section. Thus the shear stress τ, proportional to the shear strain γ, is uniform throughout the wall of the tube.

To relate this shear stress τ to the external torque T, we consider any cross-section of the tube as shown in Fig. 4.16b. For an element ds of the mean center line, the shear force is $\tau t ds$ as shown, where t is the thickness of the wall, assumed constant. The moment of this force about point O is

$$dT = \tau t ds \cdot r,$$

where r is the distance from O to the tangent to the mean center line. Summing such elemental moments over the entire length s of the mean center line, we obtain

$$T = \tau t \int_0^s r ds.$$

We see that the quantity rds under the integral sign is just double the area of the small shaded triangle of base length ds and altitude r in Fig. 4.16b.

Thus the integral of this quantity over the full length s represents double the area enclosed by the mean center line of the wall. Denoting this area by A_0, we have

$$T = \tau t \cdot 2A_0,$$

from which

$$\tau = \frac{T}{2A_0 t},$$ (4.16)

which coincides with eq. (d) above.

To calculate the angle of twist ϕ of the tube, we use the method of strain energy as discussed in Art. 4.3. From the first of eqs. (4.12) the total strain energy in the tube is

$$U = \frac{\tau^2}{2G} \times slt,$$

where slt is the volume of material in the tube and τ is the shear stress, uniform throughout the volume. Equating this strain energy to the work $T\phi/2$ of the applied torque during twist, we obtain

$$\frac{T\phi}{2} = \frac{\tau^2 slt}{2G},$$

from which

$$\phi = \frac{\tau^2}{TG} slt.$$

Since $T = 2A_0 t\tau$ from eq. (4.16), this becomes

$$\phi = \frac{\tau sl}{2A_0 G},$$ (4.17)

which agrees with eq. (e) above.

In practical problems we often encounter thin-walled tubular members of other than circular cross-section and eqs. (4.16) and (4.17) are very useful in the analysis of their behavior under torsion. The following examples will serve to illustrate several such problems.

EXAMPLE 1. Two thin-walled tubular members made of the same material have the same length, the same wall thickness, and the same total weight and are subjected to the same torque T. If their cross-sections are circular and square, respec-

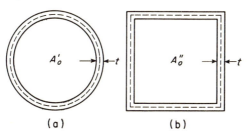

(a)   (b)

Fig. 4.17

tively, as shown in Fig. 4.17, what are the ratios of their shear stresses τ and of their angles of twist ϕ?

SOLUTION. From eq. (4.16), the ratio of shear stresses is

$$\frac{\tau'}{\tau''} = \frac{A''_0}{A'_0}, \tag{f}$$

where A'_0 and A''_0 are the areas enclosed by the mean center lines of the same length s (Fig. 4.17). For the circle $A'_0 = s^2/4\pi$ while for the square, $A''_0 = s^2/16$. Substituting these values into eq. (f), we find

$$\frac{\tau'}{\tau''} = \frac{4\pi}{16} = \frac{\pi}{4} = 0.785.$$

From eq. (4.17), the ratio of angles of twist is

$$\frac{\phi'}{\phi''} = \frac{\tau'A''_0}{\tau''A'_0} = \left(\frac{\pi}{4}\right)^2 = 0.616.$$

We conclude from this that the circular cross-section represents a more efficient use of the material.

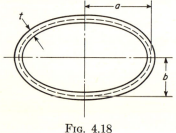

FIG. 4.18

EXAMPLE 2. A stainless steel tube having an elliptical cross-section as shown in Fig. 4.18 is subjected to torsion. If the allowable stress in shear is $\tau_w = 10,000$ psi, what is the corresponding allowable angle of twist per unit length of the tube? The following data are given: $G = 12(10)^6$ psi, $a = 3.00$ in., $b = 2.00$ in., $t = \frac{1}{8}$ in.

SOLUTION. From eq. (4.17), the angle of twist per unit length of tube will be

$$\theta = \frac{\phi}{l} = \frac{\tau s}{2A_0 G}.$$

The area enclosed by the mean center line is $A_0 = \pi ab = 6\pi$ sq in. The length s of the mean center line may be calculated with good accuracy from the approximate formula

$$s \approx \pi \left[\frac{3}{2}(a + b) - \sqrt{ab}\right],$$

which gives $s = 15.86$ in. Then the above expression for θ becomes

$$\theta = \frac{10,000 \times 15.86}{2 \times 6\pi \times 12(10)^6} = 0.000350 \text{ radian/in.}$$

PROBLEM SET 4.4

1. A tubular aluminum shaft having a square cross-section with outside dimension $a = 1$ in. must safely carry a twisting moment $T = 636$ in.-lb. Calculate the proper wall thickness t if the working stress is $\tau_w = 6000$ psi. *Ans.* $t = 0.060$ in.

2. For the aluminum tube described in the preceding problem, calculate the angle of twist per unit length of tube under the applied torque $T = 636$ in.-lb. The shear modulus $G = 4(10)^6$ psi. *Ans.* $\theta = 0.00319$ radian/in.

3. A thin-walled stainless steel tube has the rectangular cross-section shown in Fig. A. How does the angle of twist per unit length of tube due to torsion vary

with the ratio $\alpha = a/b$ if the total length of mean centerline $s = 2(a + b)$ and the applied torque T remain constant? *Ans.* θ varies as $(1 + \alpha)^4/\alpha^2$.

4. Using the result of Prob. 3, prove that the rectangular tube shown in Fig. A will have maximum torsional stiffness when $\alpha = 1$, i.e., when $a = b$.

5. A tubular shaft with circular cross section has a mean diameter $d = 4$ in. and a wall thickness $t = \frac{1}{8}$ in. This shaft must transmit 50 hp without exceeding an assigned working stress in shear of $\tau_w = 6000$ psi. What is the required rpm of the shaft? *Ans.* 167 rpm.

6. The cross-section of a thin-walled steel tube 6 ft long has the form of a hexagon with mean centerline length $s = 6a = 6$ in. and wall thickness $t = \frac{1}{16}$ in. If $\tau_w = 8000$ psi,

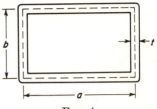

FIG. A

what is the safe twisting moment T for the tube? What is the corresponding angle of twist ϕ between the ends of the tube? *Ans.* $T = 2600$ in.-lb; $\phi = 3°10'$.

7. A slightly tapered thin-walled tube of length l has a square cross-section with mean centerline dimension a at one end and $2a$ at the other. The wall thickness is t and the applied twisting moment T. Derive an expression for the angle of twist ϕ between the two ends of the tube. *Ans.* $\phi = 3Tl/8Gta^3$.

8. The cross-section of a thin-walled tube of length l is a regular polygon with n sides. The mean centerline length is $s = na$ and the wall thickness is t. Find the angle of twist ϕ due to uniform torque T. *Ans.* $\phi = 4Tl/nGta^3\cot^2(\pi/n)$.

4.5 Shaft of Rectangular or Profile Section

The problem of twist of a shaft of rectangular cross-section is complicated, due to the warping of cross-sections during twist. This warping can be shown experimentally with a rectangular bar of rubber on whose faces a system of small squares has been traced. It is seen from Fig. 4.19* that during twist the lines originally perpendicular to the axis of the bar become curved. This indicates that the distortion of the small squares, mentioned above, varies along the sides of this cross-section, reaches a maximum value at the middle, and disappears at the corners. We therefore expect that the shearing stress will vary as this distortion: namely, it is a maximum at the middle of the sides and zero at the corners of the cross-section. Investigation of the problem† indicates that the maximum shearing stress occurs at the middle of the longer sides of the rectangular cross-section and is given by the equation

FIG. 4.19

$$\tau_{\max} = \frac{T}{\alpha bc^2},$$

(4.18)

*This figure is reproduced from Bach's *Elastizität und Festigkeit*.

†The complete solution is due to de Saint Venant, *Mém des Savants étrangers*, t. 14 (1855).

in which b is the longer and c the shorter side of the rectangular cross-section and α is a numerical factor depending upon the ratio b/c. Several values of α are given in Table 4.1.

The angle of twist per unit length in the case of a rectangular cross-section is given by the equation

$$\theta = \frac{T}{\beta b c^3 G}. \tag{4.19}$$

The values of the numerical factor β are given in the third column of Table 4.1.

Table 4.1   DATA FOR THE TWIST OF A
SHAFT OF RECTANGULAR CROSS-SECTION

b/c	α	β
1.00	.208	.141
1.50	.231	.196
1.75	.239	.214
2.00	.246	.229
2.50	.258	.249
3.00	.267	.263
4.00	.282	.281
6.00	.299	.299
8.00	.307	.307
10.00	.313	.313
∞	.333	.333

It is seen that in the case of a very narrow rectangular cross-section, such as that for a thin strip of sheet metal, α and β equal $\frac{1}{3}$ and the equations for the maximum shearing stress and the angle of twist per unit of length become

$$\tau_{\max} = \frac{3T}{bc^2}, \tag{4.20}$$

$$\theta = \frac{3T}{bc^3 G}. \tag{4.21}$$

These equations are of practical importance because they can be used not only for a narrow rectangle but also for approximate solutions in other cases in which the width of the cross-section is small. For instance, in the case of the cross-sections of uniform thickness shown in Figs. 4.20a and b, the angle of twist is obtained from eq. (4.21) by putting in this equation for b the developed length of the center line, namely, $b = \phi r$ in the case of the section represented in Fig. 4.20a, and $b = 2a - c$ in the case represented in Fig. 4.20b. The maximum stress for the first of these two sections will be obtained from eq. (4.20). For the angle section (Fig. 4.20b) the maximum

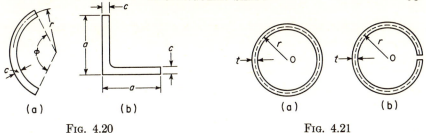

FIG. 4.20   FIG. 4.21

stress is at the reentrant corner.* This maximum stress is obtained by multiplying the stress given by eq. (4.20) by a stress concentration factor. This factor has been found to vary with the ratio of fillet radius r to thickness c: values of this factor are given in Table 4.2.

Table 4.2 STRESS CONCENTRATION FACTOR

r/c	Factor
$\frac{1}{8}$	$2\frac{1}{2}$
$\frac{1}{4}$	$2\frac{1}{4}$
$\frac{1}{2}$	2
1	$1\frac{3}{4}$

EXAMPLE 1. Two thin-walled tubes of circular cross-section are identical except that one is seamless and the other is split as shown in Fig. 4.21. If r is the radius of the mean center line and t is the wall thickness, find the ratio of their angles of twist θ per unit length of tube when they are subjected to the same torque T.

SOLUTION. For the seamless tube, we obtain from eqs. (4.16) and (4.17) of Art. 4.4,

$$\theta_1 = \frac{Ts}{4A_0^2Gt} = \frac{T}{2\pi r^3 Gt}. \tag{a}$$

Using eq. (4.21) for the split tube where $b = 2\pi r$ and $c = t$, we obtain

$$\theta_2 = \frac{3T}{2\pi rGt^3}. \tag{b}$$

Thus

$$\frac{\theta_1}{\theta_2} = \frac{1}{3}\left(\frac{t}{r}\right)^2. \tag{c}$$

Taking, for example, $t/r = 1/10$, this gives $\theta_1/\theta_2 = 1/300$, i.e., the seamless tube is 300 times as stiff in torsion as the corresponding split tube.

EXAMPLE 2. For the structural angle section shown in Fig. 4.20b, the following dimensions are given: $a = 4$ in., $c = \frac{1}{2}$ in. A steel bar having this cross-section is 4 ft long and is subjected to twisting moments $T = 2500$ in.-lb at its ends. Compute the maximum shear stress τ and the angle of twist ϕ between the two ends of the bar if the fillet radius $r = \frac{1}{4}$ in.

*The above methods apply where the cross-sections of the structural shapes are free to warp.

SOLUTION. Taking $b = 2a - c = 7.5$ in. in eq. (4.20),

$$\tau_{max} = \frac{3 \times 2500}{7.5 \times (\frac{1}{2})^2} = 4000 \text{ psi.}$$

Then with $r/c = \frac{1}{4} \div \frac{1}{2} = \frac{1}{2}$, the stress concentration factor is 2 from Table 4.2 above and the maximum shear stress at the reentrant corner becomes $\tau = 2 \times 4000 = 8000$ psi.

The total angle of twist from eq. (4.21) becomes

$$\phi = \theta l = \frac{3 \times 2500 \times 48}{7.5 \times (\frac{1}{2})^3 \times 12(10)^6} = 0.032 \text{ radian.}$$

PROBLEM SET 4.5

1. The aluminum channel section shown in Fig. A has the following dimension: $a = 1$ in., $h = 3$ in., $t = \frac{1}{16}$ in. The fillet radius at the two reentrant corners is $r = \frac{1}{32}$ in. Calculate the safe twisting moment for the channel if the working stress in shear is 8000 psi. What is the corresponding angle of twist θ per unit length of channel if $G = 4(10)^6$ psi? *Ans.* $T = 25.4$ in.-lb; $\theta = 0.0160$ rad/in.

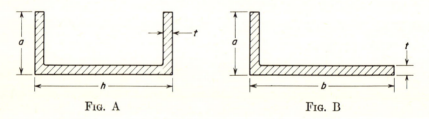

FIG. A   FIG. B

2. The structural steel angle section shown in Fig. B has the following dimensions: $a = 4$ in., $b = 9$ in., $t = \frac{1}{2}$ in. The fillet radius at the reentrant corner is $r = \frac{1}{2}$ in. Calculate the safe twisting moment for the section if the working stress in shear is 12,000 psi. What is the corresponding angle of twist θ per unit length of bar if $G = 12(10)^6$ psi? *Ans.* $T = 7140$ in.-lb; $\theta = 0.00114$ rad/in.

3. A prismatic steel bar having a 1 in. $\times$ 1 in. square cross-section is subjected to twisting moments $T = 2000$ in.-lb at its ends. Calculate the maximum shear stress τ. What is the corresponding angle of twist θ per unit length of the bar? *Ans.* $\tau_{max} = 9620$ psi; $\theta = 0.00118$ rad/in.

4. A 6-ft length of 0.8% carbon steel strap 1 in. wide by $\frac{1}{16}$ in. thick is used for a door spring as shown in Fig. C. Calculate the pull P on the door handle required to hold the door open at 90° to the plane of the opening if the handle is 32 in. from the hinge line. The steel strap was pretwisted during installation so as to sustain a torque $T_0 = 20$ in.-lb when the door is shut. What is the maximum shear stress in the strap when the door is fully open? *Ans.* $P = 1.29$ lb; $\tau_{max} = 31,700$ psi.

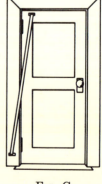

FIG. C

5

Stresses in Beams: I

5.1 Shearing Force and Bending Moment

A structural member that is reasonably long compared with its lateral dimensions when suitably supported, and subjected to transverse forces so applied as to induce *bending* of the member in an axial plane, is called a *beam*. Several examples are shown in Fig. 5.1. The beam in Fig. 5.1a, supported by a pin at A and a roller at B is called a *simple beam*. The one in Fig. 5.1b, built into a wall at B and free at A, is called a *cantilever beam*. Since, in both of these cases, the conditions of support are such that the reactions can be found from equations of statics, these beams are said to be *statically determinate*. On the other hand, the beam in Fig. 5.1c, supported by a pin at A and rollers at both C and B, is *statically indeterminate*, since the reactions cannot be found from equations of statics alone. For

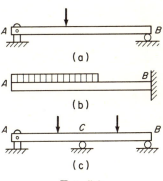

FIG. 5.1

the present, we confine our attention to statically determinate beams.*

Consider now a cantilever beam AB subjected to external loads P_1 and P_2 as shown in Fig. 5.2a. If we imagine this beam to be cut by a section mn, we see that the applied forces tend to displace the left-hand portion of the beam relative to the right-hand portion, which is anchored in the wall. This tendency is resisted by internal forces between the two parts of the beam. Thus if we isolate that portion of the beam to the left of the section mn as a free body, we represent the action of the built-in portion theron by distributed forces as shown in Fig. 5.2b. The true distribution of these internal forces on the section mn is complicated, but to maintain equilibrium of the free body, they must be statically equivalent to the equilibrant

*Methods of dealing with statically indeterminate beams will be discussed in Chapter 9.

of the applied external forces P_1 and P_2. This *stress resultant* on the section *mn* can always be represented by a force applied at the centroid of the cross-section and a couple, both in the axial plane of the applied loads. Furthermore, the force can, in turn, be resolved into rectangular components N_x, normal to the plane of the section, and V_x, lying in the plane of the section. Thus, in Fig. 5.2c, we represent the stress resultant on any

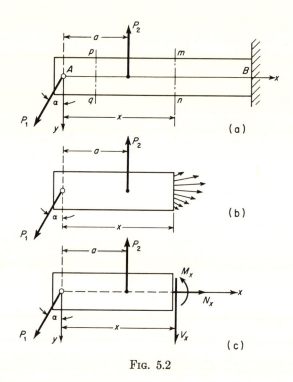

FIG. 5.2

section *mn* by the three quantities N_x, V_x, and M_x, called, respectively, the *normal force*, the *shear force*, and the *bending moment*, at that section. These quantities will be considered positive when they have the directions shown in Fig. 5.2c. Such sign conventions, although arbitrary, must be carefully observed to avoid confusion.

To pursue this question further, let us consider an element of the beam between two adjacent cross-sections as shown in Fig. 5.3a. Then normal forces N, shear forces V, and bending moments M on the two faces of this element will be considered as positive when directed as shown. We see that a positive normal force is directed away from that face of the free body on which it acts. Also a positive shear force is one that has a clockwise sense of rotation about a point inside the free body. Finally the bending moments

M are positive when they tend to bend the element concave upwards. Negative directions of N, V, and M are shown in Fig. 5.3b.

Using the three equations of equilibrium

$$\Sigma X_i = 0, \qquad \Sigma Y_i = 0, \qquad \Sigma M_i = 0 \qquad \text{(a)}$$

for the free body in Fig. 5.2c, the normal force N_x, the shear force V_x, and the bending moment M_x at any cross-section mn can readily be calculated. Using the centroid of the section mn as a moment center, these equations give

$$\left.\begin{aligned}
N_x &= P_1 \sin \alpha, \\
V_x &= P_2 - P_1 \cos \alpha, \\
M_x &= P_2(x - a) - P_1 \cos \alpha \cdot x.
\end{aligned}\right\} \qquad \text{(b)}$$

These expressions, of course, hold only for $a < x < l$. For $0 < x < a$, we consider the equilibrium conditions of that portion of the beam to the left of a section pq (Fig. 5.2a) and obtain

$$\left.\begin{aligned}
N_x &= P_1 \sin \alpha, \\
V_x &= -P_1 \cos \alpha, \\
M_x &= -P_1 \cos \alpha \cdot x.
\end{aligned}\right\} \qquad \text{(c)}$$

It should be noted from expression (b) and (c) that, in general, N_x, V_x, and M_x vary with the distance x defining the cross-section at which they occur.

In most practical problems, we encounter beams which are horizontal and are subjected only to vertical loads. In such cases, there will be no normal force N_x on any cross-section and we have to deal only with shear force V_x and bending moment M_x. The relationship between these two quantities is of special importance and will be established in the discussion that follows.

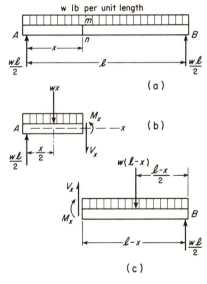

(a)

(b)

(c)

Fig. 5.4

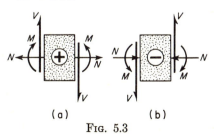

(a)   (b)

Fig. 5.3

As another example, consider the simple beam AB carrying a uniformly distributed transverse load of intensity w as shown in Fig. 5.4a. Considering the entire beam as a free body the reactions at A and B are found to be $wl/2$ each, as shown.

Then to evaluate the shearing force and bending moment at a chosen section mn, we consider the equilibrium of that portion of the beam to the left of this section (Fig. 5.4b). Acting on this free body, we have the reaction $wl/2$ at A and that part of the distributed load between A and the section mn. The resultant of this portion of the distributed load will be a vertical force wx acting at the distance $x/2$ from A as shown. On the face of the section boundary, we have a shear force V_x and a bending moment M_x, but in this case, since none of the external forces has any horizontal component, there will be no normal force N_x. Now using eqs. (a), we obtain

$$\left. \begin{aligned} V_x &= \frac{wl}{2} - wx, \\[2mm] M_x &= \frac{wl}{2}x - \frac{wx^2}{2}. \end{aligned} \right\} \tag{d}$$

If we consider the equilibrium of that portion of the beam to the right of the section mn(Fig. 5.4c), we obtain

$$\left. \begin{aligned} V_x &= w(l - x) - \frac{wl}{2}, \\[2mm] M_x &= \frac{wl}{2}(l - x) - \frac{w}{2}(l - x)^2. \end{aligned} \right\} \tag{e}$$

These expressions can readily be reduced to coincide with expressions (d) above. Since the entire beam is in equilibrium, we must obtain the same values for shear force and bending moment at a chosen section, by using the equilibrium conditions for either portion of the beam.

A significant relationship between shearing force and bending moment at any cross-section of a transversely loaded beam will now be shown. In Fig. 5.5a, consider an element of the beam cut out by two adjacent cross-sections mn and pq, distance dx apart. On the left-hand face of this element, we represent the shear force and bending moment by V_x and M_x (assumed positive). Then if no external load is applied between the cross-sections mn and pq, the shear force and bending moment on the right-hand face of the element will be V_x and $M_x + dM_x$ as shown, where dM_x is the change in bending moment between mn and pq. Since the element is in equilibrium, the algebraic sum of moments of these forces about point 0 must be zero and we find

$$-M_x + (M_x + dM_x) - V_x dx = 0,$$

from which

$$\frac{dM_x}{dx} = V_x.$$ (5.1)

Thus at any cross-section of the beam between points of application of concentrated loads, the rate of change of bending moment with respect to x is equal to the shear force.

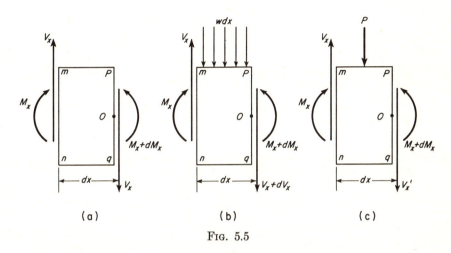

(a)   (b)   (c)

FIG. 5.5

If there is some distributed load of intensity w between mn and pq, the free-body diagram of the element will be as shown in Fig. 5.5b. Here again, we equate to zero the algebraic sum of moments of all forces with respect to point 0 and obtain

$$-M_x + (M_x + dM_x) - V_x\,dx + wdx\left(\frac{dx}{2}\right) = 0.$$

Neglecting the last term as a small quantity of second order in this expression, it reduces to

$$dM_x - V_x\,dx = 0,$$

which yields the same relationship (5.1) obtained above.

Equating to zero the algebraic sum of vertical forces on the element in Fig. 5.5b, we find

$$V_x - (V_x + dV_x) - wdx = 0,$$

from which

$$\frac{dV_x}{dx} = -w.$$ (5.2)

Thus when there is distributed load of intensity w between the cross-sections mn and pq, the shear force also changes along the beam and its rate of change with respect to x is equal to the intensity of load but with opposite sign.

If there is a concentrated load P on the beam between the cross-sections mn and pq, the free-body diagram of the element will be as shown in Fig. 5.5c. In this case, we denote the shear force on the left-hand face of the element by V_x and that on the right-hand face, by V'_x. Then equating to zero the algebraic sum of vertical forces on the element, we obtain

$$V'_x = V_x - P.$$

Thus in this case there is an abrupt change in the shear force over the length dx. Accordingly, it may be concluded from eq. (5.1) that there will be a corresponding discontinuity in the derivative dM_x/dx at the point of application of a concentrated load P on the beam.

EXAMPLE 1. A simple beam AB carries a triangular distribution of transverse load as shown in Fig. 5.6, the maximum intensity of load at B being w_0. At what cross-section (defined by x) does the maximum bending moment occur and what is its magnitude?

SOLUTION. The resultant load on the beam is $w_0l/2$ acting at a point C distance $l/3$ to the left of B. Thus, by $\Sigma M_B = 0$, for the entire beam as a free body,

$$R_A l - (w_0 l/2)(l/3) = 0,$$

from which
$$R_A = \frac{w_0 l}{6}.$$

The free-body diagram for a portion of the beam to the left of a section mn is shown in Fig. 5.6b. The resultant of that part of the distributed load which acts on this free body is $w_0 x^2/2l$ applied at point D, distance $x/3$ to the left of the section mn. Positive shear force V_x and bending moment M_x act on the section boundary as shown. From the equilibrium conditions for this free body, we find

$$\left. \begin{aligned} V_x &= \frac{w_0 l}{6} - \frac{w_0 x^2}{2l}, \\ M_x &= \frac{w_0 l x}{6} - \frac{w_0 x^3}{6l} \end{aligned} \right\} \tag{f}$$

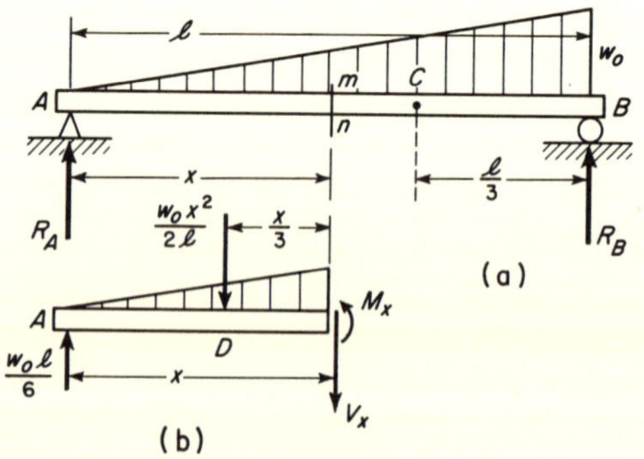

FIG. 5.6

It will be noted that these expressions satisfy the relationship expressed by eq. (5.1), i.e.,

$$\frac{dM_x}{dx} = \frac{w_0 l}{6} - \frac{w_0 x^2}{2l} = V_x.$$

Setting this expression for $dM_x/dx = 0$, we find $x = l/\sqrt{3}$, defining the location of that section for which M_x has its maximum value. Finally, substituting this value of x into the second of eqs. (f), we obtain

$$M_{max} = \frac{w_0 l^2}{9\sqrt{3}}. \tag{g}$$

EXAMPLE 2. A simply supported beam AB carries a continuously distributed load the intensity of which at any point x along the beam is $w_x = w_0 \sin(\pi x/l)$, as shown in Fig. 5.7. Find the maximum bending moment in the beam.

SOLUTION. In a case of this kind, the direct method of writing expressions for V_x and M_x, as done in Example 1, becomes cumbersome because of the difficulty in

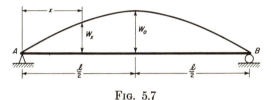

FIG. 5.7

defining the resultant load to the left of any cross-section x. Instead, we can proceed by using eqs. (5.1) and (5.2). Beginning with eq. (5.2), we have

$$\frac{dV_x}{dx} = -w_0 \sin \frac{\pi x}{l}. \tag{h}$$

Separating variables and integrating this, we obtain

$$V_x = \frac{w_0 l}{\pi} \cos \frac{\pi x}{l} + C_1, \tag{i}$$

where C_1 is a constant of integration. Now using eq. (5.1), we have

$$\frac{dM_x}{dx} = \frac{w_0 l}{\pi} \cos \frac{\pi x}{l} + C_1.$$

Again separating variables and integrating, we obtain

$$M_x = \frac{w_0 l^2}{\pi^2} \sin \frac{\pi x}{l} + C_1 x + C_2, \tag{j}$$

where C_2 is a second constant of integration.

To evaluate the constants C_1 and C_2, we note the following boundary conditions at the ends of the beam, namely,

$$\left. \begin{array}{ll} \text{when } x = 0, & M_x = 0 \\ \text{when } x = l, & M_x = 0 \end{array} \right\} \tag{k}$$

Substituting these boundary conditions into eq. (j), we find in this case that $C_1 = C_2 = 0$. Thus eq. (j) becomes

$$M_x = \frac{w_0 l^2}{\pi^2} \sin \frac{\pi x}{l}.$$ (l)

It is easy to see that this expression has its maximum value when $x = l/2$, and we obtain

$$M_{\max} = \frac{w_0 l^2}{\pi^2}$$

for the maximum bending moment at the middle of the beam.

PROBLEM SET 5.1

1. For the simple beam in Fig. A, evaluate the shear force and bending moment at a section just to the left of the point of application of the 4000-lb load. *Ans.*

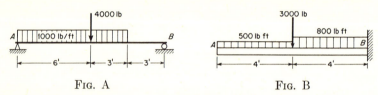

FIG. A   FIG. B

$V = 1625$ lb; $M = 27,750$ ft-lb.

2. Calculate the shear force V_B and the bending moment M_B at the section just to the left of the wall at B, for the cantilever beam in Fig. B. *Ans.* $V_B = -8200$ lb; $M_B = -30,400$ ft-lb.

3. Calculate the bending moments at cross-sections C and D of the beam shown in Fig. C. *Ans.* $M_C = 0$; $M_D = -15,000$ ft-lb.

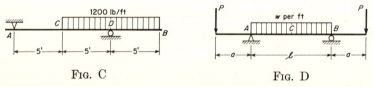

FIG. C   FIG. D

4. A simply supported beam with overhanging ends carries transverse loads as shown in Fig. D. If $wl = P$, what is the ratio a/l for which the bending moment M_c at the middle of the beam will be zero? *Ans.* $a/l = \frac{1}{8}$.

5. Referring to the simply supported beam shown in Fig. A, assume that the distributed load is replaced by its resultant as a concentrated load. What effect will this have on the reactions at A and B? What will be the shear force and the bending moment at a section just to the left of the 4000-lb load under these new conditions? *Ans.* $V = -1380$ lb; $M = 32,200$ ft-lb.

6. Referring to the simply supported beam with overhanging ends loaded as shown in Fig. D, assume that the distributed load $wl = P$ is actually replaced by a concentrated load P at C. Under these conditions, what is the ratio a/l for which the bending moment at the middle of the beam will be zero? *Ans.* $a/l = \frac{1}{4}$.

7. Referring to the simply supported beam loaded as shown in Fig. 5.6, derive general expressions for V_x and M_x by direct integration of eqs. (5.2) and (5.1). Note that the load intensity will be w_0x/l. *Ans.* See eqs. (f), page 100.

8. A simply supported beam carries a linearly varying transverse load as shown in Fig. E. The intensity of load at each end of the beam is w_0. Using eqs. (5.2) and (5.1), develop general expressions for V_x and M_x at a cross-section distance x from support A. At what value of x will the bending moment be a maximum? What is the shear force at the middle cross-section of the beam? *Ans.* max M_x at $x = 0.2114l$; $V_m = -w_0l/12$.

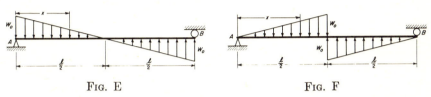

FIG. E   FIG. F

9. A simply supported beam carries linearly varying transverse load as shown in Fig. F. At what distance x from A will the shear force V_x vanish? What is the value of the shear force at the middle cross-section of the beam? *Ans.* $V_x = 0$ at $x = l/(2\sqrt{3})$; $V_m = \pm w_0l/6$.

10. Derive general expressions for normal force N_x, shear force V_x, and bending moment M_x at the cross-section mn of the beam shown in Fig. G. *Ans.* $N_x = +P$; $V_x = -0.2P$; $M_x = -0.2Px$.

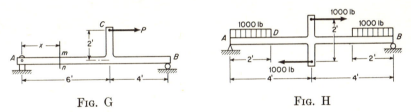

FIG. G   FIG. H

11. For the beam supported and loaded as shown in Fig. H, calculate the shear force V_D and the bending moment M_D at the cross-section D. *Ans.* $V_D = -250$ lb; $M_D = +500$ ft-lb.

12. Referring to the simply supported beam in Fig. 5.7, assume that the intensity of the transverse load varies according to a parabolic law. In such case, what is the bending moment at the middle of the beam? *Ans.* $M_c = 5w_0l^2/48$.

13. A simply supported beam is subjected to a uniformly distributed transverse load of intensity w over one-half of its span l. Calculate the maximum bending moment in the beam. *Ans.* $M_{\max} = 9wl^2/128$.

5.2 Shear and Bending Moment Diagrams

We have seen in the preceding article that the shear force V_x and bending moment M_x in a transversely loaded beam will, in general, vary with the distance x defining the location of the cross-section on which they occur. For this reason, it is often advantageous to use a graphical representation of the variation of these quantities along the axis of the beam. To do this, we let the abscissa indicate the position of the section, and the ordinate the corresponding value of shear force or bending moment. Such graphical representations are called *shearing force* and *bending moment diagrams.*

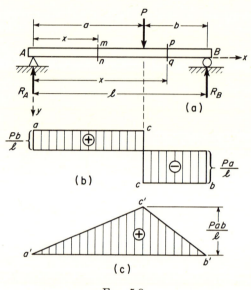

FIG. 5.8

To illustrate, let us consider as an example the case of a simple beam AB carrying a single concentrated transverse load P (Fig. 5.8a). From equilibrium considerations of the entire beam, the reactions are found to be

$$R_A = \frac{Pb}{l}, \quad R_B = \frac{Pa}{l}.$$

Then for any cross-section mn to the left of P, i.e., for $0 < x < a$, it can be concluded from the equilibrium of that portion of the beam between A and mn that

$$V_x = +\frac{Pb}{l}, \quad M_x = +\frac{Pb}{l}x. \tag{a}$$

From these expressions, we see that the shear force is constant between A and the point of application of the load P, while the bending moment varies linearly with x. For $x = 0$, the bending moment is zero and for $x = a$ it is Pab/l. The corresponding diagrams are shown by the straight lines ac and $a'c'$ in Figs. 5.8b and c.

For a cross-section pq to the right of the load P, i.e., for $a < x < l$, we obtain in the same way

$$
\left.
\begin{aligned}
V_x &= \frac{Pb}{l} - P = \frac{P(b - l)}{l} = -\frac{Pa}{l}, \\
M_x &= \frac{Pb}{l}x - P(x - a) = +Pa\left(1 - \frac{x}{l}\right).
\end{aligned}
\right\}
\tag{b}
$$

Thus again, the shear force is constant but of negative sign, while the bending moment varies linearly with x. At $x = l$, its value is zero while at $x = a$, it becomes Pab/l. The corresponding diagrams of shear force and bending moment for this part of the beam are shown by the straight lines cb and $c'b'$ in Figs. 5.8b and c.

Viewing the complete diagrams of shear force and bending moment, we see that the point of application of the load P there is an abrupt change in the shear force from $+Pb/l$ to $-Pa/l$. Correspondingly, there is a sharp discontinuity in the slope of the bending moment curve $a'c'b'$.

If several transverse forces act on a simple beam as shown in Fig. 5.9a, the same reasoning as above shows that between transverse loads, the shear force remains constant and the bending moment varies linearly. Thus the shear force diagram will have the general form shown in Fig. 5.9b, while the

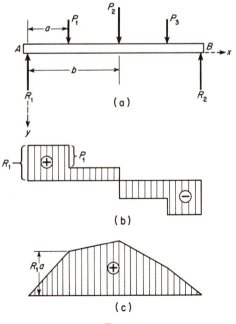

(a)

(b)

(c)

Fig. 5.9

bending moment diagram will be of the form shown in Fig. 5.9c. It should be noted that corresponding to each abrupt change in the magnitude of shear force there is a discontinuity in the slope dM_x/dx of the bending moment diagram.

The actual construction of the diagrams in Fig. 5.9, for given numerical values of the loads P_1, P_2, P_3 and distances between their points of application, is quite simple. First the reactions R_1 and R_2 will be calculated. Then on computing values of shear force and bending moment at only three cross-sections — say just to the left of each applied load — the diagrams can be constructed simply by connecting these key points by straight lines as shown. Finally it should be noted that the maximum ordinate of the bending moment diagram (under the load P_2) occurs at that section where the shear force changes sign, since here also the sign of dM/dx changes accordingly. If, going along the x-axis, the shear force changes from a positive to a negative value at a certain cross-section (as in Fig. 5.9b), we have a maximum bending moment at this section. A change in the shear force from a negative to a positive value, conversely, will indicate a minimum bending moment. In general, several changes in sign of shear force may occur along the length of the beam. If so, there will be several maxima or minima in the bending moment and each one must be investigated to ascertain the numerically largest bending moment in the beam.

Let us consider now a simple beam AB carrying a uniform distribution of transverse load of intensity w as shown in Fig. 5.10a. In this case the reactions are each equal to $wl/2$. Then at any section distance x from the left end A, we have

$$\left. \begin{aligned} V_x &= \frac{wl}{2} - wx, \\ M_x &= \frac{wlx}{2} - \frac{wx^2}{2} \end{aligned} \right\} \tag{c}$$

(see eqs. (d), p. 98). From the first of these expressions, we see that the shear force V_x has the positive value $wl/2$ at $x = 0$ and the negative value $- wl/2$ at $x = l$; furthermore, V_x varies linearly with x. Hence we may construct the shear force diagram in Fig. 5.10b simply by drawing the straight line ab as shown.

From the second of expressions (c), we see that the bending moment M_x varies quadratically with x so that the corresponding curve is a parabola. Its ordinate is zero for $x = 0$ and also for $x = l$. It has its maximum ordinate at $x = l/2$, which is $M_{max} = wl^2/8$. Thus drawing a parabola through the points a', c', b', in Fig. 5.10c, we obtain the complete bending moment diagram as shown. Again, it will be noted that the maximum

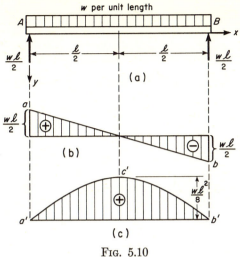

Fig. 5.10

bending moment occurs at the section for which $V_x = dM_x/dx = 0$, in accordance with eq. (5.1).

In Fig. 5.11a, we consider the case of a cantilever beam carrying a uniformly distributed load between A and C as shown. Following the sign conventions of Fig. 5.3, we see that in this case both the shear force and the bending moment are negative throughout the beam. Specifically, for a section mn between A and C,

$$\left.\begin{aligned} V_x &= -wx, \\ M_x &= -\frac{wx^2}{2}, \end{aligned}\right\} \tag{d}$$

while for a section pq between C and B,

$$\left.\begin{aligned} V_x &= -\frac{wl}{2}, \\ M_x &= -\frac{wl}{2}\left(x - \frac{l}{4}\right). \end{aligned}\right\} \tag{e}$$

From the first of expressions (d), we construct the straight line ac in Fig. 5.11b and from the first of expressions (e), the straight line cb. This completes the shear force diagram as shown. From the second of expressions (d), we construct the parabola $q'c'$ in Fig. 5.11c and from the second of expressions (e), the straight line $c'b'$. This completes the bending moment diagram as shown. Since there is no abrupt change in the shear force at section C, there is correspondingly no discontinuity in slope dM/dx of the bending moment curve at this section. Thus in Fig. 5.11c, the straight line $c'b'$ is tangent to the parabola $a'c'$ at c'.

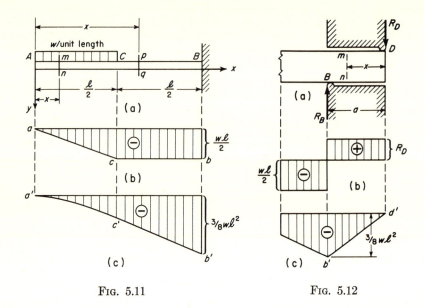

FIG. 5.11   FIG. 5.12

We note in Fig. 5.11c, that the numerically largest bending moment $M_B = -3wl^2/8$ occurs at the built-in end of the beam, but that there appears to be no corresponding change of sign in the shear force at this section. To clarify this situation, we must consider that portion of the beam which is encased in the wall (see Fig. 5.12). Assuming that the beam bears on the wall only at points B and D, we will have reactions

$$R_B = \frac{wl}{2} + \frac{3wl^2}{8a} \quad \text{and} \quad R_D = \frac{3wl^2}{8a}, \tag{f}$$

directed as shown in Fig. 5.12a. Then for any cross-section mn between D and B, we have

$$\left.\begin{aligned} V_x &= +R_D = +\frac{3wl^2}{8a}, \\ M_x &= -R_D x = -\frac{3wl^2 x}{8a}, \end{aligned}\right\} \tag{g}$$

where x is measured to the left from D, as shown. From these expressions, we may now construct diagrams showing shear force and bending moment variation along the encased portion BD of the beam (Figs. 5.12b and c). These diagrams show that there is a change in the sign of the shear force from negative to positive at B which confirms the condition of a minimum, i.e., a negative maximum of M_x at B.

It will be shown in the next article that we may expect the highest bending stresses in a beam at the section where the bending moment is numerically a maximum; therefore, the location of such sections is of practical importance. They are most easily identified by a change in sign of the shear force which can be easily detected from a study of the shear force diagram. However, there are many other useful and significant applications of both shear force and bending moment diagrams, as will be seen later.

EXAMPLE 1. Construct shear force and bending moment diagrams for the cantilever beam loaded as shown in Fig. 5.13a.

SOLUTION. In the region $0 < x < a$, we have

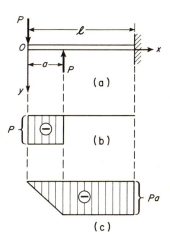

$$V_x = -P; \quad M_x = -Px. \qquad \text{(h)}$$

In the region $a < x < l$, we have

$$\left. \begin{array}{l} V_x = -P + P = 0, \\ M_x = -Px + P(x - a) = -Pa. \end{array} \right\} \qquad \text{(i)}$$

The corresponding diagrams are shown in Figs. 5.13b and c.

EXAMPLE 2. Construct shear force and bending moment diagrams for the simply supported beam with overhang, loaded as shown in Fig. 5.14a.

SOLUTION. The reactions are

$$R_A = 0, \quad R_B = wl.$$

Between A and C, we have

$$V_x = 0, \quad M_x = 0, \qquad \text{(j)}$$

FIG. 5.13

where x is measured to the right from A. Between C and B, we have

$$V_x = -wx, \quad M_x = -\frac{wx^2}{2}, \qquad \text{(k)}$$

where x is measured to the right from C. Between B and D, we have

$$V_x = w\left(\frac{l}{2} - x\right), \quad M_x = -\frac{w}{2}\left(\frac{l}{2} - x\right)^2, \qquad \text{(l)}$$

where x is measured to the right from B. The corresponding shear and moment diagrams are shown in Figs. 5.14b and c.

EXAMPLE 3. Construct shear force and bending moment diagrams for the simply supported beam with overhanging ends loaded as shown in Fig. 5.15a.

SOLUTION. From equilibrium of the entire beam, we find $R_1 = R_2 = P(1 + 2a/l)$ directed as shown. The shear force in each overhang is constant and equal to $-P$. The shear force in the middle portion is constant and equal to $-P + P(1 + 2a/l) = +2Pa/l$. The corresponding shear force diagram is shown in Fig. 5.15b.

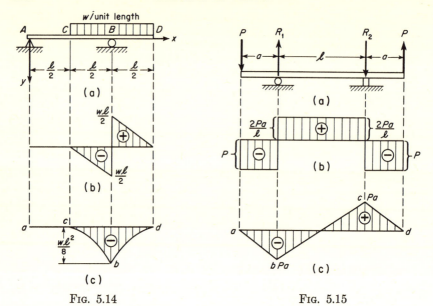

FIG. 5.14   FIG. 5.15

Since the beam is subjected only to concentrated forces, the bending moment must vary linearly between points of application of transverse loads. The bending moment over the left support is $-Pa$, while that over the right support is $+Pa$. Plotting these points in Fig. 5.15c and drawing the straight lines ab, bc, and cd, we obtain the complete bending moment diagram for the beam as shown.

PROBLEM SET 5.2

1. Construct, to scale, the complete shear force and bending moment diagrams for the beam in Fig. A on page 102.

2. Construct, to scale, the complete shear force and bending moment diagrams for the beam shown in Fig. B on page 102.

3. Construct, to scale, the complete shear force and bending moment diagrams for the beam shown in Fig. A below.

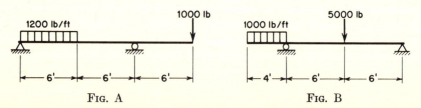

FIG. A   FIG. B

4. Construct, to scale, the complete shear force and bending moment diagrams for the beam shown in Fig. B above.

5. For the beam in Fig. D on page 102, the following numerical data are given: $P = 1000$ lb, $w = 200$ lb /ft, $l = 8$ ft, $a = 3$ ft. Using these data, construct, to scale, complete shear force and bending moment diagrams for the beam.

6. Construct shear force and bending moment diagrams for the simply supported beam AB loaded as shown in Fig. E on page 103.

7. Construct shear force and bending moment diagrams for the simply supported beam AB loaded as shown in Fig. F on page 103.

8. Construct, to scale, the complete shear force and bending moment diagrams for the beam shown in Fig. H on page 103.

9. A cantilever beam built-in at B carries a bracket at A, to the free end of which a vertical load P is applied as shown in Fig. C. Construct shear force and bending moment diagrams for the beam AB, if $a = l/3$.

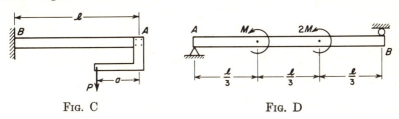

Fig. C — — — Fig. D

10. A simply supported beam AB is acted upon by two externally applied couples of moments M and $2M$ as shown in Fig. D. Neglecting the distributed weight of the beam itself, construct a complete bending moment diagram for the beam.

11. A lumberjack weighing 200 lb stands at the middle of a floating log 2 ft in diameter and 16 ft long. The specific gravity of the log is 0.50. What is the bending moment at the mid-point of the log? *Ans.* $M = 400$ ft-lb.

12. Referring to the preceding problem, assume that two men each weighing 200 lb stand at the third-points of the floating log, i.e., 5.33 ft from the two ends. Under these conditions, construct a bending moment diagram for the log and find the bending moment at the middle. *Ans.* $M_m = +267$ ft-lb.

13. Construct shear force and bending moment diagrams for the simply supported beam carrying a triangularly distributed load as shown in Fig. 5.6a.

5.3   Bending Stresses in Beams

Let us consider the beam AB transversely loaded as shown in Fig. 5.16 together with its shear force and bending moment diagrams. We note that the middle portion CD of the beam is free from shear force and that its bending moment $M_x = Pa$ is uniform between C and D. This condition is called *pure bending.*

To investigate the state of internal stress produced by pure bending, we must examine the deformation which takes place within the material. In so doing, we shall assume that the beam is prismatic and that it has an axial plane of symmetry which we take as the xy-plane. When the applied loads also act in such a plane of symmetry, bending will take place only in that plane. We assume further that the material is homogeneous and that it obeys Hooke's law, the modulus of elasticity in tension being the same as that in compression. Then since the bending moment is uniform between C and D, it is reasonable to assume that the bending deformation will also be

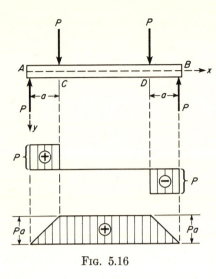

FIG. 5.16

uniform; i.e., the portion CD of the beam will take the form of a *circular arc* as shown in Fig. 5.17. In this deformed configuration, each cross-section, originally plane, is assumed to remain plane and normal to the longitudinal fibers of the beam.*

As a result of the deformation shown in Fig. 5.17, fibers on the convex side of the beam are elongated slightly while those on the concave side are shortened slightly. Somewhere in between the top and bottom of the beam, there is a layer of fibers which remain unchanged in length. This is called the *neutral surface*. The intersection of this neutral surface with the axial plane of symmetry is called the *neutral axis of the beam*. Its intersection with the plane of any cross-section is called the *neutral axis of that section*. After deformation, the planes of two adjacent cross-sections mn and pq intersect at O. We denote the angle between these planes by $d\theta$ and note that $d\theta = dx/\rho$ where $1/\rho$ is the curvature of the neutral axis of the beam.

In Fig. 5.17, we now draw through point b on the neutral axis, a line $p'q'$ parallel to mn and indicating the original orientation of the cross-section pq

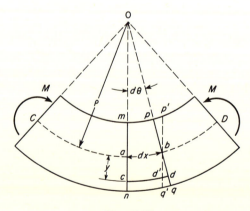

FIG. 5.17

*Careful strain measurements in the laboratory confirm this assumption for the case of pure bending.

before bending. From this construction, we see that the segment cd' of any fiber at the distance y from the neutral surface elongates by the amount $d'd = y\,d\theta$. Since its original length was $cd' = dx$, the corresponding strain is

$$\epsilon_x = \frac{y\,d\theta}{dx} = y/\rho. \tag{a}$$

If a fiber on the concave side of the neutral surface is considered, the distance y will be negative and the strain is also negative. Thus all fibers on the convex side of the neutral surface are in tension while those on the concave side are in compression. Experiments indicate that the lateral deformation of fibers is the same as in simple tension and compression. Thus the stress in each fiber will be proportional to its longitudinal strain,

$$\sigma_x = \epsilon_x E = \frac{E}{\rho}\, y. \tag{5.3}$$

This shows that the fiber stresses σ_x due to pure bending vary linearly with distance y from the neutral surface, so long as the material follows Hooke's law. Such stress distribution over the depth of the beam is shown in Fig. 5.18. The position of the neutral axis Oz of the cross-section may now be found from the condition that these stresses distributed over the section must give rise to a resisting couple M.

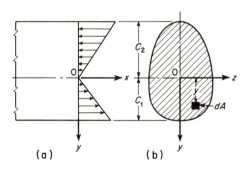

(a)   (b)

FIG. 5.18

Let dA denote an element of area of the cross-section at the distance y from the neutral axis (Fig. 5.18). Then the element of force on this area is $\sigma_x\,dA$. Using eq. (5.3), this becomes

$$\sigma_x\,dA = \frac{E}{\rho}\, y\,dA. \tag{b}$$

Now since there must be no resultant normal force N_x on the section (pure bending), the integral of $\sigma_x dA$ over the entire area of the section must vanish, i.e.,

$$\frac{E}{\rho} \int_A y \, dA = 0. \tag{c}$$

Since $E/\rho \neq 0$, we conclude from this that

$$\int_A y \, dA = A y_c = 0, \tag{d}$$

where A is the total cross-sectional area and y_c is the distance from the neutral axis to its centroid. Finally, since $A \neq 0$, we conclude that $y_c = 0$. Thus *the neutral axis of the cross-section passes through its centroid.*

The moment of the elemental force $\sigma_x dA$ about the neutral axis of the section is $dM = y\sigma_x dA$. The sum of these elemental moments over the total area of the section must produce the bending moment M on that section. Thus

$$M = \int_A y\sigma_x \, dA = \frac{E}{\rho} \int_A y^2 \, dA. \tag{e}$$

The integral in this expression in which a summation is made of each element of area dA multiplied by the square of its distance from the neutral axis (z-axis), is called the *moment of inertia* of the cross-sectional area with respect to that axis.* Introducing the notation

$$I = \int_A y^2 \, dA \tag{f}$$

for this quantity, eq. (e) may be written in the form

$$\frac{1}{\rho} = \frac{M}{EI}. \tag{5.4}$$

This shows us that the curvature $1/\rho$ of the axis of the beam is proportional to the bending moment M and inversely proportional to the quantity EI, called the *flexural rigidity* of the beam. We see that this flexural rigidity reflects both the stiffness of the material as measured by E and the proportions of the cross-sectional area as measured by I.

Substituting the value of $1/\rho$ from eq. (5.4) into eq. (5.3) above, we obtain

$$\sigma_x = \frac{My}{I}. \tag{5.5}$$

It is seen that this bending stress will be a maximum in those fibers furthest removed from the neutral surface, tension on the convex lower face of the beam and compression on the concave upper face. Denoting the distances

*See Appendix B, p. 346.

to extreme fibers in tension and compression, respectively, by c_1 and c_2 as shown in Fig. 5.18, we obtain from eq. (5.5)

$$\sigma_{max} = \frac{Mc_1}{I}, \quad \sigma_{min} = \frac{Mc_2}{I}. \tag{5.5a}$$

If the cross-section is symmetrical with respect to its centroidal axis, $c_1 = c_2 = c$ and the extreme fiber stresses in tension and compression are equal.

Introducing the notations

$$\frac{I}{c_1} = Z_1, \quad \frac{I}{c_2} = Z_2 \tag{g}$$

called the *section moduli*, eqs. (5.5a) can also be expressed in the form

$$\sigma_{max} = \frac{M}{Z_1}, \quad \sigma_{min} = \frac{M}{Z_2}. \tag{5.5b}$$

In the case of a rectangular cross-section of width b and depth h (Fig. 5.19a), $c_1 = c_2 = h/2$,

$$I = \frac{bh^3}{12}, \quad Z_1 = Z_2 = \frac{bh^2}{6}.$$

For a circular section of diameter d (Fig. 5.19b), $c_1 = c_2 = d/2$,

$$I = \frac{\pi d^4}{64}, \quad Z_1 = Z_2 = \frac{\pi d^3}{32}.$$

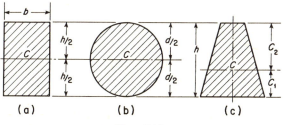

(a)   (b)   (c)

Fig. 5.19

In the case of the trapezoidal section shown in Fig. 5.19c, $c_1 < c_2$ which makes $Z_1 > Z_2$. Thus if the beam is bent concave upwards, the maximum compressive stress in the fibers of the top face will be greater than the maximum tensile stress in the fibers of the bottom face. In the case of a beam made of cast-iron this may be advantageous, since cast-iron is stronger in compression than it is in tension. Various shapes of cross-sections for beams will be discussed in more detail in the next article.

All of the foregoing theory has been developed for the case of pure bending, i.e., constant bending moment along the length of the beam. In such case, the shear force at each cross-section is zero and the normal stresses due to bending are the only ones produced. In the case of non-uniform bending of a beam where the bending moment varies from one cross-section to another, there is a shear force at each cross-section and shearing stresses are also induced in the material.* The deformation associated with these shearing stresses causes warping of the various cross-sections so that plane cross-sections before bending do not remain plane after bending. This complicates the problem, but more elaborate analysis shows that the normal stresses due to bending, as calculated from eq. (5.5), are not greatly altered by the presence of these shearing stresses. Thus it is justifiable to use the theory of pure bending for calculating the normal stresses in the case of non-uniform bending and it is accepted practice to do so.

EXAMPLE 1. A locomotive axle with outboard bearings A and B will be subjected to transverse loads as shown in Fig. 5.20. Determine the maximum bending stress σ induced in the axle if $a = 13.5$ in., $l = 59$ in., diameter $d = 10$ in., and $P = 26,000$ lb. Find, also, the deflection δ at the mid-point of the axle.

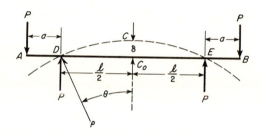

FIG. 5.20

SOLUTION. The bending moment between wheel loads at D and E is $M = Pa = 26,000 \times 13.5$ in.-lb. The section modulus is $Z = \pi d^3/32$. Thus eq. (5.5) gives

$$\sigma_{max} = \frac{M}{Z} = \frac{26,000 \times 13.5 \times 32}{\pi \times 10^3} = 3580 \text{ psi.}$$

From eq. (5.4), the radius of curvature of the circular arc DCE, in pure bending, is

$$\rho = \frac{EI}{M} = \frac{30(10)^6 \times \pi \times (10)^4}{26,000 \times 13.5 \times 64} = 41,900 \text{ in.}$$

Now, referring to Fig. 5.20, we see that the deflection $C_0C = \delta$ is

$$\delta = \rho(1 - \cos \theta), \tag{h}$$

where $\theta = l/2\rho = 0.000705$ radian. For such a small angle, we may take $\cos \theta \approx 1 - \theta^2/2$. Then eq. (h) becomes

$$\delta \approx \frac{\rho\theta^2}{2} = \frac{\rho l^2}{8\rho^2} = \frac{l^2}{8\rho} = \frac{59 \times 59}{8 \times 41,900} = 0.0104 \text{ in.}$$

*These shearing stresses are discussed in Art. 5.5.

EXAMPLE 2. Calculate the maximum bending stress that will be induced in a steel wire of diameter $d = \frac{1}{32}$ in. if it is wound on a drum of diameter $D = 20$ in.

SOLUTION. For this calculation, we use eq. (5.3). Thus

$$\sigma_x = \frac{E}{\rho}y = \frac{30(10)^6}{10.016} \times \frac{1}{64} = 46,100 \text{ psi.}$$

It must be assumed that this does not exceed the yield point of the steel, so that the material behaves elastically.

EXAMPLE 3. A simple beam 20 ft long is to carry a uniform load of 480 lb per ft, including the weight of the beam itself. (a) Select a suitable Douglas fir beam with rectangular section of depth-width ratio $h/b = 1.2$ and an allowable working stress, $\sigma_w = 1200$ psi. (b) Select a suitable steel beam of standard I-section to carry the same load, if the allowable stress is $\sigma_w = 18,000$ psi.

SOLUTION. Referring to Fig. 5.10, we see that, for a uniformly loaded simple beam, the maximum bending moment at the middle is

$$M_{\max} = \frac{wl^2}{8} = \frac{480 \times 20 \times 20}{8} = 24,000 \text{ ft-lb.} \tag{i}$$

Also, for a rectangular cross-section of depth h and width $b = h/1.2$, the section modulus is

$$Z = \frac{bh^2}{6} = \frac{h^3}{7.2}.$$

Then with $\sigma_x = \sigma_w = 1200$ psi, eq. (5.5b) becomes

$$1200 = \frac{24,000 \times 12 \times 7.2}{h^3},$$

from which

$$h = \sqrt[3]{\frac{24,000 \times 12 \times 7.2}{1200}} = \sqrt[3]{1728} = 12 \text{ in.}$$

Thus the wood beam should have a 10-in. $\times$ 12-in. section.

For the steel I-beam, eq. (5.5b) becomes

$$18,000Z = 24,000 \times 12,$$

from which the required section modulus $Z = 16$ in.[3] Now turning to Table B.3 of Appendix B, p. 364 we find that an 8I-23 standard I-beam has a section modulus $Z = 16$ in.[3] and will be satisfactory.

PROBLEMSET 5.3

1. A thin steel rule having a cross-section 0.025 in. $\times$ 1.00 in. is bent by couples applied at its ends so that a length $l = 10$ in. of the circular arc subtends a central angle $\theta = 60°$. Calculate the maximum stress induced in the rule and the magnitude of the bending moment M. *Ans.* $\sigma = 39,300$ psi; $M = 4.09$ in.-lb.

2. A wood beam 6 ft long is simply supported at its ends, has a cross-section 6 in. wide by 24 in. deep and carries a uniformly distributed load of intensity $w = 5000$ lb/ft over the full span. Calculate the bending stress at a point 8 in. above the bottom of the beam and 24 in. from the left support. *Ans.* $\sigma = 139$ psi.

3. A simply supported beam having a span $l = 12$ ft is to carry a uniformly distributed load of intensity $w = 1000$ lb/ft. The cross-section is to be rectangular with depth h and width $b = h/2$. If the allowable bending stress in tension or compression is $\sigma_w = 1200$ psi, what is the required depth h for the cross-section? *Ans.* $h = 13$ in.

4. A railroad tie (Fig. A) is 8 ft long and has a 12 in. $\times$ 10 in. rectangular cross-section with the 12-in. faces horizontal. The maximum loads transmitted to the tie by the rails are $P = 50,000$ lb. each, and the ballast is assumed to exert a uniformly distributed reactive load on the bottom of the tie as shown. Calculate the maximum bending stress in the tie if $l = 4$ ft 9 in. and $a = 1$ ft $7\frac{1}{2}$ in. *Ans.* $\sigma_{max} = 1125$ psi.

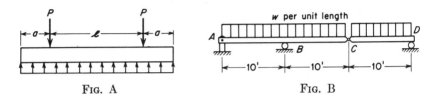

FIG. A   FIG. B

5. A compound beam made of two bars AC and CD hinged together at C is supported and loaded as shown in Fig. B. Each portion of the beam is a standard 5-I-10 steel section for which the section modulus $Z = 4.8$ in.[3] What is the safe value of the intensity w of the uniformly distributed load if the allowable working stress in bending is $\sigma_w = 20,000$ psi? *Ans.* $w = 80$ lb/ft.

6. A simply supported beam with overhang is loaded as shown in Fig. C. The beam is made of cast iron and has a T-section the flange of which is 4 in. wide and the stem of which is 3 in. deep. Flange and stem are each 1 in. thick. Find the safe value of the load P if the allowable working stresses for cast iron are $\sigma_t = 6000$ psi in tension and $\sigma_c = 10,000$ psi in compression. *Ans.* $P = 992$ lb.

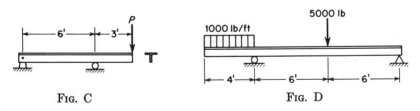

FIG. C   FIG. D

7. A simply supported beam with overhang is loaded as shown in Fig. D. The beam has a heavy T-section the flange of which is 9 in. wide and the stem of which is 6 in. deep. Flange and stem are each 4 in. thick. Calculate the maximum tensile stress and the maximum compressive stress due to bending. *Ans.* $(\sigma_t)_{max} = 1650$ psi; $(\sigma_c)_{max} = 1200$ psi.

8. A simply supported beam with overhang is loaded as shown in Fig. E. The cross-section of the beam is a standard 10-U-30 channel section with the open side up. The allowable bending stresses in tension and compression, respectively, are $\sigma_t = 20{,}000$ psi and $\sigma_c = 16{,}000$ psi. Find the safe value of the intensity w of the distributed load on the overhang. *Ans.* $w = 348$ lb/ft.

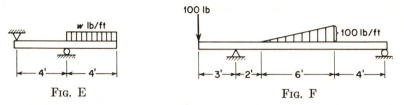

FIG. E   FIG. F

9. What is the required section modulus for the beam shown in Fig. F if the allowable bending stress is $\sigma_w = 1200$ psi? *Ans.* $Z = 5.85$ in.[3]

10. A temporary dam in a water channel is formed by setting vertical 3×12-in. planks between guide rails AA and BB as shown in Fig. G. Assuming no support at D, calculate the maximum bending stress induced in each plank when the water depth on the left side of the dam is 6 ft as shown. *Ans.* $\sigma_{max} = 3000$ psi.

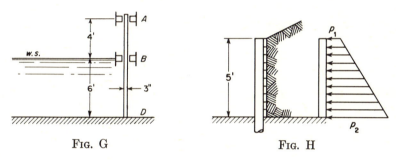

FIG. G   FIG. H

11. A retaining wall for an earth embankment is made with 12-in.-diameter wood piles backed by 3-in.-thick planks as shown in Fig. H. The earth fill behind the retaining wall is assumed to exert the pressure diagram shown where $p_1 = 100$ psf and $p_2 = 400$ psf. The planks are assumed to act as simply supported beams between piles and the working stress in bending for both planks and piles is $\sigma_w = 1200$ psi. What is the proper spacing of the piles? *Ans.* 6 ft 0 in.

12. A six-ton tractor is to cross a bridge consisting of two parallel beams as shown in Fig. I. What is the required section modulus Z for each beam if the allowable working stress in bending is $\sigma_w = 16{,}000$ psi? *Hint:* Define the position of the tractor on the span by the distance x and then choose x so as to maximize the bending moment under the rear wheel. *Ans.* $Z = 14.4$ in.[3]

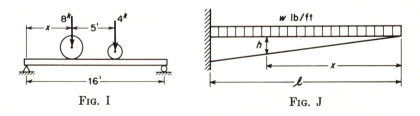

FIG. I   FIG. J

13. Prove that the depth h of a cantilever beam of rectangular cross-section should vary linearly with x to attain a beam of constant strength under uniformly distributed load (Fig. J). Assume that the flexure formula for prismatic beams applies.

5.4 Various Shapes of Cross-Sections of Beams

From the discussion in Art. 5.3 it follows that the maximum tensile or compressive stress in a bent bar is proportional to the distance of the most remote fibers from the neutral axis. Hence if the material has the same strength in tension and compression, it will be logical to choose shapes of cross-section in which the centroid is at the middle of the depth of the beam. In this manner the same factor of safety for fibers in tension and for those in compression will be obtained. This is the underlying idea in the choice of sections symmetrical with respect to the neutral axis for materials such as structural steel, which have about the same yield point in tension and compression. Such cross-sections are shown in Fig. 5.21a, b, and e. If the section is not symmetrical with respect to the neutral axis, for example a rail section, the material is frequently distributed between the head and the base so as to have the centroid near the middle of its height.

For a material of low strength in tension and high strength in compression — for example, cast-iron or concrete — the best cross-section for a beam is not symmetrical with respect to the neutral axis but is such that the dis-

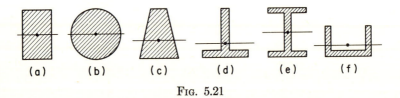

$$(a) \quad (b) \quad (c) \quad (d) \quad (e) \quad (f)$$

FIG. 5.21

tances c_1 and c_2 from the neutral axis to the most remote fibers in tension and compression are in the same proportion as the strengths of the material in tension and in compression. In this manner equal strength in tension and compression is obtained. For example, with a T-section (Fig. 5.21d), the position of the centroid of the section may be adjusted along the height of the section by properly proportioning its flange and web.

In designing a beam to undergo bending, not only the conditions of strength should be satisfied but also the condition of economy in the weight of the beam. Of two cross-sections having the same section modulus, that is, satisfying the condition of strength with the same factor of safety, the one with the smaller area is more economical.

In comparing various shapes of cross-sections, consider first the rectangular section of depth h and width b (Fig. 5.21a). The section modulus is

$$Z = \frac{bh^2}{6} = \frac{1}{6} Ah, \tag{a}$$

where A denotes the cross-sectional area. It is seen that the rectangular cross-section becomes more and more economical with increase in its depth h. However there is a certain limit to this increase and the question of lateral instability of the beam arises as the section becomes too narrow. The collapse of a beam of very narrow rectangular section may be due not to overcoming the strength of the material but to sidewise buckling.*

In the case of a circular cross-section of diameter d (Fig. 5.21b),

$$Z = \frac{\pi d^3}{32} = \frac{1}{8} Ad. \tag{b}$$

Comparing circular and square cross-sections of the same area, the side h of the square will be $h = d\sqrt{\pi}/2$, for which eq. (a) gives

$$Z = 0.147 \, Ad.$$

Comparison of this with (b) shows a square cross-section to be more economical than a circular one.

Consideration of the stress distribution along the depth of the cross-section (Fig. 5.18) leads to the conclusion that for economical design most of the material of the beam should be put as far as possible from the neutral axis. The theoretically ideal case for a given cross-sectional area A and depth h would be to distribute each half of the area at a distance $h/2$ from the neutral axis as shown in Fig. 5.22. Then

$$I = 2 \times \frac{A}{2} \times \left(\frac{h}{2}\right)^2 = \frac{1}{4} Ah^2$$

and the section modulus becomes

$$Z = \frac{1}{2} Ah. \tag{c}$$

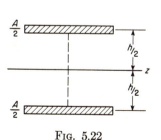

FIG. 5.22

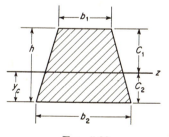

FIG. 5.23

*For a discussion of lateral buckling of beams, see Timoshenko and Gere, *Elastic Stability*, McGraw-Hill Book Co., Inc., New York, 1961.

This ideal limit, although unattainable, may be approached in practice by using an I-section or wide-flange section with most of the material in the flanges (Fig. 5.21e). Due to the necessity of putting part of the material in the web, the limiting condition (c) can never be realized; but for standard wide-flange sections, we have

$$Z \approx \frac{1}{3} Ah. \tag{d}$$

Comparison of (d) with (a) shows that an I-section is more economical than a rectangular section of the same depth. Furthermore, due to its wide flanges, an I-beam will be more stable with respect to sidewise buckling than a rectangular section having the same depth and section modulus.

In the design and selection of beam sections, reference is usually made to a handbook* giving various properties such as moment of inertia, section modulus, etc., of standard structural shapes. These shapes such as I-sections (Fig. 5.21e) and ⊔-sections (Fig. 5.21f) are usually specified by the nominal depth in inches and the weight in pounds per lineal foot of beam. An 8-in., 18.4-lb I-beam, for instance, is 8 in. deep, weighs 18.4 lb per ft, and is designated simply by 8I-18.4. Such beams having extra wide flanges are specified in a separate class. For example, the notation 12WF65 denotes a wide-flange section having a nominal depth of 12 in. and weighing 65 lb per lineal foot.

EXAMPLE 1. A prismatic beam in pure bending has a trapezoidal cross-section as shown in Fig. 5.23, the top fibers being in compression. If allowable working stresses in tension and compression are $\sigma_t = 5000$ psi and $\sigma_c = 8000$ psi, calculate the ratio of bases b_1/b_2 for maximum economy.

SOLUTION. Let c_1 and c_2 denote distances to extreme fibers from the neutral (centroidal) axis as shown. Then from eq. (5.5a) we have

$$8000 = \frac{M}{I}c_1,$$

$$5000 = \frac{M}{I}c_2,$$

from which $c_1/c_2 = 8/5$. Also $c_1 + c_2 = h$. Hence

$$c_1 = \frac{8}{13}h, \quad c_2 = \frac{5}{13}h. \tag{e}$$

The distance y_c to the centroid of the trapezoid is

$$y_c = \frac{h}{3}\left(\frac{b_2 + 2b_1}{b_2 + b_1}\right). \tag{f}$$

*See, for example, *Steel Construction* by the American Institute of Steel Construction, New York, 1959; also Appendix B-1, p. 357.

Equating this to the value of c_2 from (e), we find

$$\frac{b_1}{b_2} = \frac{2}{11}.$$

EXAMPLE 2. A beam of square cross-section $a \times a$ is bent in the vertical plane of one of its diagonals as shown in Fig. 5.24. Show that, for a given moment, the maximum fiber stresses can be reduced by cutting off the shaded corners to a depth $\alpha a/\sqrt{2}$ and calculate the optimum value of α.

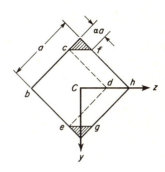

FIG. 5.24

SOLUTION. The moment of inertia of the complete cross-section is $I = a^4/12$ and the corresponding section modulus is $Z = \sqrt{2}\, a^3/12$ (see Appendix B). The moment of inertia of the reduced section will be obtained by adding to the moment of inertia of the square $bcde$ about its diagonal bd that of the two parallelograms $cfdh$ and $gedh$ about their common base dh. Thus

$$I = \frac{a^4(1-\alpha)^4}{12} + 2\frac{\sqrt{2}\,\alpha a}{3}\left[\frac{a(1-\alpha)}{\sqrt{2}}\right]^3 = \frac{a^4(1-\alpha)^3}{12}(1+3\alpha)$$

and the corresponding section modulus becomes

$$Z = \frac{I\sqrt{2}}{a(1-\alpha)} = \frac{\sqrt{2}}{12}a^3(1-\alpha)^2(1+3\alpha). \tag{g}$$

This section modulus is a maximum for that value of α which makes $dZ/d\alpha = 0$. Making this differentiation, we find $\alpha = \frac{1}{9}$ and the corresponding maximum section modulus becomes, from (g),

$$(Z)_{\max} = 1.053\,\frac{\sqrt{2}a^3}{12}. \tag{h}$$

Thus, by cutting off the corners, the section modulus is increased by about 5 per cent and the maximum bending stresses will be reduced by this percentage.

To understand how this removal of material actually increases the strength of the section, we must observe that removal of the small corner areas simply reduces the distance to extreme fibers in a greater proportion than it reduces the moment of inertia of the section. Thus the section modulus $Z = I/c$ is actually increased.

PROBLEM SET 5.4

1. A simply supported cast-iron beam is to have the inverted T-section shown in Fig. A. If the allowable stresses for cast-iron in tension and compression are $\sigma_t = 4000$ psi and $\sigma_c = 8000$ psi, calculate the proper stem thickness t of the section. Ans. $t = 1.2$ in.

2. For the channel beam shown in Fig. B, it is desired to have the ratio of extreme fiber bending stresses $\sigma_t : \sigma_c = 3 : 7$. What is the proper wall thickness t to realize this condition? Ans. $t = 2$ in.

3. Calculate the proper width b of the upper flange of the beam section shown in Fig. C in order that extreme fiber stresses in bending will be in the ratio 4:3. *Ans.* $b = 2.24$ in.

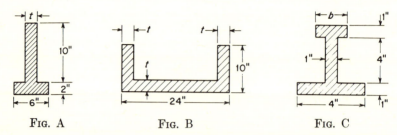

FIG. A   FIG. B   FIG. C

4. Calculate the two section moduli Z_1 and Z_2 for the T-section shown in Fig. A when $t = 1.2$ in. *Ans.* $Z_1 = 40$ in.³; $Z_2 = 80$ in.³

5. Calculate the two section moduli Z_1 and Z_2 for the channel section shown in Fig. B when $t = 2$ in. *Ans.* $Z_1 = 95.2$ in.³; $Z_2 = 222$ in.³

6. Calculate the moment of inertia of the section shown in Fig. C with respect to the centroidal axis parallel to the flanges if $b = 2.24$ in. *Ans.* $I = 43.0$ in.⁴

7. A prismatic bar having a semicircular cross-section is subjected to pure bending in its longitudinal plane of symmetry. Calculate the ratio of the extreme fiber stresses in bending. *Ans.* $\sigma_1/\sigma_2 = 0.736$.

8. A rectangular wood beam is to be cut from a circular log as shown in Fig. D. Calculate the ratio b/h required to attain a beam of maximum strength in bending. *Ans.* $b/h = 1/\sqrt{2}$.

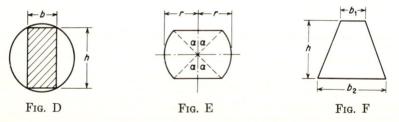

FIG. D   FIG. E   FIG. F

9. The top and bottom of a log of diameter $d = 12$ in. are adzed to form the cross-section shown in Fig. E. Calculate the section modulus for the case where $\alpha = 45°$. *Ans.* $Z = 120$ in.³

10. Calculate the two-section moduli for the trapezoidal cross-section shown in Fig. F if $h = 12$ in., $b_2 = 10$ in., $b_1 = 6$ in. *Ans.* $Z_1 = 173$ in.³; $Z_2 = 205$ in.³

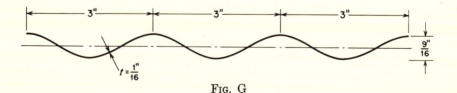

FIG. G

11. In cross-section, a sheet of 16-gage corrugated sheet metal has the dimensions shown in Fig. G. Assuming that the corrugations have a sine-wave form, calculate the section modulus per foot of width of sheet. The $\frac{9}{16}$-in. dimension is from outside surface to outside surface. *Ans.* $Z = 0.086$ in.^{3}/ft.

5.5   Shear Stresses in Bending

In Art. 5.1 we have seen that in the case of a beam bent by transverse loads there is, in general, both a bending moment M_x and a shear force V_x at each cross-section. In Art. 5.3 we have seen how the bending moment represents the resultant of a certain linear distribution of normal stresses σ_x over the cross-section. Similarly, the shear force V_x on any cross-section must be the resultant of a certain distribution of shear stresses τ over that section. The question to be considered now is just what that distribution must be to satisfy equilibrium conditions of the various elements of the beam.

We begin with the simplest case of a beam of rectangular cross-section (Fig. 5.25) and assume that V_x is the resultant shear force on a chosen cross-section. Dividing the cross-section into infinitesimal strips parallel to the z-axis, it is reasonable to assume that on any one such strip, the shear stress τ will be uniform across the width b of the beam and parallel to the y-axis as shown. We also know from the discussion of Art 2.1 that such shear stress on one side of the prismatic element mn must be accompanied by equal shear stresses on each of the other three sides of the element (see Fig. 5.25). Thus, we must observe at the outset that there will be *horizontal shear stresses* between layers of the beam parallel to the neutral plane as well as *transverse shear stresses* between cross-sections and that at any point in

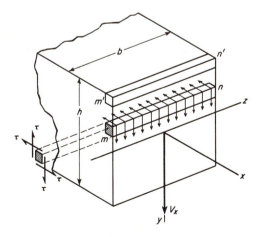

FIG. 5.25

the beam these complementary shear stresses are equal in magnitude. This at once leads us to conclude that the shear stress τ must vary as we go from top to bottom of the beam. For if we consider the prismatic element $m'n'$ whose upper face coincides with the free top surface of the beam, we see that the shear stress τ must vanish. The same conclusion holds for an element one face of which coincides with the bottom surface of the beam. Thus the shear stress τ must vary with y and must vanish at $y = \pm h/2$.

The key to the true law of variation of τ with y comes from examining the equilibrium of an elemental block pnp_1n_1 of the beam between two adjacent cross-sections mn and m_1n_1 distance dx apart as shown in Fig. 5.26a. The bottom face bdx of this block coincides with the bottom surface of the beam and is free from stress. Its upper face is parallel to the neutral plane and at the arbitrary distance y_1 therefrom and is acted upon by the horizontal shear stress τ existing at this level. The end faces of the block are represented by the shaded portion of the cross-section of the beam as shown in Fig. 5.26b. These end faces are acted upon by the normal bending stresses

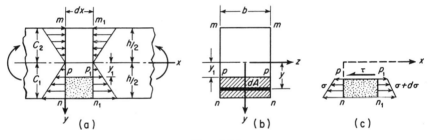

Fig. 5.26

σ. The complete elemental block, isolated as a free body, is shown in Fig. 5.26c. There are, of course, also shear stresses on the end faces pn and p_1n_1, but we will be interested only in the equilibrium of this block in the x-direction; hence, these shear stresses will not enter in the equation of equilibrium and are not shown on the free body.

Before going further, we may observe at once that if the bending moment in the beam has the same magnitude at sections mn and m_1n_1, the normal stress distributions on the ends pn and p_1n_1 of the elemental block (Fig. 5.26c) will be identical. Then for equilibrium of the block in the x direction, we conclude that $\tau = 0$. This simply verifies the fact that pure bending can induce no shear stresses in the beam.

Consider now the more general case of varying bending moment, denoting by M and $M + dM$ the moments at the cross-sections mn and m_1n_1, respectively. Then the normal force acting to the left on an elemental area dA of the end face pn of the block will be, from eq. (5.5),

$$\sigma dA = \frac{My}{I} dA.$$

The sum of these forces over the face pn of the block will be

$$\int_{y_1}^{c_1} \frac{My}{I} dA. \tag{a}$$

In the same manner, the sum of the normal forces over the face $p_1 n_1$ of the block becomes

$$\int_{y_1}^{c_1} \frac{(M + dM)y}{I} dA. \tag{b}$$

The shear force acting on the upper face of the block is

$$\tau b \, dx, \tag{c}$$

where τ is the shear stress at the arbitrary level y_1. The forces (a), (b), (c), must be in equilibrium; hence

$$\tau b \, dx = \int_{y_1}^{c_1} \frac{(M + dM)y}{I} dA - \int_{y_1}^{c_1} \frac{My}{I} dA,$$

from which

$$\tau = \frac{dM}{dx} \frac{1}{Ib} \int_{y_1}^{c_1} y \, dA,$$

or, by using eq. (5.1),

$$\tau = \frac{V}{Ib} \int_{y_1}^{c_1} y \, dA. \tag{5.6}$$

The integral in this expression is seen to represent the *statical moment*, about the neutral axis of the cross-section, of the shaded portion $ppnn$ of the cross-section, i.e., of that portion of the cross-section below (or above) the arbitrary level y_1 at which the shear stress τ is required. Denoting this statical moment by Q, eq. (5.6) takes the form

$$\tau = \frac{VQ}{Ib}. \tag{5.7}$$

To see how this shear stress varies with the distance y_1 from the neutral axis, we must now examine the variation of Q with y_1.

Referring to Fig. 5.26b, we see that for a rectangular cross-section $dA = b\,dy$ and $c_1 = h/2$, so that

$$Q = \int_{y_1}^{c_1} y \, dA = b \int_{y_1}^{h/2} y \, dy = \frac{b}{2}\left(\frac{h^2}{4} - y_1^2\right). \tag{d}$$

This same statical moment will be obtained as the product of the shaded area

$$b\left(\frac{h}{2} - y_1\right)$$

and the distance

$$\frac{1}{2}\left(\frac{h}{2} + y_1\right)$$

of its centroid from the neutral axis of the cross-section.

Substituting (d) into eq. (5.7), we obtain

$$\tau = \frac{V}{2I}\left(\frac{h^2}{4} - y_1^2\right). \tag{5.8}$$

This shows that the shear stress τ varies parabolically with y_1 as shown in Fig. 5.27b. When $y_1 = \pm h/2$, $\tau = 0$, and when $y_1 = 0$, $\tau_{max} = Vh^2/8I$. Noting that $I = bh^3/12$, we have

$$\tau_{max} = \frac{3}{2}\frac{V}{A}, \tag{5.9}$$

where $A = bh$ is the total area of the cross-section. Thus the maximum shear stress (horizontal or vertical) occurs at the neutral axis ($y_1 = 0$) and is 50 per cent larger than the average shear stress V/A.

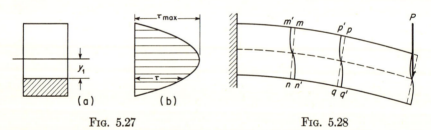

FIG. 5.27   FIG. 5.28

Since the shear stress τ varies from top to bottom of the beam, it follows that the shear strain $\gamma = \tau/G$ must vary in a similar way. Thus originally plane cross-sections of the beam become warped after bending. This warping can be demonstrated by bending a rubber beam as shown in Fig. 5.28. If straight lines mn and pq are scribed on the side of the beam before bending, they will become curved lines $m'n'$ and $p'q'$ after bending as shown in the figure. At the points m', p', n', q', the shearing strain is zero so that the curves $m'n'$ and $p'q'$ remain normal to the upper and lower surface of the beam after bending. At the neutral surface, the angles between the tangents to these curves and the normal sections mn and pq

are equal to the shear strain τ_{max}/G. As long as the shear force V_z remains constant along the beam, the warping of all cross-sections is the same; that is $mm' = pp'$, $nn' = qq'$, etc. Thus the shear stresses τ do not contribute to the longitudinal strains in the fibers and the distribution of normal stresses σ is the same as in the case of pure bending.

A more elaborate theoretical investigation of this problem shows that warping due to shear strain does not substantially affect the longitudinal fiber strains, even when V_z varies along the beam, provided this variation is continuous as in the case of a beam carrying a uniformly distributed load. In the case of concentrated loads on the beam, however, the stress distribution near points of application of external loads becomes more complicated, but these irregularities are very localized and do not appreciably affect the over-all stress distribution in the beam. Thus it is usually justifiable to use flexure formulas derived for pure bending in the case of non-uniform bending, as we have already done in the preceding articles.

In discussing the distribution of shear stress in the web of an I-beam, we can proceed in about the same manner as for a beam of rectangular cross-section. Referring to Fig. 5.29a, let y_1 denote the level at which we wish to evaluate the shear stress τ. Then with the dimensions of the section as shown, the statical moment of the shaded portion of the cross-section about the neutral axis is

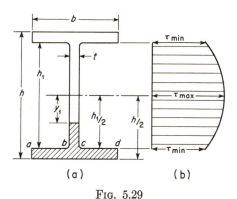

(a)   (b)

Fig. 5.29

$$Q = b\left(\frac{h}{2} - \frac{h_1}{2}\right)\frac{1}{2}\left(\frac{h}{2} + \frac{h_1}{2}\right) + t\left(\frac{h_1}{2} - y_1\right)\frac{1}{2}\left(\frac{h_1}{2} + y_1\right),$$

and eq. (5.7) becomes

$$\tau = \frac{V}{It}\left[\frac{b}{2}\left(\frac{h^2}{4} - \frac{h_1^2}{4}\right) + \frac{t}{2}\left(\frac{h_1^2}{4} - y_1^2\right)\right]$$

(5.7′)

Taking $y_1 = 0$, this gives for the maximum shear stress at the neutral axis,

$$\tau_{\max} = \frac{V}{It}\left[\frac{b}{2}\left(\frac{h^2}{4} - \frac{h_1{}^2}{4}\right) + \frac{th_1{}^2}{8}\right]. \tag{e}$$

Taking $y_1 = h_1/2$, the minimum shear stress in the web at the junction with the flange becomes

$$\tau_{\min} = \frac{V}{It}\left[\frac{b}{2}\left(\frac{h^2}{4} - \frac{h_1{}^2}{4}\right)\right]. \tag{f}$$

From eq. (5,7'), it is seen that between $y_1 = 0$ and $y_1 = h_1/2$, the shear stress varies parabolically as represented by the horizontal ordinates of the diagram in Fig. 5.29b. The area of this diagram multiplied by the thickness t of the web represents the total shear force V_w carried by the web. Calculations show that for the usual proportions of standard I-sections and WF-sections, $V_w \approx V$, i.e., the flanges contribute very little to the total shear force V carried by the cross-section, and need not be considered.

In this regard it will only be pointed out that eq. (5.7) is not applicable in studying shear stresses in the flanges anyway, since it is not justifiable to assume that the shear stress will be uniform across the width b of a flange. For example, we see at once that for $y_1 = h_1/2$, the shear stress over the free surfaces ab and cd of the flange must be zero, while across the junction bc, it has the value given by eq. (f). This non-uniformity across the width b of the flange prevails to a considerable extent throughout the flange, and is too complex to be analyzed by elementary methods.

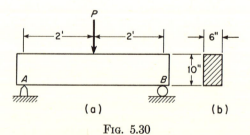

$$(a) \qquad\qquad (b)$$

FIG. 5.30

EXAMPLE 1. A simply supported wood beam of rectangular cross-section carries a concentrated load P at its mid-section as shown in Fig. 5.30. Allowable working stresses in tension or compression and in shear parallel to the grain are given as follows: $\sigma_w = 1000$ psi, $\tau_w = 160$ psi. What is the safe value of the load P?

SOLUTION. The cross-sectional area A and section modulus Z are

$$A = bh = 60 \text{ in.}^2; \quad Z = \frac{bh^2}{6} = 100 \text{ in.}^3$$

The maximum shear force V and maximum bending moment M are

$$V = \frac{P}{2}; \quad M = \frac{Pl}{4} = 12P \text{ in.-lb.}$$

Then from eqs. (5.9) and (5.5), we have

$$\tau_{max} = \frac{3}{2}\frac{V}{A} = \frac{P}{80}; \quad \sigma_{max} = \frac{M}{Z} = \frac{12P}{100}.$$

Setting these maximum stresses equal to the given working stresses and solving for the load P in each case gives

$$P_1 = 80\tau_w = 12{,}800 \text{ lb, based on shear stress,}$$

$$P_2 = 8.33\sigma_w = 8333 \text{ lb, based on bending stress.}$$

Thus the bending stress governs and the safe load is $P = 8333$ lb.

EXAMPLE 2. A beam AB supported at its ends has a span $l = 4$ ft and carries a uniform load of intensity $w = 2000$ lb/ft over the full span. The cross-section of the beam is a T-section having the dimensions shown in Fig. 5.31. Calculate the maximum shear stress τ, induced in the beam.

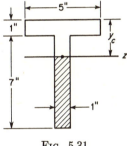

FIG. 5.31

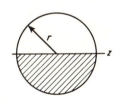

FIG. 5.32

SOLUTION. The maximum shear force occurs on a section just inside one of the supports and has the value equal to the reaction, i.e.,

$$V_{max} = \frac{wl}{2} = 4000 \text{ lb.}$$

The centroid of the section is located by the distance

$$y_c = \frac{4(\tfrac{1}{2}) + 8 \times 4}{12} = 2.83 \text{ in.}$$

The moment of inertia of the cross-section is

$$I_z = \frac{5(1)^3}{3} + \frac{1(7)^3}{3} - 12(1.83)^2 = 75.4 \text{ in.}^4$$

The maximum shear stress occurs at the neutral axis of the cross-section; the statical moment of the shaded area of the stem, with respect to this axis is

$$Q = 5.17 \times 1 \times \tfrac{1}{2}(5.17) = 13.34 \text{ in.}^3$$

Substituting these quantities into eq. (5.7), we obtain

$$\tau = \frac{VQ}{It} = \frac{4000 \times 13.34}{75.4 \times 1} = 708 \text{ psi.}$$

EXAMPLE 3. Calculate the maximum shear stress induced at the neutral axis of a beam of circular cross-section (Fig. 5.32) if the total shear force at the section is V.

SOLUTION. The moment of inertia of the cross-section is

$$I_z = \frac{\pi r^4}{4}.$$

The statical moment about the neutral axis of the shaded semicircular area is

$$Q = \frac{\pi r^2}{2} \times \frac{4r}{3\pi} = \frac{2}{3} r^3.$$

With these values, eq. (5.7) becomes

$$\tau_{\max} = \frac{VQ}{Ib} = \frac{4V \times 2r^3}{\pi r^4 \times 3 \times 2r} = \frac{4}{3} \frac{V}{A}, \tag{g}$$

where $A = \pi r^2$ is the area of the circular cross-section.

PROBLEM SET 5.5

1. If the load P on the beam in Fig. 5.30 is replaced by uniformly distributed load of intensity w lb/ft and the working stresses in shear and bending are as given in Example 1, what is the safe value of w? *Ans. $w = 3200$ lb/ft.*

2. Referring again to the beam in Fig. 5.30, assume that the load P is replaced by a uniformly distributed load of intensity w lb/ft. Then using the same data as given in Example 1, calculate the critical span length l of the beam below which the shear stress will govern and above which the bending stress will govern the safe intensity of load w. *Ans. $l = 5.21$ ft.*

3. A simply supported beam of span $l = 10$ ft carries a uniform load of intensity w. The cross-section of the beam is a rectangle 8 in. wide $\times$ 12 in. deep. The maximum flexure stress due to bending is 1200 psi. Calculate the maximum shear stress. *Ans. $\tau_{\max} = 120$ psi.*

4. A laminated wood beam is made up of three 2 $\times$ 4-in. planks glued together to form a solid cross-section 4 $\times$ 6 in. as shown in Fig. A. The allowable shear stress in the glued joints is $\tau_w = 50$ psi. If the beam is 6 ft long and simply supported at

FIG. A

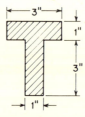

FIG. B

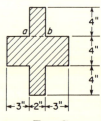

FIG. C

its ends, what is the safe load P that can be carried at mid-span? What is the corresponding maximum flexure stress? *Ans.* $P = 1800$ lb; $\sigma_{max} = 1350$ psi.

5. A simply supported beam of span $l = 9$ ft has the T-section shown in Fig. B. The beam carries a load $P = 1000$ lb at a point 3 ft from the right-hand support. Calculate the maximum shear stress in the beam. *Ans.* $\tau_{max} = 245$ psi.

6. A wood beam 18 ft long is simply supported with a span of 12 ft and an overhang at one end of 6 ft. The beam carries a uniformly distributed load of intensity $w = 200$ lb/ft over its entire length. The rectangular cross-section is 6 in. wide by 12 in. deep. Calculate the maximum shear stress in the beam. *Ans.* $\tau_{max} = 31.2$ psi.

7. The cross-section of a wood beam is an equilateral triangle with sides of length $a = 10$ in. Assuming that the horizontal shear stress at the neutral surface is uniformly distributed across the width of the beam, calculate its magnitude. The shear force at the cross-section is $V = 5000$ lb. *Ans.* $\tau_{max} = 153$ psi.

8. A beam of extruded magnesium has the cruciform cross-section shown in Fig. C. Calculate the shear stress at the junction ab if the total shear force on the cross-section is $V = 50,000$ lb. *Ans.* $\tau = 2500$ psi.

5.6 Stresses in Built-up Beams

Fabricated or built-up beams and girders are frequently found in engineering practice. Several examples of such beams are shown in Fig. 5.33. Figure 5.33a represents the cross-section of a simple box beam built up of four wood planks held together by nails or screws spaced at intervals along the length of the beam. Figure 5.33b represents the cross-section of a wood girder formed simply by bolting two rectangular timbers

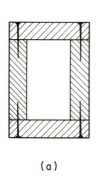 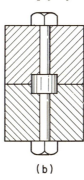 

(a) (b) (c)

Fɪɢ. 5.33

together and Fig. 5.33c represents a typical cross-section of a steel girder formed by riveting angle sections and cover plates to the edges of a webplate.

The stresses in such built-up beams are usually calculated on the assumption that the parts are rigidly connected so that the beam behaves like one of solid section. The computations will then involve: (a) designing the beam as a solid beam and (b) designing and spacing the connections so as to fulfill the requirements of (a). In the first step, the formulas for solid beams are used, making allowance for the effect of rivet holes, slots, etc., by using reduced cross-sectional areas. The calculations required to design the connecting elements can best be discussed in connection with particular cases.

Let us consider first the case of a box beam having the cross-sectional dimensions shown in Fig. 5.34a. We assume that the beam is simply supported at its ends with a span $l = 6$ ft and that it is to carry a vertical

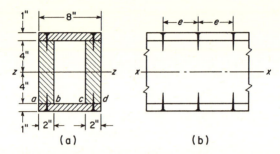

FIG. 5.34

load P at mid-span. The allowable working stress in bending is $\sigma_w = 1000$ psi. It is required to find the proper spacing e of lag screws, each of which can transmit a shear force $F = 600$ lb.

The moment of inertia of the cross-section with respect to its neutral axis zz is

$$I = \frac{4 \times (8)^3}{12} + 2\left[\frac{8 \times (1)^3}{12} + 8 \times (4.5)^2\right] = 495 \text{ in.}^4$$

and the corresponding section modulus

$$Z = \frac{I}{c} = \frac{495}{5} = 99.0 \text{ in.}^3$$

Then from eq. (5.5b)

$$\sigma_w = \frac{M}{Z} = \frac{P \times 72}{4 \times 99} = 1000 \text{ psi,}$$

from which $P = 5500$ lb. Thus the shear force at each cross-section of the beam becomes

$$V = \frac{P}{2} = 2750 \text{ lb.}$$

The shear stress across the junctions ab and cd of the section, from eq. (5.7), is

$$\tau = \frac{VQ}{Ib} = \frac{2750 \times 36}{495 \times 4} = 50 \text{ psi.}$$

The corresponding horizontal shear force to be transmitted by each screw is

$$F = \tfrac{1}{2}(4e\tau)$$

from which

$$e = \frac{F}{2\tau} = \frac{600}{2 \times 50} = 6 \text{ in.}$$

Built-up I-beams are more frequently used in practice than built-up wood beams. Such a beam (Fig. 5.35) is first designed as one of solid section, making suitable deductions in cross-sectional area to allow for the weakening effect of the rivet holes, and using a suitable working stress σ_w. The size and spacing of rivets will then be selected so as to develop the nominal flexure strength of the beam. To illustrate, let us consider the required spacing of the rivets which connect the flange angles to the web.

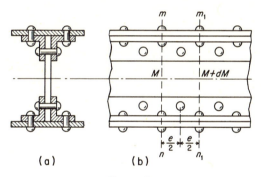

FIG. 5.35

Referring to Fig. 5.35, let e denote the spacing of these rivets and consider corresponding cross-sections mn and m_1n_1 at which the bending moments are M and $M + \Delta M$, respectively. Due to the difference ΔM in these bending moments, the normal stresses σ_x on the sections mn and m_1n_1 will be unbalanced and there is a tendency for the flange, represented by the shaded portion of the cross-section in Fig. 5.35a, to slide with respect to the web. This sliding is prevented by friction and by the shearing strength of the rivet. Neglecting friction, the force on the rivet becomes equal to the difference in normal stress resultants in the flange at sections mn and m_1n_1. This difference is

$$F = \frac{\Delta M}{I} \int y \, dA, \qquad\qquad (a)$$

where $\int y\, dA$ is the statical moment of the shaded area of the flange about the neutral axis of the section. Now using the relationship $dM/dx = V$, eq. (5.1), and replacing dM by ΔM and dx by e, we have

$$\Delta M = Ve. \tag{b}$$

Substituting this in eq. (a) and also denoting $\int y\, dA$ by Q_F, we obtain

$$F = \frac{Ve}{I} Q_F. \tag{5.10}$$

This represents the total shear force that must be carried by each rivet. Observing that each rivet is in double shear and assuming an allowable average shear stress for the rivet, the force F is readily computed for any given rivet size. The corresponding spacing e is then calculated from eq. (5.10).

EXAMPLE 1. A box beam is made up of two 10⌴20 channel sections connected to 16-in. × $\frac{1}{2}$-in. side plates with $\frac{3}{4}$-in.-diameter rivets spaced at 5 in. as shown in Fig. 5.36. Also, to allow for riveting and painting, access holes are cut in the side plates as shown. The beam has a span $l = 20$ ft and is to carry a concentrated load P at the middle. To allow for the weakening effect of rivet holes, the working stress for bending is chosen as $\sigma_w = 16,000$ psi. Calculate the safe load P, the average shear stress τ in the side plates along bc between access holes, and the shear stress in the rivets.

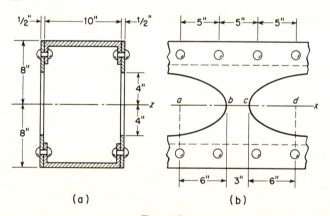

(a)   (b)

FIG. 5.36

SOLUTION. Referring to Table B.4 of Appendix B, we find that the smaller moment of inertia of one channel about its own centroidal axis is 2.8 in.[4] and the cross-sectional area of one channel is 5.86 in.[2] Then the moment of inertia of the entire cross-section, ignoring rivet holes, is

$$I_z = 2[2.8 + 5.86\,(7.39)^2] + \frac{1 \times 16^3}{12} - \frac{1 \times 8^3}{12} = 944.4 \text{ in.}[4]$$

Now using eq. (5.5a), and noting that the maximum bending moment is $M = Pl/4$, we obtain

$$\sigma_{max} = \frac{Mc}{I} = \frac{P \times 240 \times 8}{4 \times 944.4} = 16,000 \text{ psi,}$$

from which the safe load $P = 31,480$ lb. The corresponding shear force at any cross-section in $V = \frac{1}{2}P = 15,740$ lb.

To calculate the average shear stress in the side plates between b and c, we must first calculate the statical moment Q of half the cross-sectional area in Fig. 5.36a, with respect to the neutral axis of the cross-section. This becomes

$$Q = 5.86 \times 7.39 + 1 \times 4 \times 6 = 67.3 \text{ in.}^3$$

The total shear force F to be carried between cross-sections through a and d can now be found from eq. (5.10) by taking $e = 15$ in., the distance between centers of access holes. Thus

$$F = \frac{15,740 \times 15}{944.4} \times 67.3 = 16,800 \text{ lb.}$$

The web area available to transmit this force is $1 \times 3 = 3$ in.² Hence the average shear stress in the side plates along bc is

$$\tau_{av} = \frac{F}{A} = \frac{16,800}{3} = 5600 \text{ psi.}$$

To calculate the shear stress in each rivet, we must first find the statical moment of one channel section about the neutral axis of the cross-section. This becomes

$$Q_F = 5.86 \times 7.39 = 43.3 \text{ in.}^3$$

Then, again using eq. (5.10) and taking $e = 5$ in., we find

$$F = \frac{15,740 \times 5}{944.4} \times 43.3 = 3610 \text{ lb.}$$

There are two rivets each of cross-sectional area $A = 0.442$ in.² to transmit this force. Hence the average shear stress in one rivet becomes

$$\tau_{av} = \frac{3610}{2 \times 0.442} = 4090 \text{ psi.}$$

PROBLEM SET 5.6

1. A built-up beam having the cross-section shown in Fig. A consists of two standard 4I-7.7 steel beams connected by $\frac{3}{4}$-in.-diameter rivets spaced on 4-in. centers along the length of the beam. The beam is 5 ft long, simply supported, and uniformly loaded so as to develop a maximum bending stress of 16,000 psi. Calculate the average shear stress induced in the rivets nearest the ends of the beam. *Ans.* $\tau = 5330$ psi.

2. A cantilever beam is composed of two 6 × 6-in. timbers held together by bolts and connector rings as shown in Fig. B. The bolt holes are $\frac{3}{4}$ in. in diameter and each connector ring can safely transmit a force of 6000 lb in shear. If the load $P = 5000$ lb, what is the required spacing of the bolts? *Ans.* $e = 9.6$ in.

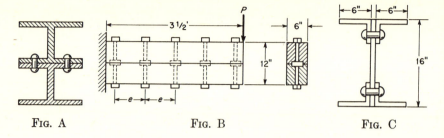

FIG. A   FIG. B   FIG. C

3. Calculate the maximum bending stress σ in the beam of the preceding problem, assuming that the first bolt on the left is 2 in. from the wall. *Ans.* $\sigma = 1590$ psi.

4. A box beam like that shown in Fig. 5.33a is made of four 6×1-in. wood planks connected by screws, each of which can safely transmit a shear force of 250 lb. Calculate the minimum spacing of screws along the length of the beam if the maximum shear force $V = 1000$ lb. *Ans.* $e = 4.38$ in.

5. A built-up girder is made by riveting four $6 \times 4 \times \frac{1}{2}$-in. angles to the edges of a 16×1-in. plate, using 1-in.-diameter rivets, as shown in Fig. C. The girder is 20 ft long, simply supported at the ends, and carries a concentrated load P at the middle. Calculate the rivet spacing e to develop the full flexural strength of the girder if $\sigma_w = 16,000$ psi and the allowable average shear stress in the rivets is $\tau_w = 6000$ psi. *Ans.* $e = 8.5$ in.

6. A built-up box girder consists of two 12-U-25 channel sections connected by two 12-in. $\times \frac{1}{2}$-in. flange plates as shown in Fig. D. This girder has a span of 16 ft and carries a concentrated load P at midspan. The allowable working stress in flexure is $\sigma_w = 16,000$ psi and the allowable shear stress for the $\frac{3}{4}$-in.-diameter rivets is $\tau_w = 6000$ psi. Calculate the safe value of the load P and the proper rivet spacing e. Neglect rivet holes. *Ans.* $P = 38,800$ lb; $e = 5\frac{1}{2}$ in.

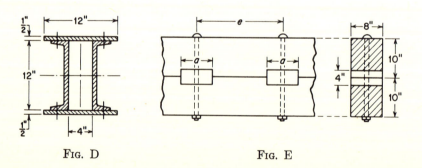

FIG. D   FIG. E

7. A wood girder is built up out of two 8-in. $\times$ 10-in. timbers connected by bolts and oak blocks as shown in Fig. E. The girder has a span of 12 ft and is to carry a concentrated load $P = 8$ tons at mid-span. For the oak blocks, the allowable bearing stress parallel to the grain is $\sigma_w = 1000$ psi and the allowable shear stress parallel to the grain is $\tau_w = 225$ psi. Calculate the proper values for dimensions a and e. Neglect the shear resistance offered by the bolts. *Ans.* $e = 27.5$ in.; $a = 8.9$ in.

6

Stresses in Beams: II

6.1 Plastic Bending of Beams

The preceding theory of bending has been developed on the assumptions that plane cross-sections of a beam before bending remain plane after bending and that the material follows Hooke's law. These assumptions lead to the condition of a linear distribution of stresses over the depth of the beam. This theory applies with good accuracy for many materials, so long as the deformations remain within the elastic limit. However, for the bending of a steel beam stressed beyond the yield point, it is no longer permissible to assume proportionality between stress and strain in the longitudinal fibers. Such bending of a steel beam beyond the elastic range of strain is called *plastic bending*.

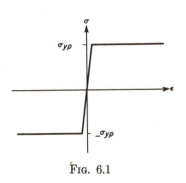

FIG. 6.1

To develop a theory of plastic bending of steel beams, we begin with a consideration of the stress-strain diagram for structural steel as shown in Fig. 2.8a, p. 33. Since the elastic limit stress and the yield stress are close together and since the plastic strain during yielding may be many times greater than the elastic strain before yielding, it is customary to idealize this stress-strain diagram by three straight lines as shown in Fig. 6.1. This diagram assumes that proportionality between stress and strain holds up to the yield stress $\sigma_{y.p.}$ and that for any strain beyond this, the stress remains constant and equal to $\sigma_{y.p.}$. It is also assumed that the matereial has the same yield point in tension and compression.

Now let us consider a prismatic beam of arbitrary symmetrical cross-section subjected to pure bending in the plane of symmetry as shown in Fig. 6.2. If the bending moment M is not too large, the beam will be elastic, the distribution of bending stresses over the depth of the beam will be *linear*

(Fig. 6.2a), and the maximum stress in the most remote fiber from the neutral axis will be less than the yield stress. This represents the case of elastic bending as discussed in the preceding chapter.

If the bending of the beam is increased beyond the elastic range, experiments show that plane cross-sections before bending continue to remain plane after bending. Thus we always have the strain in each fiber proportional to its distance from the neutral axis. Some of the fibers more remote from the neutral axis will acquire strains beyond the elastic limit. However, we see from the stress-strain diagram, Fig. 6.1, that the stresses in these overstrained fibers remain constant and equal to the yield stress $\sigma_{y.p.}$. This condition is represented in Fig. 6.2b. The central unshaded portion of the beam is still elastic while the shaded outer regions have become plastic.

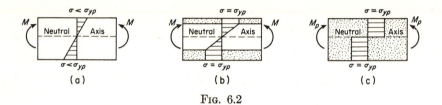

FIG. 6.2

With further increase in bending, more and more of the inner fibers reach the yield condition until finally the entire beam, with the exception of a very thin layer at the neutral axis, becomes plastic. This condition is represented in Fig. 6.2c. The corresponding bending moment is called the *plastic moment* for the section and is denoted by M_p. Neglecting strain hardening in the outer fibers, no further increase in the bending moment can be attained. Thus the plastic moment M_p represents the *limiting strength* of the beam in bending.

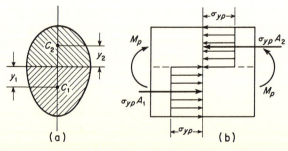

FIG. 6.3

To evaluate the plastic moment M_p for a given cross-section, Fig. 6.3a, we assume that in the fully plastic condition every fiber of the beam is stressed to the yield point $\sigma_{y.p.}$, those above the neutral axis being in compression and those below the neutral axis being in tension, as shown in Fig. 6.3b. Then since the sum of the internal forces over the entire cross-section must reduce to a couple equal to the bending moment M_p, we conclude that

$$\sigma_{y.p.}A_1 - \sigma_{y.p.}A_2 = 0, \qquad \text{(a)}$$

and

$$\sigma_{y.p.}A_1y_1 + \sigma_{y.p.}A_2y_2 = M_p, \qquad \text{(b)}$$

where A_1, A_2 are the areas below and above the neutral axis, respectively, and y_1 and y_2 are the distances to their centroids. For locating the neutral axis, we obtain from eq. (a)

$$A_1 = A_2 = \frac{1}{2}A, \qquad \text{(6.1)}$$

where $A = A_1 + A_2$ is the total area of the cross-section. Thus the neutral axis divides the total area of the cross-section into equal parts. It will be noted that the location of the neutral axis for plastic bending is, in general, different from that for elastic bending (see p. 114). Replacing A_1 and A_2 by $A/2$ in eq. (b), we obtain

$$M_p = \sigma_{y.p.} \frac{A}{2}(y_1 + y_2). \qquad \text{(6.2)}$$

For the particular case of a rectangular cross-section, $A = bh$, $y_1 = y_2 = h/4$ and eq. (6.2) becomes

$$M_p = \sigma_{y.p.}\left(\frac{bh^2}{4}\right). \qquad \text{(c)}$$

The maximum elastic moment M_e for the same section is

$$M_e = \sigma_{y.p.} Z = \sigma_{y.p.}\left(\frac{bh^2}{6}\right). \qquad \text{(d)}$$

Thus, for the rectangular cross-section, we have the ratio

$$\frac{M_p}{M_e} = \frac{3}{2}$$

and conclude that the plastic moment is 50 per cent greater than the maximum elastic moment for the same rectangular section.

For a circular cross-section of radius r, $A = \pi r^2$, $y_1 = y_2 = 4r/3\pi$ and eq. 6.2 becomes

$$M_p = \sigma_{y.p.}\left(\frac{4r^3}{3}\right). \qquad \text{(e)}$$

The maximum elastic moment for the same section is

$$M_e = \sigma_{\text{y.p.}} Z = \sigma_{\text{y.p.}} \left(\frac{\pi r^3}{4}\right). \tag{f}$$

Thus, in this case, the ratio

$$\frac{M_p}{M_e} = \frac{16}{3\pi} \approx 1.70,$$

from which the plastic moment is seen to be 70 per cent greater than the maximum elastic moment.

If we write eq. (6.2) in the form

$$\sigma_{\text{y.p.}} = \frac{M_p}{\frac{1}{2} A (y_1 + y_2)}, \tag{g}$$

we see then the quantity $\frac{1}{2} A (y_1 + y_2)$ may be considered as a section modulus for the case of plastic bending. Denoting this quantity by Z_p, we have

$$Z_p = \frac{1}{2} A (y_1 + y_2). \tag{6.3}$$

Applying this formula to the particular case of a 12WF50 wide flange section, we have, from Table B.2 of Appendix B, $A = 14.71$ in.^{2}, and

$$y_1 = y_2 = \frac{8.08 \times 0.641 \times 5.78 + 5.46 \times 0.371 \times 2.73}{7.36} = 4.82 \text{ in.}$$

Thus

$$Z_p = \tfrac{1}{2} \times 14.71 \times 9.64 = 70.9 \text{ in.}^3,$$

while the elastic section modulus is $Z = 64.7$ in.^{3} For such a section the ratio

$$\frac{M_p}{M_e} = \frac{Z_p}{Z} = \frac{70.9}{64.7} = 1.10$$

and the plastic moment is only 10 per cent greater than the maximum elastic moment.

In the practical design of beams, the allowable loads are sometimes selected on the basis of the plastic moment that can be developed by the section rather than the maximum elastic moment. This procedure, analogous to that of Art. 2.3, often results in a more efficient use of the material and is called *limit design*. To illustrate, let us consider the case of a simply supported beam loaded at the middle as shown in Fig. 6.4.

If the load P is not too large, the
entire beam will be elastic and we
have the deflection curve ACB as
shown in Fig. 6.4a. Increasing the
load gradually, a value will soon be
reached for which the maximum
extreme fiber stress at C will be equal
to $\sigma_{y.p.}$ although the beam is still
completely elastic. This load, denoted by P_E, represents the elastic
limit load for the beam. As P is
increased beyond this value, the
fibers in the neighborhood of point C

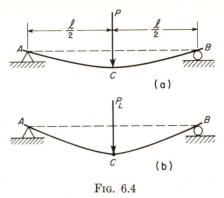

Fig. 6.4

(maximum bending moment) begin
to yield and this yielding will progress until, when the bending moment at C
reaches the value M_p, a *plastic hinge* will be formed at C. Most of the beam
will still be elastic but with the plastic hinge at C, it continues to deflect
without further increase in the load. The value of P which produces this
condition is called the *limit load* and is denoted by P_L. To calculate this
limit load, we have

$$\frac{P_L}{2} \cdot \frac{l}{2} = M_p,$$

from which $P_L = 4M_p/l$. A safe working load taken as $P_w = P_E/n$
will have a factor of safety n based on the elastic limit load. A safe working
load taken as $P_w = P_L/n_1$ will have a factor of safety n_1 based on the
plastic limit load. Using this second procedure for selecting the allowable
working load represents the idea of limit design.

EXAMPLE 1. A stepped steel shaft of circular cross-section is built-in at C and
loaded at the free end A as shown in Fig. 6.5. Calculate the ratio of diameters
$d_1 : d_2$ in order that plastic hinge conditions will develop simultaneously at B and C.

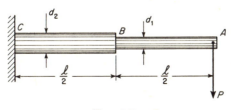

Fig. 6.5

SOLUTION. The bending moment at B is $-Pl/2$ while that at C is $-Pl$; thus
they are in the ratio $1 : 2$, and the plastic section moduli at B and C must be in
the same ratio. From this requirement, we have (see eq. e)

$$d_1{}^3/d_2{}^3 = \tfrac{1}{2}$$

from which $d_1/d_2 = 1/\sqrt[3]{2} = 0.794$.

EXAMPLE 2. Calculate the plastic moment M_P for the unbalanced I-section shown in Fig. 6.6 if the material is structural steel with a yield stress of 40,000 psi.

SOLUTION. For fully developed plastic bending, the neutral axis divides the total area of the section into two equal parts. Hence we have

$$(3 \times 1) + a_1(1) = (4 \times 1) + a_2(1).$$

FIG. 6.6

Then, since $a_1 + a_2 = 4$ in., we find $a_1 = 2.5$ in., $a_2 = 1.5$ in. We may now find the distances y_1 and y_2 to the lower and upper centroids as follows:

$$y_1 = \frac{2.5 \times 1.25 + 3 \times 3.0}{2.5 + 3} = 2.21 \text{ in.}$$

$$y_2 = \frac{1.5 \times 0.75 + 4 \times 2.0}{1.5 + 4} = 1.66 \text{ in.}$$

Finally, using eq. (6.2), we obtain

$$M_P = 40,000 \times 5.5(2.21 + 1.66) = 852,000 \text{ in.-lb.}$$

PROBLEM SET 6.1

1. A steel beam having a cross-section in the form of an isosceles triangle as shown in Fig. A is subjected to pure bending in its longitudinal plane of symmetry. Locate the neutral axis NN for fully plastic bending and calculate the plastic section modulus. *Ans.* $h_1 = 0.707h$; $Z_P = 0.0977bh^2$.

2. A prismatic steel bar has the trapezoidal cross-section shown in Fig. B. Locate the neutral axis NN for fully plastic bending and calculate the corresponding plastic section modulus. *Ans.* $Z_P = 18.2$ in.^{3}

3. Locate the neutral axis NN for fully plastic bending of a steel beam of semi-circular cross-section as shown in Fig. C. *Ans.* $h_2 = 0.404r$.

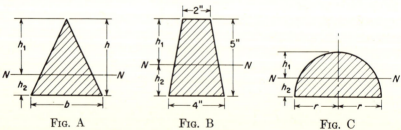

FIG. A   FIG. B   FIG. C

4. A simply supported steel beam has a span $l = 4$ ft and a rectangular cross-section 1 in. wide by 2 in. deep. If the beam is uniformly loaded over the full span, what intensity of load w will induce a plastic hinge at the mid-point? The yield stress is $\sigma_{y.p.} = 36,000$ psi. *Ans.* $w = 1500$ lb/ft.

5. Calculate the maximum plastic moment M_P for a 12WF72 steel girder if the yield stress is 40,000 psi. *Ans.* $M_P = 354,000$ ft-lb.

6.2 Beams of Two Materials

Beams made of two or more different materials are frequently encountered in engineering practice. Common examples are wood beams reinforced by steel plates and concrete beams reinforced by imbedded steel rods. The theory of bending of such beams within the elastic range of the materials is quite simple. For a beam of two materials, the procedure is to transform the composite beam into an equivalent beam of one material.

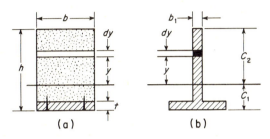

FIG. 6.7

Consider, for example, the case of a wood beam of rectangular cross-section reinforced on the bottom by a steel plate as shown in Fig. 6.7. Assuming that no slip takes place between the wood and steel during bending, the theory of solid beams will still apply. That is, plane cross-sections before bending remain plane after bending and the strains in the longitudinal fibers are proportional to their distances from the neutral axis. Thus for any bending curvature $1/\rho$ within the elastic range of the material, the normal fiber stress in the wood, at the distance y from the neutral axis, is

$$\sigma = E_w y/\rho$$

and the corresponding element of normal force on the area $dA = b\,dy$ is

$$dF_w = \frac{E_w}{\rho}\,by\,dy, \tag{a}$$

where E_w is the modulus of elasticity of wood. Similarly, for an elemental area of the steel,

$$dF_s = \frac{E_s}{\rho}\,by\,dy, \tag{b}$$

where E_s is the modulus of elasticity of steel. In the case of pure bending these elemental forces, summed over the total areas of wood and steel, respectively, must have a net resultant force equal to zero, and the sum of their moments about the neutral axis must be equal to the resisting moment developed by the section. Without actually making these summations, we note that the results will be unchanged if we write eq. (a) in the equivalent form

$$dF_w = \frac{E_s}{\rho} \left(\frac{E_w}{E_s} b \right) y \, dy. \tag{a'}$$

This shows that we may regard the wood portion of the cross-section of width b as equivalent to a steel stem of reduced width

$$b_1 = \frac{E_w}{E_s} b \tag{c}$$

as shown in Fig. 6.7b. Under a given loading, the composite section in Fig. 6.7a and the *transformed section* in Fig. 6.7b will both develop the same resisting moment.

As soon as we have the transformed section, the problem of the bending of the beam of two materials is reduced to that of the bending of a steel beam of T-section which can be treated on the basis of the theory previously developed in Art. 5.3. Thus for a given bending moment M on the section, the extreme tensile stress at the bottom will be

$$\sigma_1 = \frac{Mc_1}{I},$$

and at the top the compressive stress will be

$$\sigma_2 = \frac{Mc_2}{I},$$

where I is the moment of inertia of the transformed section, Fig. 6.7b. These bending stresses are, of course, for the equivalent steel section. To obtain the true maximum compressive stress in the wood, the stress σ_2 must be reduced in the ratio E_w/E_s.

In the above discussion, the less stiff material (wood) was transformed

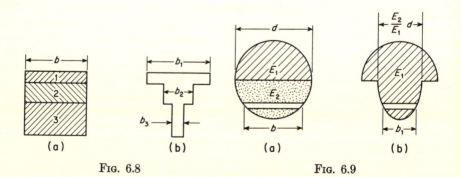

(a)   (b)   (a)   (b)

FIG. 6.8   FIG. 6.9

into an equivalent width b_1 of the stiffer material (steel). The reverse procedure will obviously be equally satisfactory.

If the composite beam consists of more that two materials, the transformed section may, in general, be obtained in terms of any one of them. For example, Fig. 6.8a represents the rectangular cross-section of a laminated beam of three materials having moduli of elasticity $E_1 > E_2 > E_3$. Then a suitable transformed section of the stiffest material will be that shown in Fig. 6.8b, where $b_1 = b$, $b_2 = E_2 b/E_1$, and $b_3 = E_3 b/E_1$.

If the composite beam does not have a rectangular cross-section, the problem of finding an equivalent transformed section may be more complicated. Consider, for example, a beam of two materials having a circular cross-section as shown in Fig. 6.9a and assume $E_1 > E_2$. Then in the transformed section, Fig. 6.9b, each elemental strip of width b in the lower half of the cross-section must be reduced to width $b_1 = (E_2/E_1)b$. This will be accomplished by making the lower half of the transformed section elliptical as shown.

EXAMPLE 1. A timber beam having an 8×12-in. rectangular cross-section is reinforced top and bottom by steel plates 8 in. wide $\times \frac{1}{2}$ in. thick as shown in Fig. 6.10a. The moduli of elasticity are $E_s = 30(10)^6$ psi and $E_w = 1.5(10)^6$ psi. The allowable working stresses are $\sigma_s = 16,000$ psi and $\sigma_w = 1200$ psi. Find the maximum allowable bending moment for the section.

SOLUTION. The ratio $E_w : E_s = 1 : 20$. Hence the transformed section will be an I-section with web thickness $b_1 = b/20 = 0.4$ in. as shown in Fig. 6.10b.

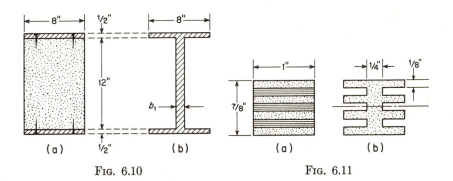

FIG. 6.10   FIG. 6.11

To determine the safe bending moment, we must first decide which allowable working stress will govern. The allowable stress in the web of the equivalent beam will be $\sigma'_s = 20 \times 1200 = 24,000$ psi. The allowable stress in the flanges is $\sigma_s = 16,000$ psi, hence the flange stress governs.

The moment of inertia of the transformed section is

$$I = 2\left[\frac{1}{12} + 4(6.25)^2 + \frac{0.4(6)^3}{3}\right] = 370 \text{ in.}^4$$

Then, from eq. (5.4)

$$M_{max} = \frac{I}{c}\sigma_s = \frac{370}{6.5} \times 16{,}000 = 911{,}000 \text{ in.-lb.}$$

EXAMPLE 2. The cross-section of a small beam cut from a sheet of 7-ply plywood has the dimensions shown in Fig. 6.11a. Alternate layers of the plywood have the grain parallel to the length of the beam. The beam is 4 ft long, simply supported, and loaded at the middle. The modulus of elasticity parallel to the grain is $E_1 = 1.6(10)^6$ psi, that perpendicular to the grain is $E_2 = 0.4(10)^6$ psi. The corresponding working stresses are $\sigma_1 = 1200$ psi and $\sigma_2 = 300$ psi. Calculate the safe value of the load P.

SOLUTION. The ratio of the smaller modulus to the larger is $E_2 : E_1 = 1 : 4$. Hence the width of layers having the grain across the axis of the beam should be reduced in the ratio 1 : 4 and the transformed section will be as shown in Fig. 6.11b. The moment of inertia of this transformed section is

$$I = \frac{1(0.875)^3}{12} - 3\left[\frac{0.75(0.125)^3}{12}\right] - 2[0.09375(0.25)^2] = 0.0439 \text{ in.}^4$$

The corresponding safe bending moment for the section is

$$M_{max} = \frac{I}{c}\sigma_1 = \frac{0.0439}{0.437} \times 1200 = 120 \text{ in.-lb.}$$

and the safe load P at the middle of the beam is

$$P = \frac{4M_{max}}{l} = \frac{480}{48} = 10 \text{ lb.}$$

If a laminated beam of two materials has many layers, it will be justifiable to express the moment of inertia of the transformed section by the approximate formula

$$I \approx \frac{bh^3}{12} - \frac{1}{2}\frac{(b - b_1)\, h^3}{12}, \tag{d}$$

where $b_1 = E_2 b/E_1$ is the reduced width of the softer layers. Substituting this value of b_1 into eq. (d), we obtain

$$I \approx \frac{bh^3}{24}\left(1 + \frac{E_2}{E_1}\right). \tag{e}$$

PROBLEMSET 6.2

1. A simply supported beam 10 ft long and carrying a concentrated load P at the middle has the cross-section shown in Fig. 6.7a. Taking $P = 2000$ lb, $E_w/E_s = 1/20$, $b = 4$ in., $h = 6$ in., and $t = 1/2$ in., calculate the maximum tensile stress in the steel and the maximum compressive stress in the wood. *Ans.* $\sigma_s = 8580$ psi; $\sigma_w = 1530$ psi.

2. A wood beam having a rectangular cross-section 8 in. wide by 12 in. deep is reinforced top and bottom by two plates as shown in Fig. 6.10a. The top plate is 1 in. thick and is made of cast iron for which $E_c = 15(10)^6$ psi and the bottom plate is $\frac{1}{4}$ in. thick and is made of steel for which $E_s = 30(10)^6$ psi. The modulus of elasticity for the wood is $E_w = 1.5(10)^6$ psi. If the beam is to be loaded so that the

maximum tensile stress in the steel is 16,000 psi, what is the corresponding maximum compressive stress in the cast iron? *Ans.* $\sigma_c = 6090$ psi.

3. Referring to the circular cross-section of two materials as shown in Fig. 6.9 and denoting the ratio of moduli of elasticity E_2/E_1 by α, calculate the distance y_c from the horizontal diameter to the centroid of the transformed section. *Ans.*

$$y_c = \frac{2d}{3\pi} \frac{1-\alpha}{1+\alpha}.$$

4. A laminated wood beam has a rectangular cross-section 8 in. wide by 12 in. deep. The layers are each $\frac{1}{16}$ in. thick, and alternate layers have their grains crossed at right angles, the two outside layers being grained along the axis of the beam. The ratio $E_2 : E_1 = 1 : 4$, and the corresponding working stresses are $\sigma_1 = 1200$ psi parallel to the grain and $\sigma_2 = 250$ psi across the grain. Calculate the safe bending moment for the section. *Ans.* $M = 10,100$ ft-lb.

5. A prismatic rod of hexagonal cross-section is made up of six equilateral triangular segments as shown in Fig. A. The segments are alternately Redwood ($E_1 = 1.2 (10)^6$ psi) and soft White Pine ($E_2 = 1.0(10)^6$ psi) glued together to form a solid section. The dimension $a = 1$ in. If the bar is subjected to pure bending in its longitudinal plane of symmetry, calculate the maximum bending stress σ, assuming that $M = 1000$ in.-lb. *Ans.* $\sigma = 1750$ psi.

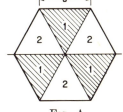

Fig. A

6. If the rod described in the preceding problem is supported at its ends as a simple beam and loaded at the middle by a transverse force $P = 300$ lb, calculate the shear stress in the glue between the upper and lower halves of the section. *Ans.* $\tau = 69$ psi.

6.3 Reinforced Concrete Beams

As mentioned in the preceding article, concrete beams reinforced with steel represent an important example of beams of two different materials. A complete treatment of the theory of reinforced concrete is beyond the scope of this book; we give here only a brief discussion of this important problem.*

It is well known that the strength of concrete is much greater in compression than in tension. Hence a rectangular beam of concrete will fail from the tensile stresses on the convex side. The beam can be greatly strengthened by imbedding steel bars on the convex side as shown in Fig. 6.12. Since concrete grips the steel strongly, there will be no sliding of the steel bars with respect to the concrete during bending and the methods developed in the previous article can also be used here for calculating bending stresses. In practice, the cross-sectional area of the steel bars is usually such that the tensile strength of the concrete on the convex

---

*For a more complete treatment of reinforced concrete, see C. W. Dunham, *The Theory and Practice of Reinforced Concrete*, McGraw-Hill Book Co., Inc., New York, 1953

side is overcome before yielding of the steel begins, and at larger loads
the steel alone takes practically all the tension. Hence it is established
practice in calculating bending stresses in reinforced-concrete beams of the
type shown in Fig. 6.12 to assume that all the tension is taken by the steel
and all the compression by the concrete.

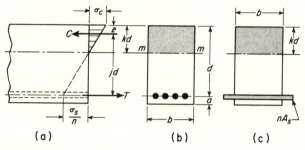

FIG. 6.12

Concrete does not follow Hooke's law, and a compression test diagram
for this material has a shape similar to that for cast-iron (see Fig. 2.8b).
As the compressive stress increases, the slope of the tangent to the diagram
decreases, that is, the modulus of concrete decreases with increase in stress.
In calculating stresses in reinforced-concrete beams, it is usual practice
to assume that Hooke's law holds for concrete and to compensate for the
variable modulus by taking a lower value for this modulus than that
obtained from compression tests at small compressive stresses. In specifications for reinforced concrete it is usually assumed that n, the ratio of
the modulus of elasticity of the steel to that of the concrete, is

$$n = \frac{E_s}{E_c} = 15.$$

Transforming the area of the steel A_s into an area of concrete nA_s,
equivalent as far as elastic properties are concerned, the transformed
cross-section is that shown in Fig. 6.12c. The stress distribution follows
a linear law, since plane cross-sections remain plane during bending and
since Hooke's law is assumed to hold for the concrete. Under these
conditions, the neutral axis will lie at the centroid of the shaded cross-
section (Fig. 6.12c). This requires that the first moment of the shaded
area above the neutral axis mm with respect to mm must equal that of
the shaded area below; that is,

$$(bkd) \frac{kd}{2} = nA_s (d - kd).$$

This equation is quadratic in terms of k, the value of which defines the
position of the neutral axis.

If the ratio of the area of the steel A_s to that of the area of. the concrete above the center of the steel be denoted by

$$p = \frac{A_s}{bd},$$ (a)

the solution of the above equation gives

$$k = \sqrt{2pn + (pn)^2} - pn.$$ (6.4)

Instead of using the moment of inertia of the transformed section as in the preceding article, it is more direct to proceed as follows: The average stress acting over the area above the neutral axis is $\sigma_c/2$ where σ_c is the maximum compressive stress in the concrete. The total compressive force C is $(\sigma_c/2)$ bkd. For a linear stress distribution in bending, the center of pressure of C is at the centroid of the triangle which represents the stress distribution; that is, at $z = \frac{1}{3}kd$. With no axial forces acting on the beam, the tension T must equal C and these two forces constitute a couple whose moment is the moment of resistance and is equal to the product of C and the moment arm jd, where $j = 1 - k/3$. Hence the bending moment M_c based on a maximum stress σ_c in the concrete is

$$M_c = \frac{\sigma_c}{2} bkd(jd) = \frac{1}{2} \sigma_c kjbd^2.$$ (6.5)

As the maximum allowable fiber stress in the steel will probably not be produced by the same value of bending moment as produced σ_c, it is necessary to compute also the bending moment M_s based on σ_s. The couple used for this purpose is $T(jd)$ where $T = \sigma_s A_s$ is the force in the steel. Then

$$M_s = \sigma_s A_s jd.$$ (6.6)

In determining the safe bending moment that may be applied, the smaller of the two values M_c and M_s should be used.

If the bending moment be given rather than the maximum allowable fiber stresses, the value of the maximum fiber stresses caused by the given moment can be computed in the same manner as above and then examined to see if either or both exceed permissible values.

To protect the steel from damage by fire, the reinforcement in beams, girders, and columns should not be placed nearer the exposed surface than $1\frac{1}{2}$ in. This specification fixes a lower limit for the depth of concrete below the reinforcing bars designated as a in Fig. 6.12b.

EXAMPLE 1. If $n = 15$ and $\sigma_c = 650$ psi, determine the safe load at the middle of a reinforced-concrete beam 10 ft long, freely supported at the ends, and having $b = 10$ in., $d = 12$ in., $A_s = 1.17$ sq in. What is the stress in the steel? Consider two cases: (a) if the weight of the concrete is neglected; (b) if the concrete weighs 150 lb per cu ft.

SOLUTION. (a) The transformed area of the reinforcing rods is $nA_s = 15(1.17) = 17.55$ sq in. The neutral axis 1-1 must be so located that it is a centroidal axis for the shaded areas shown in Fig. 6.13a. This requires that $\Sigma M_{1-1} = 0$, i.e.,

$$(10y)\,\frac{y}{2} = 17.55\,(12 - y),$$

from which $y = 4.965$ in.

The compressive force C is the product of the compression area and the average compressive stress acting thereon; namely, $C = (650/2)(10 \times 4.965) = 16,100$ lb. Then since $T = C$, and the moment arm of the couple is $(12 - 4.965/3) = 10.35$ in., the moment based on a safe stress of 650 psi in the concrete is

$$M_c = 16,100(10.35) = 166,500 \text{ in.-lb.}$$

The maximum bending moment is $Pl/4$; equating this to 166,500 in.-lb, one finds $P = 5570$ lb.

Since $T = C = 16,100$ lb, the stress in the steel is $T/A = 16,100/1.17 = 13,750$ psi.

(b) The concrete weighs $150(10)(14)/144 = 146$ lb per ft of length of the 10×14-in. cross-section, allowing 2 in. of concrete below the reinforcing rods. The maximum bending moment caused by this uniform load is $wl^2/8 = [146(10)^2/8]12 = 21,900$ in.-lb. Hence the moment available for carrying the concentrated load P is $166,500 - 21,900 = 144,600$ in.-lb. Then $Pl/4 = 144,600$ and $P = 4820$ lb.

From the foregoing discussion it is seen that the bending moment which produces the maximum permissible stress in the concrete of a reinforced-concrete beam will not necessarily stress the steel to its safe limit. Or it may, on the other hand, overstress the steel, and if the bending moment is reduced so as to lower the steel stress to a safe value, then the concrete will be understressed. Both of these conditions are unfavorable from the standpoint of economy, since the full strength of either the steel or the concrete is not being utilized. When a reinforced concrete beam is being designed, however, it is possible to so proportion the ratio of the steel to the concrete in the cross-section that the maximum permissible stresses in the concrete and the steel are realized simultaneously. The beam then is said to have *balanced reinforcement*. In most practical cases, balanced reinforcement can

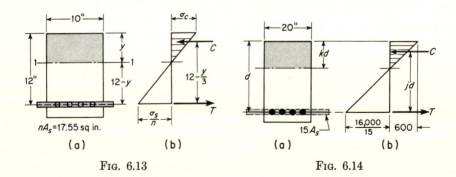

FIG. 6.13   FIG. 6.14

only be approximated on account of design conditions, such as size of reinforcing rods available, etc.

In ideal balanced reinforcement, the tensile stress in the equivalent concrete, which replaces the steel and whose area is n times as great as that of the steel, is to be σ_s/n at the same time as the compressive stress in the outside fiber of the actual concrete is σ_c. Since these stresses are proportional to their distances from the neutral axis,

$$\frac{\sigma_c}{kd} = \frac{\sigma_s/n}{d - kd} \quad \text{or} \quad \frac{1 - k}{k} = \frac{\sigma_s}{n\sigma_c}. \tag{6.7}$$

Once k is determined, the remaining procedure is exactly the same as before and is illustrated in the following numerical example.

EXAMPLE 2. Design a concrete beam with balanced reinforcement to sustain a bending moment of 2,000,000 in.-lb. Width of beam to be 20 in., $\sigma_s = 16,000$ psi; $\sigma_c = 600$ psi; $n = 15$.

SOLUTION. The transformed area and the stress distribution are shown in Fig. 6.14. The tensile stress in the transformed area of concrete equivalent to the original steel is one-fifteenth of that in the steel since its area has been multiplied by 15. The distribution of stress intensities is assumed to be a straight line as shown in Fig. 6.14b. The neutral axis is located by using eq. (6.7), i.e.,

$$\frac{1 - k}{k} = \frac{16,000}{15 \times 600},$$

from which

$$k = 0.360.$$

The moment arm of the couple TC is

$$jd = d - \frac{kd}{3} = 0.88d.$$

The compressive force in the concrete above the neutral axis is $(300)(0.36d)(20) = 2160d$. The resisting moment is $(2160d)(0.88d)$ and is equal to the given bending moment of 2,000,000 in.-lb. Therefore $d^2 = 1052$ sq in., from which $d = 32.5$ in.

In balanced reinforcement, $T = C$, and hence

$$16,000(A_s) = 2160\,(32.5)$$

from which

$$A_s = 4.39 \text{ sq in.}$$

This area of steel can be supplied by six 1-inch-diameter reinforcing bars.

PROBLEM SET 6.3

1. For a reinforced-concrete beam, $b = 6$ in., $d = 10$ in., and the total area of the reinforcing rods is $\frac{2}{3}$ sq in. Assume $n = 15$, $\sigma_c = 600$ psi, and $\sigma_s = 18,000$ psi. What is the maximum bending moment in in.-lb that the beam can carry? *Ans.* $M = 66,800$ in.-lb.

2. For a reinforced-concrete beam, $b = 15$ in., $d = 28$ in., and the total area of the reinforcing rods is 3.02 sq in. Assume $n = 20$, $\sigma_c = 450$ psi, $\sigma_s = 12,500$ psi. What is the maximum resisting moment in ft-lb? *Ans.* 76,000 ft-lb.

3. For a reinforced-concrete beam, $b = 10$ in., $d = 18$ in., total area of reinforcing rods is 1 sq in., $n = 15$. If this beam is subjected to a bending moment of 10,000 ft-lb, what will be the maximum stresses set up in the concrete and the steel? *Ans.* $\sigma_c = 250$ psi; $\sigma_s = 7500$ psi.

4. In a certain beam, the stresses produced in the concrete and steel are to be determined. $b = 20$ in., $d = 36$ in., area of reinforcing rods = 7.2 sq in., $n = 18$ and the applied bending moment is 2,500,000 in.-lb. *Ans.* $\sigma_s = 11,330$ psi; $\sigma_c = 508$ psi.

5. For a reinforced-concrete beam, $b = 18$ in., $d = 27$ in., $n = 15$, $\sigma_c = 500$ psi. If the beam is to resist a bending moment of 800,000 in.-lb, how much area of steel reinforcement must be used? What is the stress set up in the steel? *Ans.* 1.593 sq in.; 20,450 psi.

6. Solve Problem 5 if the bending moment is 1,200,000 in.-lb. *Ans.* 5.12 sq in.; 10,100 psi.

7. A cantilever beam, 6 ft long and 8 in. wide, is to be made of reinforced concrete. It is to be designed to carry a total uniformly distributed load of 6000 lb, including its own weight. What should be its effective depth at the wall? 3 ft from the wall? Indicate the position of the steel reinforcement. Assume $n = 15$, $\sigma_c = 650$ psi, $\sigma_s = 16,000$ psi. *Ans.* 15.84 in.; 7.92 in.

8. A reinforced-concrete beam is to carry a bending moment of 102,400 in.-lb. If $d = 1.5b$, what are b, d, and the total area of reinforcing steel needed? Assume $\sigma_s = 18,000$ psi and $\sigma_c = 600$ psi; $E_s = 30 \times 10^6$ psi and $E_c = 2 \times 10^6$ psi. *Ans.* $d = 12$ in.; $A = 0.533$ sq in.

9. A simply supported beam 15 ft long carries a center load of 6500 lb. The economic percentage of steel is to be used and $d = 20$ in. Find b and A_s, assuming $\sigma_s = 16,000$ psi, $\sigma_c = 500$ psi, and $n = 15$. Take account of the weight of the beam, assuming concrete to weigh 150 lb per cu ft. Allow 2 in. for depth a below the reinforcing bars. *Ans.* $b = 14.1$ in.; $A_s = 1.41$ sq in.

10. In the beam of Problem 9 three $\frac{5}{8}$-in.-diameter round bars and two $\frac{1}{2}$-in. square bars are used for reinforcement. The neutral axis may be assumed to be 6.39 in. below the top. What is the maximum shearing stress set up in the concrete and what is the maximum bond stress between reinforcement and concrete? *Ans.* 22.5 psi; 32 psi.

6.4   Bending of Beams of Arbitrary Cross-Section

In developing the theory of bending of prismatic beams, Art. 5.3, it was emphasized that the beam was assumed to have an axial plane of symmetry and that the applied loads were acting in this plane. In general, if the applied loads do not act in a plane of symmetry, bending of the beam will not take place in the plane of loading and the theory developed in Art. 5.3 becomes invalid.

Consider, for example, the case of a beam of Z-section subjected to bending couples M applied in the xy-plane as shown in Fig. 6.15. Let it be assumed that the beam also bends only in this plane, i.e., that Oz is the neutral axis of the cross-section. Then tensile and compressive stresses in the flanges will be uniformly distributed over the widths b, and resultant internal forces T and C will act at the mid-points of the flanges as shown. Clearly these equal and opposite forces constitute an internal bending

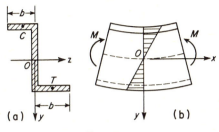

Fig. 6.15

moment about the y-axis which is not balanced by any corresponding applied external moment, the applied couples M being in the xy-plane. Thus the assumed deformation (curvature only in the xy-plane) is inconsistant with the conditions of equilibrium, and the assumption is proved invalid. Unless external constraint against bending in the xz-plane is provided, we must conclude that bending couples in the xy-plane produce also some bending in the xz-plane.

We now proceed to develop a more general theory of bending of prismatic beams of *arbitrary cross-section*. In Fig. 6.16, let such a beam be subjected to pure bending and assume that plane cross-sections before bending remain plane after bending. Then for the conditions shown, all longitudinal fibers above a certain line nn in the cross-section will be in compression, and all fibers below this line will be in tension. The line nn represents the neutral axis of the section. Through any point O on this line as origin, we take coordinate axes x, y, z, such that Ox is parallel to the axis of the beam and the plane xy is the plane in which the external couples M_z are acting. We have already observed that such couples will, in general, produce bending of the beam both in the xy-plane of loading and

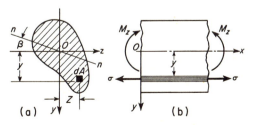

Fig. 6.16

in the perpendicular xz-plane. Let $1/\rho_z$ denote the curvature in the xy-plane and $1/\rho_y$, that in the xz-plane. Then plane cross-sections before bending remaining plane after bending, the strain in any longitudinal fiber, defined by coordinates y, z, will be

$$\epsilon = \frac{y}{\rho_z} + \frac{z}{\rho_y}.$$

For elastic behavior, stress is proportional to strain so that

$$\sigma = \frac{Ey}{\rho_z} + \frac{Ez}{\rho_y}. \tag{a}$$

This defines the stress distribution over the cross-section and since the stress resultant must reduce to a couple M_z about the z-axis, we have

$$\int_A \sigma dA = 0, \quad \int_A \sigma y \, dA = M_z, \quad \int_A \sigma z \, dA = 0. \tag{b}$$

Substituting the value of σ from eq. (a) into these equilibrium equations, we obtain.

$$\left. \begin{array}{l} \dfrac{E}{\rho_z} Ay_c + \dfrac{E}{\rho_y} Az_c = 0, \\[2mm] \dfrac{EI_z}{\rho_z} + \dfrac{EI_{yz}}{\rho_y} = M_z, \\[2mm] \dfrac{EI_{yz}}{\rho_z} + \dfrac{EI_y}{\rho_y} = 0, \end{array} \right\} \tag{c}$$

where

$$I_z = \int y^2 dA, \quad I_y = \int z^2 dA, \quad I_{yz} = \int yz dA,$$

are moments of inertia and product of inertia of the cross-section and y_c, z_c, are the coordinates of the centroid of its area A.

The first of these three equations is satisfied by taking $y_c = z_c = 0$; hence we conclude that the neutral axis passes through the centroid of the cross-section. From the last two equations, we obtain

$$\frac{1}{\rho_z} = \frac{M_z I_y}{E(I_y I_z - I_{yz}^2)}; \quad \frac{1}{\rho_y} = \frac{M_z I_{yz}}{E(I_{yz}^2 - I_y I_z)}. \tag{6.8}$$

These expressions define the curvatures in the xy and xz planes, respectively, due to bending couples applied in the xy-plane as shown in Fig. 6.16 b. We note that when xy is a plane of symmetry, $I_{yz} = 0$ and $1/\rho_y = 0$, while $1/\rho_z = M_z/EI_z$ which coincides with eq. (5.4) of Art. 5.3.

Substituting expressions (6.8) for curvatures into eq. (a), we obtain

$$\sigma = \frac{M_z(I_y y - I_{yz} z)}{I_z I_y - I_{yz}^2}, \tag{6.9}$$

which gives the bending stress in any fiber the location of which is defined by the coordinates y, z. Again if xy is a plane of symmetry, $I_{yz} = 0$ and eq. (6.9) reduces to $\sigma = M_z y / I_z$, which coincides with eq. (5.5) of Art. 5.3.

Setting $\sigma = 0$ from eq. (6.9), we obtain, for the equation of the neutral axis,

$$I_y y - I_{yz} z = 0$$

or

$$y = \frac{I_{yz}}{I_y} z = z \tan \beta. \tag{6.10}$$

Thus the neutral axis of the cross-section makes the angle $\beta = \tan^{-1} (I_{yz}/I_y)$ with the z-axis as shown in Fig. 6.16a.

For any cross-section, it can be shown that there are always two orthogonal centroidal axes in its plane for which the product of inertia $I_{yz} = 0$. These are called the *principal axes* of the cross-section (see Appendix B). The corresponding axial planes of the beam are called the principal planes of bending. From the foregoing discussion, we conclude that for bending moments applied in such a principal plane, only bending in this plane will take place and the usual theory of bending is valid. Thus for a beam subjected to bending moment that is not in a principal plane, we may always resolve this bending moment into components coinciding with the two principal planes of the beam. Then, by superposition, the total bending stress in any fiber will be obtained by adding algebraically the two stresses produced separately by these components. This procedure is illustrated in the first of the examples following.

All of the above discussion holds rigorously only for pure bending. If a beam is bent by transverse loads, there will be shearing stresses and deformations as well as bending stresses. In the case of beams of solid section, these shearing stresses do not greatly effect the bending action and it is satisfactory to calculate the bending stresses as above without consideration of the shear. In the case of beams of thin-walled profile section, such as the Z-section shown in Fig. 6.15, the shearing stresses are of especial interest, and this question will be considered further in Art. 6.5.

EXAMPLE 1. A simply supported wood beam (roof purlin) of rectangular cross-section carries a uniform load of intensity w as shown in Fig. 6.17. The plane of symmetry $xO2$ of the beam is inclined to the vertical xy-plane of loading by an angle α as shown. Calculate the maximum bending stress σ_{max} if $l = 10$ ft, $w = 200$ lb per ft, $b = 6$ in., $h = 8$ in., and $\tan \alpha = \frac{1}{3}$.

STRESSES IN BEAMS: II

SOLUTION. The maximum bending moment at the middle of the span is

$$M_{\max} = \frac{wl^2}{8} = \frac{200 \times 120^2}{12 \times 8} = 30{,}000 \text{ in.-lb.}$$

The components of this bending moment in the two planes of symmetry of the beam are $M_1 = M_{\max} \cos \alpha = 30{,}000 \times 0.949 = 28{,}500$ in.-lb and $M_2 = M_{\max} \sin \alpha = 30{,}000 \times 0.316 = 9500$ in.-lb. Thus the maximum bending stress at A or B is

$$\sigma_{\max} = \frac{28{,}500 \times 6}{6 \times 8^2} + \frac{9{,}500 \times 6}{8 \times 6^2} = 643 \text{ psi.}$$

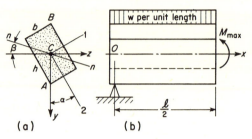

FIG. 6.17

EXAMPLE 2. Locate the neutral axis of the cross-section for the beam loaded as shown in Fig. 6.17.

SOLUTION. The principal moments of inertia of the cross-section with respect to axes of symmetry 1 and 2 are

$$I_1 = \frac{bh^3}{12}, \quad I_2 = \frac{hb^3}{12}, \quad I_{12} = 0. \tag{e}$$

From Appendix B, we have then

$$\left. \begin{aligned} I_y &= \frac{I_1 + I_2}{2} - \frac{I_1 - I_2}{2} \cos 2\alpha + I_{12} \sin 2\alpha, \\ I_{yz} &= \frac{I_1 - I_2}{2} \sin 2\alpha + I_{12} \cos 2\alpha. \end{aligned} \right\} \tag{f}$$

Substituting the values of I_1, I_2, I_{12} for $b = 6$ in. and $h = 8$ in. into eqs. (f), we find, for $\alpha = \tan^{-1}(\tfrac{1}{3}) = 18°\ 26'$, $I_y = 155.2$ in.^{4}, and $I_{yz} = 33.6$ in.^{4} Then from eq. (6.10)

$$\tan \beta = \frac{I_{yz}}{I_y} = \frac{33.6}{155.2} = 0.2164$$

and $\beta = 12°\ 13'$, measured as shown in Fig. 6.17a.

EXAMPLE 3. A cantilever beam of Z-section (Fig. 6.18) is 100 in. long and is loaded at the free end by a force $P = 400$ lb which acts in the vertical plane of the web. Find the maximum bending stress $\sigma_{\max}$.

SOLUTION. From a handbook of Z-sections, we find for this section: $I_z = 34.4$ in.^{4}, $I_y = 12.9$ in.^{4}, $I_{yz} = 14.4$ in.^{4} The maximum bending moment is $(M_z)_{\max} = -Pl = -40{,}000$ in.-lb. The maximum bending stress may occur at either point A

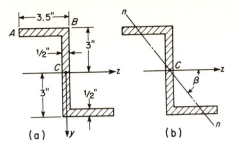

FIG. 6.18

or point B. For point A, $y_a = -3.00$ in., $z_a = -3.25$ in.; for point B, $y_b = -3.00$ in., $z_b = +0.25$ in. Using all these data in eq. (6.9), we obtain

$$\sigma_a = \frac{-40,000(-12.9 \times 3 + 14.4 \times 3.25)}{34.4 \times 12.9 - (14.4)^2} = -1380 \text{ psi},$$

$$\sigma_b = \frac{-40,000(-12.9 \times 3 - 14.4 \times 0.25)}{34.4 \times 12.9 - (14.4)^2} = +7150 \text{ psi}.$$

Thus the largest bending stress occurs at point B. This indicates that the neutral axis of the section has approximately the position shown in Fig. 6.18b. To actually locate this axis, we use eq. (6.10) and find

$$\beta = \tan^{-1}\left(\frac{I_{yz}}{I_y}\right) = \tan^{-1}\left(\frac{14.4}{12.9}\right) = \tan^{-1}(1.116) = 48° \ 08'.$$

PROBLEM SET 6.4

1. A cast-iron beam of triangular cross-section, Fig. A, is to be subjected to pure bending by couples applied at its ends and acting in the xy-plane, so as to induce compression at A. Find the safe bending moment if the working stresses in tension and compression are, respectively, $\sigma_t = 5,000$ psi, $\sigma_c = 8000$ psi. The dimensions of the cross-section are $b = 3$ in., $h = 6$ in. *Ans.* $M = 22,500$ in.-lb.

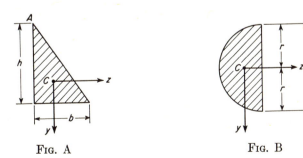

FIG. A   FIG. B

2. A prismatic steel bar of semicircular cross-section, Fig. B, is used as a cantilever beam loaded parallel to the y-axis at its free end. The length $l = 72$ in., $r = 1$ in., and $\sigma_w = 18,000$ psi. Calculate the safe load P. *Ans.* $P = 98.2$ lb.

3. A simply supported beam of span $l = 8$ ft carries a uniformly distributed vertical load of intensity $w = 1000$ lb/ft over the full span. The beam has a 4 in. $\times$ 6 in. $\times$ $\frac{1}{2}$ in. angle section as shown in Fig. C. (a) Ignoring the fact that the beam is not loaded in a plane of symmetry, calculate the bending stress σ at point A. (b) Calculate the same stress, taking into consideration the fact that the xy-plane of loading is not a plane of symmetry. *Ans.* (a) $\sigma = 22{,}300$ psi; (b) $\sigma = 25{,}440$ psi.

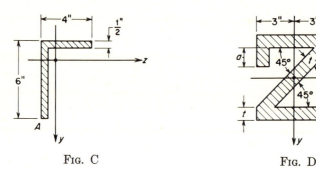

FIG. C   FIG. D

4. It is desired to make a beam of Z-section, as shown in Fig D, of such proportions that the z- and y-axes will be principal axes for the section. If all elements of the section have the same thickness $t = 1$ in. and other dimensions are as shown in the figure, what is the proper dimension a to realize the required condition? *Ans.* $a = 1.45$ in.

6.5   Shearing Stresses in Beams of Thin-Walled Profile Section

In the preceding article we have seen that pure bending couples acting in a *principal plane* of a beam produces bending only in that plane. For such pure bending, there are no shearing stresses and the internal stress resultant on any section is a couple which equilibrates the externally applied couple at either end of the beam. When bending of a beam in a principal plane is produced by transverse loads, there will also be shearing stresses to consider, as was done in Art. 5.5. In general, the resultant of these shear stresses on any section will be a force parallel to the plane of loading but not necessarily in this plane. This stiuation causes some twisting action on the beam so that cross-sections rotate about the longitudinal axis during bending. To attain *simple bending*, i.e., bending without twist, it is necessary to apply the external loads in the same axial plane as that in which the shear stress resultants act. The determination of the location of this plane will, of course, require a careful examination of the distribution of shear stresses over the cross-section.

Let us begin with the case of a beam of singly symmetric cross-section loaded in a plane perpendicular to the plane of symmetry, as shown in

Fig. 6.19a, and assume that the external load P is applied at such distance e from the principal plane xy that simple bending without torsion occurs. The neutral axis of the cross-section then coincides with the z-axis and the normal stress σ at any point in the cross-section is proportional to the distance y from the neutral axis. From the discussion of Art. 5.5, we know that the shear stresses τ will be distributed according to a parabolic law

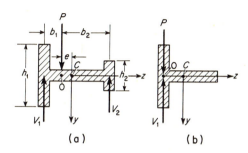

Fig. 6.19

over the depths h_1 and h_2 of the flanges and that the shear stress in the thin horizontal web will be negligible. Thus practically all of the shear will be carried by the two flanges. If we consider the two flanges as separate beams having moments of inertia I_1 and I_2, then their curvatures in bending will be equal if the load P is distributed between them in the ratio $I_1:I_2$, and the shearing forces V_1 and V_2 will then be in the same ratio. This condition will be satisfied if the transverse load P acts in the vertical plane such that

$$\frac{b_1}{b_2} = \frac{I_2}{I_1}. \tag{a}$$

Thus, to obtain simple bending, the load P must be applied in a plane that lies between the centroid C and the stiffer flange. In the limiting case, Fig. 6.19b, where there is only one flange, we may take $I_2 = 0$, and conclude that $b_1 = 0$. Thus, for simple bending of a T-section perpendicular to its plane of symmetry, the load P must be applied in the middle plane of the flange as shown. That point O in the cross-section, in each case, representing the point of application of the shear stress resultant, is called the *shear center* of the cross-section.

In general, the location of the shear center for a solid cross-section of arbitrary shape (Fig. 6.20a) is a complicated problem. Fortunately, it is not so important for beams of such solid sections because they have considerable torsional rigidity and twist very little during bending by loads

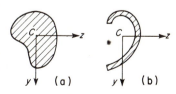

Fig. 6.20

acting through the centroid C. Hence in our further discussion, we shall consider only the case of an arbitrary thin-walled profile section as shown in Fig. 6.20b.

We begin with a beam of such cross-section loaded parallel to one of its two principal planes, say xy, so as to undergo simple bending in this plane, Fig. 6.21. Then the z-axis becomes the neutral axis and the normal stress σ in any longitudinal fiber, the location of which is defined by coordinates y, z, is simply

$$\sigma = \frac{M_x y}{I_z}. \tag{b}$$

Isolating an element A of the wall between cross-sections x and $x + dx$, and of arc length s, we see that normal stress resultants N_x and $N_x + dN_x$ will act on its transverse edges as shown. The bending moment at $x + dx$ being larger than that at x, there will be a net force dN_x in the positive direction of the x-axis, which induces shear stress τ along the inner edge of the element. Since the wall thickness t is assumed small, this shear stress can have no

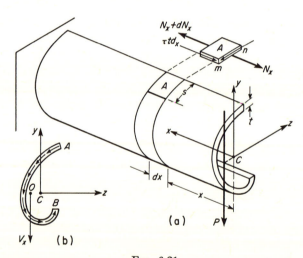

FIG. 6.21

transverse component and must act parallel to the x-axis. The outer edge of the element is a free surface and carries no shear stress. Thus for equilibrium of the element in the x direction, we must have

$$\tau \, t \, dx = dN_x. \tag{c}$$

Now using eq. (b), we conclude that

$$N_x = \frac{M_x}{I_z} \int_0^s y \, dA$$

and

$$N_x + dN_x = \frac{M_x + \dfrac{dM_x}{dx}\,dx}{I_z} \int_0^s y\,dA,$$

so that

$$dN_x = \frac{dM_x}{dx}\frac{dx}{I_z}\int_0^s y\,dA = \frac{V_x dx}{I_z}\int_0^s y\,dA, \qquad (d)$$

where $V_x = dM_x/dx$ is the shear force at the cross-section defined by x. Substituting expression (c) for dN_x into eq. (d), we obtain

$$\tau t\,dx = \frac{V_x dx}{I_z}\int_0^s y\,dA,$$

from which

$$\tau = \frac{V_x}{I_z t}\int_0^s y\,dA. \qquad (6.11)$$

This expression, analogous to eq. (5.6), gives the shear stress at any point in the wall distance s from the free edge. Since the integral therein is a function of s, we conclude that this horizontal shear stress between adjacent fibers varies with the distance s, being a maximum at the neutral plane and zero at each free edge.

From the requirement for equality of complementary shear stresses (see p. 30), we may now conclude that there must be the same shear stress distribution in the plane of the cross-section as shown in Fig. 6.21b. Again, since the wall thickness t is assumed small, these shear stresses in the plane of the cross-section must act in the tangential direction at each point. The shearing stress per unit length of the center line AB of the section is

$$\tau t = \frac{V_x}{I_z}\int_0^s y\,dA, \qquad (6.12)$$

and this is sometimes called the *shear flow*.

Considering now the equilibrium of that portion of the beam between any cross-section x and the free end, it is seen that the elemental shear forces $\tau t\,ds$ over any cross-section must reduce to a force V_x parallel to the y-axis. This shear force will act through a point O such that its moment about the centroid C is equal to the sum of moments of the elemental forces about this same point. This requirement will enable us to locate the shear center O for any specific cross-section. The beam undergoes simple bending without twist only if the load P at the free end acts through the shear center of the end cross-section.

EXAMPLE 1. A beam having a thin-walled channel section as shown in Fig. 6.22a is loaded in a vertical plane parallel to the web so as to produce simple bending in this plane. Find the distance e defining the location of the shear center O of the section.

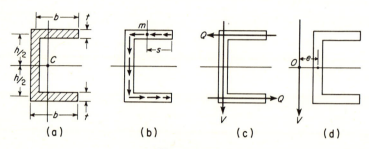

FIG. 6.22

SOLUTION. Since the wall thickness t is small, the shear stresses in the flanges will be horizontal, and those in the web vertical, as shown in Fig. 6.22b. The resultant shear force for the web alone will be a vertical force coinciding with the middle line of the web, and its magnitude must be equal to the total shear force V on the entire section. This is shown in Fig. 6.22c. To find the resultant shear force for either flange, we must first find the intensity of shear stress at any point s of a flange. Referring to Fig. 6.22b and using eq. (6.11) this becomes

$$\tau_s = \frac{V}{It}\int_0^s \frac{h}{2}tds = \frac{Vhs}{2I}.$$

Then the resultant shear force Q for either flange becomes

$$Q = \int_0^b \tau_s tds = \frac{Vht}{2I}\int_0^b sds = \frac{Vhtb^2}{4I}. \tag{e}$$

These horizontal shear stress resultants for the flanges act along their middle lines as shown in Fig. 6.22c, and constitute a counter-clockwise couple of moment Qh. This couple, together with the vertical shear force V coinciding with the web, are statically equivalent to a vertical force V acting through a point O in the plane of the cross-section as shown in Fig. 6.22d. This point O is the required shear center and its distance e from the middle line of the web is found from the relation

$$Ve = Qh,$$

from which

$$e = \frac{Qh}{V} = \frac{b^2h^2t}{4I}. \tag{f}$$

The beam will undergo simple bending in the principal plane parallel to the web only if the plane of loading is removed from the plane of the web by this distance e.

As a specific example, consider the case of a standard $10\lfloor 20$ channel section. Then from Table B.4 of Appendix B, we find $h = 9.56$ in., $b = 2.55$ in., $t = 0.436$ in., $I = 78.5$ in.[4] Substituting these data into eq. (f),

$$e = \frac{(2.55)^2(9.56)^2(0.436)}{4 \times 78.5} = 0.825 \text{ in.}$$

EXAMPLE 2. A beam having the thin-walled semicircular cross-section shown in Fig. 6.23 is loaded in a principal plane xy so as to produce simple bending in this plane. Find the distance e defining the location of the shear center O.

SOLUTION. The shear stress τ at each point $s = r\phi$ along the middle line of radius r will be in the direction of the tangent to this line as shown. From eq. (6.11), its magnitude is

$$\tau = \frac{V}{I_z t}\int_0^s y \, dA = \frac{V}{I_z t}\int_0^\phi r\cos\psi \cdot tr d\psi = \frac{Vr^2}{I_z}\sin\phi.$$

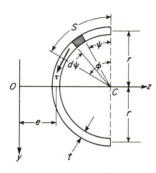

FIG. 6.23

We note that this shear stress is a maximum when $\phi = \pi/2$ and zero for $\phi = 0$ and for $\phi = \pi$. The corresponding elemental shear force is $\tau t ds = \tau tr d\phi$ and its moment about point C is $\tau tr^2 d\phi$. The sum of these moments over the entire cross-section becomes

$$T = \int_0^\pi \tau tr^2 d\phi = \frac{Vr^4 t}{I_z}\int_0^\pi \sin\phi d\phi = \frac{2Vr^4 t}{I_z}.$$

The horizontal components of the elemental shear forces $\tau tr d\phi$ above the neutral axis cancel the horizontal components of those below the neutral axis; hence, the shear stress resultant is a vertical force equal to the shear V at the section. To produce the twisting moment T, calculated above, this force must act through a point O such that

$$V(r + e) = T = \frac{2Vr^4 t}{I_z},$$

from which,

$$e = \frac{2r^4 t}{I_z} - r, \tag{g}$$

where

$$I_z = 2\int_0^{\frac{\pi}{2}} r^3 t\cos^2\psi d\psi = \frac{\pi r^3 t}{2}.$$

Substituting this value of I_z into eq. (g), we find

$$e = r\left(\frac{4}{\pi} - 1\right) = 0.272r. \tag{h}$$

PROBLEM SET 6.5

1. Prove that the shear center O for a thin-walled angle section lies at the intersection of the two legs.

2. Prove that the shear center O for a thin-walled balanced Z-section such as that shown in Fig. 6.18 coincides with its centroid C.

3. Calculate the distance e from the plane of the web to the shear center O of the section shown in Fig. 6.22 if it is a standard 6-U-8.2 channel section. The average flange thickness is $t = 0.343$ in. *Ans. e = 0.699 in.*

4. Locate the shear center O for the unbalanced I-section shown in Fig. A, for simple bending in the plane of the web. *Ans. e = 0.41 in.*

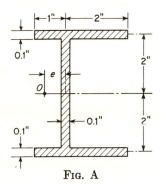

FIG. A

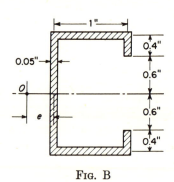

FIG. B

5. Locate the shear center O for the C-section shown in Fig. B, for simple bending in the plane of the web. *Ans. e = 0.55 in.*

6. A simply supported girder of span $l = 20$ ft has a 24WF120 section and carries a concentrated vertical load $P = 200,000$ lb at mid-span. (a) Calculate the shear stress on the vertical junction plane between a half flange and the web. (b) Calculate the shear stress on the horizontal junction plane between a full flange and the web. *Ans.* (a) $\tau = 1850$ psi; (b) $\tau = 6500$ psi.

7. Referring to the C-section shown in Fig. B, assume that the lips on the two flanges extend out away from the neutral axis instead of in toward it thus producing what is commonly called a hat-section. Will this change the distance e from the plane of the web to the shear center and, if so, what is the value of e in this case? *Ans. e = 0.459 in.*

8. A cantilever beam has the Z-section shown in Fig. 6.18 and is loaded as described in Example 3, page 158. Calculate the intensity of shear stress on the junction between the web and the lower flange. *Hint:* Note that the beam is not loaded in a principal plane; this means that eq. (6.9) must be used to calculate the bending stress at any point in the section and that eq. (6.11) is not directly applicable. *Ans.* $\tau = 66$ psi.

6.6   Bending Stresses in Curved Beams

In this section we consider briefly the theory of pure bending of an initially curved bar within the elastic range. Referring to Fig. 6.24a, consider a short portion of a curved bar acted upon by couples of moment M in the plane of initial curvature. Such bending moment which tends to decrease the initial curvature will be considered as *positive*. Each cross-section of the bar is assumed to have an axis of symmetry which lies in the plane of initial curvature. The locus of the centroids then is a plane curve called the *center line* of the bar and its radius of curvature is denoted by R.

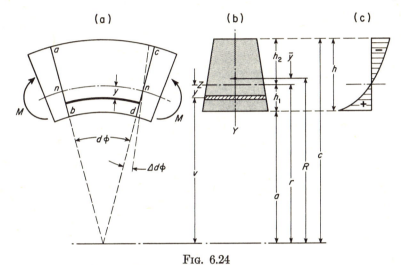

FIG. 6.24

In discussing the stress distribution produced by pure bending of such a curved bar, we make the same assumptions as in the case of straight bars, namely, that transverse cross-sections of the bar, originally plane and normal to the center line, remain so after bending. Let ab and cd denote two neighboring cross-sections of the bar and let $d\phi$ denote the small angle between them before bending. As a result of bending, the cross-section cd rotates with respect to ab. Let $\Delta d\phi$ denote the small angle of rotation. Due to this rotation, the longitudinal fibers on the convex side of the bar are compressed and the fibers on the concave side are extended. If n-n denotes the neutral surface, the extension of any fiber at the distance y from this surface is $y(\Delta d\phi)$ and the corresponding unit elongation is

$$\epsilon = \frac{y(\Delta d\phi)}{(r - y)d\phi},\tag{a}$$

where r denotes the radius of the neutral surface and the denominator in eq. (a) is the length of the fiber between the adjacent cross-sections before bending. Assuming that there is no lateral pressure between the longitudinal fibers,* the bending stress at a distance y from the neutral surface is

$$\sigma = E\epsilon = \frac{Ey(\Delta d\phi)}{(r - y)d\phi}. \tag{b}$$

Eq. (b) shows that the stress distribution is no longer linear as in the case of straight bars, but that it follows a hyperbolic law as shown in Fig. 6.24c. From the condition that the sum of the normal forces distributed over the cross-section is zero, it can be concluded that the neutral axis is displaced from the centroid of the cross-section towards the center of curvature of the bar.

In the case of a *rectangular* cross-section, the shaded area (Fig. 6.24c) in tension must equal that in compression; hence the greatest bending stress acts on the concave side. In order to make the stresses in the most remote fibers in tension and in compression equal, it is necessary to use sectional shapes which have the centroid nearer the concave side of the bar.

Eq. (b) contains two unknowns, the radius r of the neutral surface and the angle $\Delta d\phi$ which represents the angular displacement due to bending. To determine them, we must use two equations of statics. The first equation is based on the condition that the sum of the normal forces distributed over a cross-section is equal to zero. The second equation is based on the condition that the moment of these normal forces is equal to the bending moment M. Thus

$$\int \sigma dA = \frac{E(\Delta d\phi)}{d\phi} \int \frac{y dA}{r - y} = 0, \tag{c}$$

$$\int \sigma y \, dA = \frac{E(\Delta d\phi)}{d\phi} \int \frac{y^2 dA}{r - y} = M. \tag{d}$$

The integration in both equations is extended over the total area of the cross-section.

Eq. (c) enables one to determine r and, in turn, the distance $\bar{y}$ (considered a positive quantity) from the centroidal axis to the neutral axis of the cross-section. Let v represent the distance from the center of curvature to any element dA; then $y = r - v$, and eq. (c) can be written

$$\int \frac{(r - v)dA}{v} = 0,$$

from which

$$r = \frac{A}{\int \frac{dA}{v}}, \tag{6.13}$$

---

*The exact theory shows that there is a certain radial pressure but that it has no substantial effect on the stress σ and can be neglected.

or

$$\bar{y} = R - \frac{A}{\int \frac{dA}{v}},$$ (6.14)

where R is the initial radius of curvature of the center line of the bar.

Eq. (d) may be used to obtain a formula for the fiber stresses in terms of the bending moment. The integral in eq. (d) is first simplified as follows:

$$\int \frac{y^2 dA}{r - y} = -\int \left(y - \frac{ry}{r - y} \right) dA = -\int y\, dA + r \int \frac{y dA}{r - y}.$$ (e)

The first integral on the right side of eq. (e) represents the moment of the cross-sectional area with respect to the neutral axis, and the second, as is seen from eq. (c), is equal to zero. Hence

$$\int \frac{y^2 dA}{r - y} = -[A(-\bar{y})] = A\bar{y}.$$ (f)

Eq. (d) then becomes

$$\frac{E(\Delta d\phi)}{d\phi} = \frac{M}{A\bar{y}}.$$

Substituting this in eq. (b),

$$\sigma = \frac{My}{A\bar{y}(r - y)}.$$ (g)

The stresses in the most remote fibers which are the maximum stresses in the bar are

$$\sigma_{max} = \frac{Mh_1}{A\bar{y}a} \quad \text{and} \quad \sigma_{min} = -\frac{Mh_2}{A\bar{y}c},$$ (6.15)

in which h_1 and h_2 are the distances from the neutral axis to the most remote fibers, and a and c are the inner and outer radii of the bar.

So far we have considered the case of pure bending where the bar is subjected to end couples only. In a more general case when a curved bar is bent by transverse forces acting in its plane of symmetry, the forces acting upon the portion of the bar to one side of any cross-section may be reduced to a couple and a force applied at the centroid of the cross-section. The moment of this couple equals that of the external forces with respect to the centroidal axis of the cross-section. The stresses produced by the couple are then obtained as explained above. The force is resolved into two components, a longitudinal force N in the direction of the tangent to the center line of the bar and a shearing force V in the plane of the cross-section. The longitudinal force produces tensile or compressive stresses uniformly distributed over the cross-section and equal to N/A. To get the total axial stress acting in any fiber, this uniform stress is added algebraically to the stress caused by the couple. The transverse force V produces shearing stresses and the distribution of these stresses over the cross-section can be taken the same as for a straight bar.

EXAMPLE 1. Determine the numerical value of the ratio $\sigma_{max}/\sigma_{min}$ for the case of a curved beam of rectangular cross-section in pure bending if $R = 5$ in. and $h = 4$ in.

SOLUTION. From eqs. (6.15)

$$\frac{\sigma_{max}}{\sigma_{min}} = \frac{h_1 c}{h_2 a}, \tag{h}$$

where $h_1 = h/2 - \bar{y}$ and $h_2 = h/2 + \bar{y}$ (see Fig. 6.24b). To calculate $\bar{y}$, we use eq. (6.14) in which

$$\int \frac{dA}{v} = \int_a^c \frac{b\,dv}{v} = b \ln\left(\frac{c}{a}\right) = b \ln\left(\frac{7}{3}\right) = 0.847\,b.$$

Thus

$$\bar{y} = R - \frac{bh}{0.847b} = 5 - 4.72 = 0.28 \text{ in.}$$

Then $h_1 = 2 - 0.28 = 1.72$ in. and $h_2 = 2 + 0.28 = 2.28$ in. With these values of h_1 and h_2, eq. (h) becomes

$$\frac{\sigma_{max}}{\sigma_{min}} = \frac{1.72 \times 7}{2.28 \times 3} = 1.76.$$

EXAMPLE 2. A curved beam with a circular center line has the inverted T-section shown in Fig. 6.25, and is subjected to pure bending in its plane of symmetry. Find the dimension b_1 in order to have equal tensile and compressive stresses in extreme fibers.

SOLUTION. Since we require $|\sigma_{max}| = |\sigma_{min}|$, it follows from eqs. (6.15) that

$$\frac{h_1}{a} = \frac{h_2}{c},$$

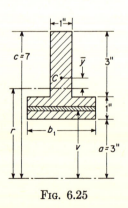

FIG. 6.25

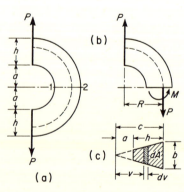

FIG. 6.26

from which

$$\frac{h_1}{h_2} = \frac{a}{c} = \frac{3}{7}.$$

Then since $h_1 + h_2 = 4$ in., we find $h_1 = 1.2$ in., $h_2 = 2.8$ in. This locates the neutral axis of the cross-section and its radius $r = a + h_1 = 3 + 1.2 = 4.2$ in. This radius of the neutral axis is also defined by eq. (6.13) wherein

$$\int_A \frac{dA}{v} = \int_3^4 \frac{b_1 dv}{v} + \int_4^7 \frac{dv}{v} = b_1 \ln\left(\frac{4}{3}\right) + \ln\left(\frac{7}{4}\right) = 0.288 b_1 + 0.560,$$

so that

$$r = \frac{A}{\int \frac{dA}{v}} = \frac{b_1 + 3}{0.288 b_1 + 0.560} = 4.2 \text{ in.}$$

This gives $b_1 = 3.09$ in.

EXAMPLE 3. A semicircular curved bar is loaded as shown in Fig. 6.26a and has the trapezoidal cross-section shown in Fig. 6.26c. Calculate the tensile stress σ_1 at point 1 if $a = b = h = 1$ in. and $P = 1000$ lb.

SOLUTION. Considering that portion of the bar above the section 1-2 (Fig. 6.26b), we see that the stress resultant on this section consists of a force P acting at the centroid of the section and a bending moment $M = PR$ where R is the radius of the centroidal axis. The extreme fiber stress at point 1 due to the moment M will be given by the first of eqs. (6.15), and that due to the force P will be simply P/A. Thus the total stress at point 1 is

$$\sigma_1 = \frac{PRh_1}{A\bar{y}a} + \frac{P}{A}. \tag{i}$$

For the given cross-section (see Fig. 6.26c) we find

$$A = \frac{b + b/2}{2} \times h = \frac{3}{4}bh = \tfrac{3}{4} \text{ in.}^2,$$

$$R = \frac{\int_A v\,dA}{A} = \frac{\frac{b}{c}\int_a^c v^2 dv}{\frac{3}{4}bh} = \frac{4(c^3 - a^3)}{9ch} = \frac{14}{9} \text{ in.,}$$

$$\int_A \frac{dA}{v} = \frac{b}{c}\int_a^c dv = \frac{b}{c}(c - a) = \frac{1}{2} \text{ in.,}$$

$$\bar{y} = R - \frac{A}{\int \frac{dA}{v}} = \frac{14}{9} - \frac{\frac{3}{4}}{\frac{1}{2}} = \frac{14}{9} - \frac{3}{2} = \frac{1}{18} \text{ in.,}$$

$$h_1 = R - a - \bar{y} = \frac{14}{9} - 1 - \frac{1}{18} = \frac{1}{2} \text{ in.}$$

Substituting these numerical values into eq. (i), we obtain

$$\sigma_1 = \frac{1000 \times \frac{14}{9} \times \frac{1}{2}}{\frac{3}{4} \times \frac{1}{18} \times 1} + \frac{1000}{\frac{3}{4}} = 20{,}000 \text{ psi.}$$

1. Determine the numerical value of the ratio $\sigma_{max}/\sigma_{min}$ for the case of pure bending of a curved beam having a 1 in. $\times$ 1 in. square cross-section if the radius of curvature of the centroidal axis is $R = 1.5$ in. *Ans.* $\sigma_{max}/\sigma_{min} = 1.58$.

2. A curved beam with a circular centerline has the T-section shown in Fig. A and is subjected to pure bending in its plane of symmetry. The radius of curvature of the concave face is $a = 3$ in. All dimensions of the cross-section are fixed as shown except the thickness t of the stem. Find the proper value of t so that the extreme fiber stresses in bending will be numerically equal. *Ans.* $t = 1.19$ in.

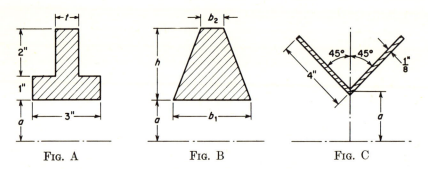

Fig. A   Fig. B   Fig. C

3. A curved beam with a circular centerline has the trapezoidal cross-section shown in Fig. B and is subjected to pure bending in its plane of symmetry. The face b_1 is the concave side of the beam. If $b_1 = 4$ in., $h = 4$ in., and $a = 4$ in., find the proper value of b_2 to make extreme fiber stresses in tension and compression numerically equal. *Ans.* $b_2 = 0.66$ in.

4. A circular ring has the V-section shown in Fig. C, which is a 4 in. $\times$ 4 in. $\times \frac{1}{8}$ in. angle section. What is the allowable bending moment in the ring if the compressive stress at the outer edges of the legs is not to exceed 16,000 psi and $a = 3$ in.? *Ans.* $M = 8900$ in.-lb.

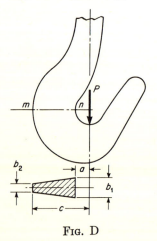

Fig. D

5. A steel crane hook carries a load $P = 4500$ lb as shown in Fig. D. The cross-section mn of the hook is trapezoidal as shown in the figure. Find the total stresses σ_m and σ_n at points m and n. The following numerical data are given: $b_1 = 1\frac{5}{8}$ in., $b_2 = \frac{3}{8}$ in., $a = 1\frac{1}{4}$ in., and $c = 5$ in. *Ans.* $\sigma_m = -3560$ psi; $\sigma_n = 9340$ psi.

6. A circular torus of mean radius R has a thin-walled circular cross-section of mean diameter d. Calculate the radius r of the neutral axis for the particular case where $R = d$. *Ans.* $r = 0.866d$.

7. Determine the numerical value of the ratio $\sigma_{max}/\sigma_{min}$ for the case of a curved beam of solid circular cross-section in pure bending if $R = 5$ in. and the diameter d of the cross-section is 4 in. *Ans.* $\sigma_{max}/\sigma_{min} = 1.91$.

7

Analysis of Plane Stress
and Plane Strain

7.1 General Case of Plane Stress

In preceding discussions of beams bent by transverse loads, we have
seen how an element of material in the beam can be subjected to both
normal and shearing stresses on its edges as shown in Fig. 7.1a. A similar
situation will occur in the case of an element of a shaft subjected to axial
loads and twisting moments as shown in Fig. 7.1b. Such a state of stress

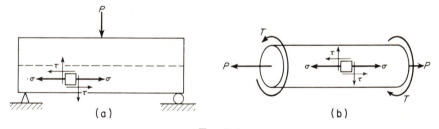

(a) (b)

Fig. 7.1

on the edges of a rectangular element in which there are no stresses normal
to its face is called *plane stress*. After such normal and shearing stresses
as those shown in Fig. 7.1 have been found, it is frequently necessary to
examine further the state of stress within the element to find the magnitudes
and directions of maximum stresses.

Let us consider now the general case of an element under plane stress
as shown in Fig. 7.2a. The normal stress in the x direction is denoted
by σ_x, that in the y direction by σ_y, and tension is considered positive.
The shear stresses on the edges of the element that are normal to the
x-axis are denoted by τ_{xy} while those on the edges normal to the y-axis are
denoted by τ_{yx}. The shear streseses τ_{xy}, having a *clockwise* sense of rotation
about a point inside the element, are to be considered *positive* in accordance
with our previous rule (see p. 28). The shear stresses τ_{yx}, having a *counter-*

clockwise sense of rotation, are *negative*. From the requirement of equality of complementary shear stresses (see p. 30), we have $\tau_{xy} = -\tau_{yx}$. Because of this equality of orthogonal shear stresses, it is customary to use only one notation τ_{xy} for these stresses without regard to order of subscripts, but in so doing, it is necessary to remember that shear stresses giving counterclockwise rotation are to be treated as negative.

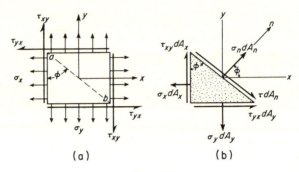

(a)   (b)

FIG. 7.2

Given the state of plane stress shown in Fig. 7.2a, the normal stress σ_n and the shear stress τ on any plane whose normal n makes the angle ϕ with the x-axis can easily be found from the equilibrium conditions of the triangular element shown in Fig. 7.2b. Let the area of the inclined face of this element be denoted by dA_n; then the areas of the other two faces are $dA_x = dA_n \cos\phi$ and $dA_y = dA_n \sin\phi$. Multiplying the various stresses by the areas of the faces on which they act, the total forces on the triangular element will be as shown in Fig. 7.2b. Then for equilibrium in the n direction, we must have

$$\sigma_n dA_n = \sigma_x dA_n \cos^2\phi + \sigma_y dA_n \sin^2\phi - 2\tau_{xy} dA_n \cos\phi \sin\phi. \qquad \text{(a)}$$

Similarly, for equilibrium in the direction perpendicular to n, we must have

$$\tau dA_n = \sigma_x dA_n \cos\phi \sin\phi - \sigma_y dA_n \sin\phi \cos\phi + \tau_{xy} dA_n(\cos^2\phi - \sin^2\phi). \qquad \text{(b)}$$

Equations (a) and (b) are readily reduced to

$$\left.\begin{aligned}
\sigma_n &= \sigma_x \cos^2\phi + \sigma_y \sin^2\phi - 2\tau_{xy} \sin\phi \cos\phi \\
&= \tfrac{1}{2}(\sigma_x + \sigma_y) + \tfrac{1}{2}(\sigma_x - \sigma_y)\cos 2\phi - \tau_{xy}\sin 2\phi, \\
\tau &= (\sigma_x - \sigma_y)\sin\phi \cos\phi + \tau_{xy}(\cos^2\phi - \sin^2\phi) \\
&= \tfrac{1}{2}(\sigma_x - \sigma_y)\sin 2\phi + \tau_{xy}\cos 2\phi,
\end{aligned}\right\} \qquad (7.1)$$

which are analogous to eqs. (3.2) in Art. 3.2.

To find the location of the plane of maximum normal stress σ_n, we make the derivitive $d\sigma_n/d\phi = 0$ from the first of eqs. (7.1) and obtain

$$-(\sigma_x - \sigma_y) \sin 2\phi - 2\tau_{xy} \cos 2\phi = 0, \qquad (c)$$

from which

$$\tan 2\phi = -\frac{2\tau_{xy}}{\sigma_x - \sigma_y}. \qquad (7.2)$$

This condition defines two values of 2ϕ differing by 180° and hence two values of ϕ differing by 90°. For one of these values, σ_n is a maximum and for the other, a minimum.

Considering the second of eqs. (7.1) and setting the shear stress τ equal to zero, we again obtain eq. (c): From this, it may be concluded that on those planes where σ_n is a maximum or a minimum the shear stress τ vanishes. The corresponding normal stresses $(\sigma_n)_{max}$ and $(\sigma_n)_{min}$ are called *principal stresses*, and the planes on which they act are called *principal planes* of stress.

Referring to the second of eqs. (7.1) and setting $d\tau/d\phi = 0$, we obtain

$$(\sigma_x - \sigma_y) \cos 2\phi - 2\tau_{xy} \sin 2\phi = 0, \qquad (d)$$

from which

$$\cot 2\phi = \frac{2\tau_{xy}}{\sigma_x - \sigma_y}. \qquad (7.3)$$

Comparing this with eq. (7.2), we see that the maximum shear stresses occur on orthogonal planes bisecting the angle between principal planes, i.e., at 45° to the planes of principal stress as already concluded in Art. 3.2.

To evaluate the maximum normal and shear stresses in the element, it is necessary to replace ϕ in eqs. (7.1) by the values defined by eqs. (7.2) and (7.3). Because of the transcendental character of the equations, this becomes somewhat involved and, for this purpose, it will be simpler to use Mohr's circle.

The general case of plane stress is shown again in Fig. 7.3a. To construct Mohr's circle for this case, one proceeds as follows: Lay out first the coordinate axes σ and τ with origin at O as shown in Fig. 7.3.b Then to locate the point D representing the stress conditions on the x-plane, i.e., the plane normal to the x-axis, lay off the value of σ_x as abscissa OF and the shear stress τ_{xy} as a positive ordinate FD. Next, locate the point D_1, representing the state of stress on the y-plane, by laying off the abscissa OF_1 to represent the normal stress σ_y and the negative ordinate F_1D_1 to represent the shear stress $-\tau_{xy}$. Since the x- and y-planes are orthogonal, the corresponding points D and D_1 on Mohr's circle are 180° apart and represent the ends of a diameter. Connecting these points with a straight line locates the center C on the σ-axis and the circle can be drawn as shown.

The maximum and minimum normal stresses are represented in Fig. 7.3b by OA and OB, respectively. These principal stresses are denoted by $(\sigma_n)_{max} = \sigma_1$ and $(\sigma_n)_{min} = \sigma_2$, as shown. To locate their directions in Fig. 7.3a, we start with point D on the circle, corresponding to the x-plane of the element, and label this point $\phi = 0$, as shown. Then to reach point A on the circle, corresponding to the plane of maximum prinicpal stress, it is necessary to pass through the *clockwise* angle $2\phi_a$. Hence, the direction of σ_1 in Fig. 7.3a is found by laying out, also clockwise, the angle ϕ_a from the x-axis, as shown. The direction of σ_2 is then at right angles to that of σ_1 and the principal planes are located as shown.

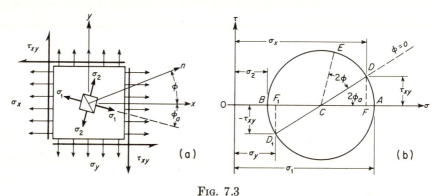

FIG. 7.3

In general, any plane through the element whose normal n makes the angle ϕ with the x-axis and the corresponding point E on the circle, representing the state of stress on this plane, are related in the same way, namely: the angle DCE in Fig. 7.3b is always double the angle ϕ in Fig. 7.3a and is to be measured in the same direction.

Expressions for the principal stresses σ_1 and σ_2 are easily found in terms of σ_x, σ_y, and τ_{xy}, from the geometry of Mohr's circle, Fig. 7.3b, as follows:

$$\left.\begin{aligned}\sigma_1 = OA = OC + CD = \frac{\sigma_x + \sigma_y}{2} + \sqrt{\left(\frac{\sigma_x - \sigma_y}{2}\right)^2 + \tau_{xy}{}^2},\\[2mm]\sigma_2 = OB = OC - CD = \frac{\sigma_x + \sigma_y}{2} - \sqrt{\left(\frac{\sigma_x - \sigma_y}{2}\right)^2 + \tau_{xy}{}^2}.\end{aligned}\right\} \quad (7.4)$$

These are the same values which would be found by substituting the value of 2ϕ from eq. (7.2) into the first of eqs. (7.1) on p. 174.

EXAMPLE 1. A square element of a thin plate subjected to plane stress is shown in Fig. 7.4a. The given stresses on its mutually perpendicular faces are $\sigma_x = -500$

psi, $\sigma_y = +1500$ psi, $\tau_{xy} = -\tau_{yx} = +1000$ psi. Construct Mohr's circle for this element and find:

(a) the magnitudes and directions of principal stresses σ_1 and σ_2.

(b) the normal stress σ_n and shear stress τ on the diagonal plane ab.

(c) the aspect of the plane of maximum positive shear stress τ_{max}.

SOLUTION. From the origin O of the $\sigma\tau$-plane in Fig. 7.4b, lay off $OD' = -500$ psi and $D'D = +1000$ psi to locate point D, representing the given state of stress on the x-plane. Also from O, lay off $OD'_1 = +1500$ psi and $D'_1D_1 = -1000$ psi to locate point D_1, representing the given state of stress on the y-plane. Connect points D and D_1 by a straight line to locate the center C of Mohr's circle and draw the circle as shown. Also label CD with $\phi = 0$ to serve as a reference line. Then the

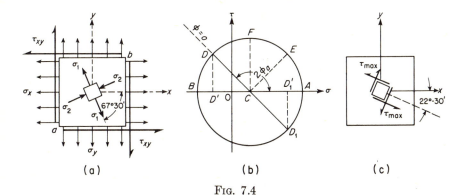

(a)   (b)   (c)

FIG. 7.4

measured angle $DCA = 2\phi_a = -135°$, negative to show that it is *clockwise*. Hence $\phi_a = -67°30'$ and the direction of the principal stress σ_1 is as shown in Fig. 7.4a; the direction of σ_2 is, of course, at right angles to this. The magnitudes of σ_1 and σ_2 are represented, respectively, by $OA = \sigma_1 = +1910$ psi and by $OB = \sigma_2 = -910$ psi, scaled directly from the drawing. The same values can be obtained from eqs. (7.4) by substituting therein the given numerical values of σ_x, σ_y, and τ_{xy}, with due regard to signs.

The normal to the plane ab which bisects the angle between the x- and y-planes is seen to make an angle $\phi = -45°$ with the x-axis. Hence from the reference point D in Fig. 7.4b, we measure off the angle $DCE = 2\phi = -90°$ to locate point E, representing the state of stress on this plane. Scaling the coordinates of point E, we find $\sigma_n = +1500$ psi and $\tau = +1000$ psi.

Reference to Fig. 7.4b shows that point F on Mohr's circle represents the plane of maximum positive shear stress in the element. Then since the angle DCF measures $2\phi = -45°$, the plane of this maximum positive shear stress is defined by $\phi = -22°30'$ as shown in Fig. 7.4c. The magnitude of this maximum shear stress is $\tau_{max} = CF = 1410$ psi, scaled from the drawing. As a check, it will be noted that the angle between planes of principal stress and planes of maximum shear stress is $67°30' - 22°30' = 45°$ as it should be.

EXAMPLE 2. Fig. 7.5a represents one element of a plate in a state of plane stress and Fig. 7.5b, the corresponding Mohr's circle. Point D on this circle corresponds to the x-plane, point D_1 to the y-plane, and points A and B, to the two principal planes. Point D' is vertically below D and point D'_1, vertically above D_1. Show

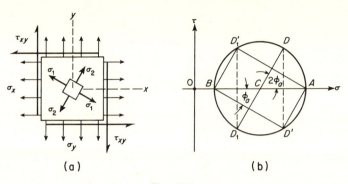

FIG. 7.5

that the sides of the rectangle AD'_1BD' represent the directions of principal stress in the element.

SOLUTION. Since AB is a diameter of the circle, it follows that $AD'B$ and $D'BD'_1$ are right angles. Also, since the arc AD' is equal to the arc DA, the angle ABD' is equal to half the angle DCA. Thus the angle $ABD' = \phi_a$, which defines the direction of the principal stress σ_1, as shown. This simple construction furnishes a convenient way to find the directions of principal stresses from Mohr's circle.

PROBLEM SET 7.1

1. For the rectangular element shown in Fig. 7.3, the following data are given: $\sigma_x = 5000$ psi, $\sigma_y = 3000$ psi, and $\tau_{xy} = 750$ psi. Using eqs. (7.4) and (7.2), determine the magnitudes of the two principal stresses σ_1 and σ_2 and the angle ϕ_a between the directions of σ_x and σ_1. Ans. $\sigma_1 = 5250$ psi; $\sigma_2 = 2750$ psi; $\phi_a = -18°\ 26'$.

2. Construct Mohr's circle for the case of plane stress described in the preceding problem and determine therefrom the values of σ_1, σ_2, and ϕ_a.

3. Find the magnitudes and directions of principal stresses for the element shown in Fig. 7.3 if $\sigma_x = -5000$ psi, $\sigma_y = 3000$ psi, and $\tau_{xy} = 3000$ psi. Ans. $\sigma_1 = 4000$ psi; $\sigma_2 = -6000$ psi; $\phi_a = -71°\ 34'$.

4. Construct Mohr's circle for the case of plane stress described in the preceding problem and determine therefrom the values of σ_1, σ_2, and ϕ_a.

5. For the rectangular element shown in Fig. 7.3, the following numerical data are given: $\sigma_x = 12,000$ psi, $\sigma_y = 4000$ psi, $\tau_{xy} = 3000$ psi. Using eqs. (7.1), calculate the normal stress σ_n and the shear stress τ on the plane defined by the angle $\phi = 30°$. Ans. $\sigma_n = 7400$ psi; $\tau = 4960$ psi.

6. Construct Mohr's circle for the state of plane stress described in the preceding problem and determine therefrom the values of σ_n and τ on the plane whose normal is defined by $\phi = 30°$.

7. Construct Mohr's circle for the element in Fig. 7.3 if $\sigma_x = \sigma_y = 0$ and $\tau_{xy} = 5000$ psi. From this construction, find the principal stresses σ_1 and σ_2 and also the angle ϕ_a between the x-axis and the direction of the principal stress σ_1. Ans. $\sigma_1 = -\sigma_2 = 5000$ psi; $\phi_a = -45°$.

8. For the square element shown in Fig. 7.4, both the normal stress and the shear

stress on the diagonal plane ab are zero. If $\sigma_x = \sigma_y = -5000$ psi, what is the value of the shear stress τ_{xy}? What are the magnitudes of the principal stresses σ_1 and σ_2? *Ans.* $\tau_{xy} = 5000$ psi; $\sigma_1 = 0$; $\sigma_2 = -10,000$ psi.

9. For the element shown in Fig. 7.3, $\sigma_x = 6000$ psi and $\sigma_y = 0$. What is the maximum permissible magnitude of the shear stress τ_{xy} if the larger principal stress σ_1 is not to exceed 9000 psi? *Ans.* $\tau_{xy} = 5196$ psi.

7.2   Principal Stresses in Beams

Consider in Fig. 7.6 the case of a simply supported beam of rectangular cross-section subjected to uniformly distributed transverse loading. Then for any cross-section defined by the distance x from the left support, the shear force will be

$$V_x = \frac{wl}{2} - wx$$

and the bending moment will be

$$M_x = \frac{wlx}{2} - \frac{wx^2}{2},$$

both of which vary with the distance x defining the location of the cross-section. The normal stress σ_x and the shear stress τ_{xy} at any distance y from the neutral surface can be readily calculated from the formulas developed in Chapter V. Thus, from eq. (5.5)

$$\sigma_x = \frac{M_x y}{I}, \tag{a}$$

and, from eq. (5.8),

$$\tau_{xy} = \frac{V_x}{2I}\left(\frac{h^2}{4} - y^2\right). \tag{b}$$

Since M_x and V_x both vary with x, these stresses are seen to vary continuously with both x and y throughout the beam. For any particular cross-section, the bending stress σ_x varies linearly with y and the shear stress τ_{xy} varies parabolically as shown in Fig. 7.6a. For a rectangular element such as A or B, Fig. 7.6b, the shear stress vanishes and the corresponding tensile or compressive stress due to bending represents a principal stress. For an element C situated at the neutral surface, the normal stress vanishes and the element is in a state of pure shear. For such conditions the directions of principal stress are inclined by $\pm 45°$ to the x-axis. For any intermediate element D, there are both normal and shearing stresses to consider. These stress conditions for the element D are shown separately in Fig. 7.6c.

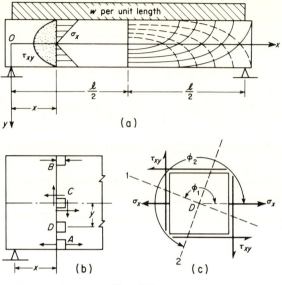

FIG. 7.6

The state of plane stress shown in Fig. 7.6c is simply a particular case of plane stress as discussed in Art. 7.1, one of the normal stresses in this case being $\sigma_y = 0$. Thus to find the magnitudes of principal stress for this element we use eqs. (7.4), which, with $\sigma_y = 0$, become

$$\sigma_{1,2} = \frac{\sigma_x}{2} \pm \sqrt{\left(\frac{\sigma_x}{2}\right)^2 + \tau_{xy}{}^2}. \tag{7.5}$$

It is seen that in this case the maximum principal stress σ_1 will always be positive, i.e., tension, while the minimum principal stress σ_2 will always be negative, i.e., compression.

The directions of these principal stresses will be found by using eq. (7.2) which, with $\sigma_y = 0$, becomes

$$\tan 2\phi = -\frac{2\tau_{xy}}{\sigma_x}. \tag{7.6}$$

For known values of σ_x and τ_{xy}, this gives two values of ϕ which locate the directions of principal stresses σ_1 and σ_2 as shown in Fig. 7.6c.

Applying eq. (7.6) to elements A or B, we have $\tau_{xy} = 0$ and obtain $\tan 2\phi = 0$, yielding $\phi_1 = 0$, $\phi_2 = 90°$. Thus at the lower and upper free surfaces of the beam, the directions of principal stress coincide with the x- and y-axes as anticipated above. For the element C at the neutral surface, $\sigma_x = 0$, and eq. (7.6) gives $\tan 2\phi = -\infty$, yielding $\phi = \pm 45°$.

Thus, for points in the neutral surface of the beam, the directions of principal stress cross the x-axis at $\pm 45°$ as already anticipated.

By calculating ϕ from eq. (7.6) for a number of points defined by coordinates x and y, it is possible to construct two families of orthogonal curves whose tangents at each point coincide with the directions of principal stresses at these points. Such curves for the right half of the beam under uniform load will be as shown in Fig. 7.6a. They are called *principal stress trajectories*. The solid line curves represent tensile stress trajectories and the dotted line curves, compressive stress trajectories. Both sets of curves cross the x-axis at 45° and always cross each other at right angles. They terminate in the upper and lower free surfaces normal to the axis of the beam.

In beam design, one usually wants the numerically maximum values of normal stress σ_n. From eq. (7.5) it can be seen that for the most remote fibers in tension, where the shear stress is zero, the longitudinal normal stress σ_x becomes the principal stress. For fibers nearer to the neutral axis, the longitudinal fiber stress σ_x is less than at the extreme fiber; however, we now have a shear stress τ_{xy} also and the stresses σ_x and τ_{xy} acting together at this point may produce a principal stress, given by eq. (7.5), which will be numerically larger than that at the extreme fiber. In the case of beams of rectangular or circular cross-section, in which the shearing stress τ_{xy} varies continuously over the depth of the beam, this is not usually the case, that is, the stress $(\sigma_x)_{\max}$ calculated for the most remote fiber at the section of maximum bending moment is the maximum stress acting in the beam. However, in the case of an I-beam, where a sudden change occurs in the magnitude of shearing stress at the junction of flange and web (see p. 129), the maximum stress calculated at this junction from eq. (7.5) may be larger than the tensile stress $(\sigma_x)_{\max}$ in the most remote fiber and should be taken into account in design.

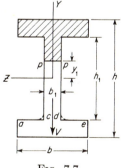

Fig. 7.7

To illustrate, consider the case of a simple beam carrying a concentrated load P at the middle, assuming $l = 2$ ft and $P = 60,000$ lb. The cross-section is an I-section as shown in Fig. 7.7; $h = 12$ in., $h_1 = 10.5$ in., $b = 5$ in., and $b_1 = 0.5$ in., $I_z = 286$ in.[4] Then $M_{\max} = 30,000$ ft-lb, $V_{\max} = 30,000$ lb. From eq. (a), the tensile stress in the most remote fiber is

$$(\sigma_x)_{\max} = \frac{30,000 \times 12 \times 6}{286} = 7550 \text{ psi.}$$

Now for a point at the junction of flange and web, we obtain the following values of normal and shearing stresses:

$$\sigma_x = 7550 \times \frac{10.5}{12} = 6610 \text{ psi};$$

$$\tau_{xy} = 4430 \text{ psi.}$$

Then from eq. (7.5), the principal stress is

$$(\sigma_n)_{\max} = 8830 \text{ psi.}$$

It is seen that $(\sigma_n)_{\max}$ at the junction between flange and web is larger than the tensile stress in the most remote fiber and therefore should be considered in design.

PROBLEM SET 7.2

1. For the simply supported I-beam discussed on p. 181, determine the span length l such that $(\sigma_n)_{\max}$ at the junction between flange and web will be equal to $(\sigma_x)_{\max}$ in an extreme fiber. *Ans.* $l = 39.8$ in.

2. A simply supported beam has an 8WF17 section and carries a concentrated load P at mid-span. Determine the span length l such that $(\sigma_n)_{\max}$ at the junction between flange and web will be equal to $(\sigma_x)_{\max}$ in an extreme fiber. *Ans.* $l = 48.6$ in.

3. A cantilever beam 1 ft long has a rectangular cross-section 4 in. wide by 8 in. deep and carries a load P at the free end. Considering a cross-section at the built-in end, what is the ratio of the principal tension halfway between the neutral axis and the upper surface to that at the upper surface? *Ans.* 0.53.

4. A short cast iron beam has a span of 1 ft and carries a load $P = 4000$ lb at the middle as shown in Fig. A. The cross-section of the beam is an inverted T-section with dimensions as shown, both flange and stem being $\frac{1}{4}$ in. thick. What is the maximum tensile stress induced in the material and at what point does it occur? *Ans.* $\sigma_{\max} = 4390$ psi.

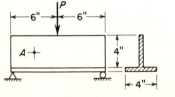
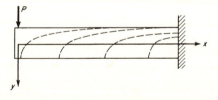

FIG. A   FIG. B

5. Referring to the beam supported and loaded as shown in Fig. A, construct Mohr's circle for the state of stress at a point A which is midway between the left-hand support and the load P and also midway between the top and bottom surfaces. Using this circle, find the magnitude and direction of the principal tensile stress σ_1 at this point. *Ans.* $\sigma_1 = 1700$ psi; $\phi = -54°$.

6. Write the differential equation for the family of principal stress trajectories for the cantilever beam of rectangular cross-section shown in Fig. B.

$$Ans. \quad \left(\frac{h^2}{4} - y^2\right)\left(\frac{dy}{dx}\right)^2 + 2xy \frac{dy}{dx} - \left(\frac{h^2}{4} - y^2\right) = 0.$$

7.3 Stresses Due to Combined Bending and Torsion

One of the most important applications of the theory of combined stresses arises in the case of shafts subjected to both bending and torsion. Consider, in Fig. 7.8, a portion of a shaft of circular cross-section acted upon at its ends by bending couples M in the xy-plane and by twisting couples T about the x-axis. Under such loading, an element A in the top

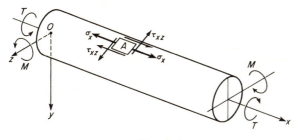

Fig. 7.8

surface of the shaft will carry normal stress σ_x due to bending, and shearing stresses τ_{xz} due to torsion. These stresses are readily computed from eq. (5.5) and from eq. (4.5). Thus, we have

$$\sigma_x = \frac{Mc}{I} = \frac{32M}{\pi d^3} \qquad (a)$$

and

$$\tau_{xz} = \frac{Tr}{J} = \frac{16T}{\pi d^3}, \qquad (b)$$

where d is the diameter of the shaft.

The element A is seen to be in a state of plane stress and the principal stresses will be found from eqs. (7.4) of Art. 7.1. Thus the principal normal stresses are

$$\sigma_{1,2} = \frac{\sigma_x}{2} \pm \sqrt{\left(\frac{\sigma_x}{2}\right)^2 + \tau_{xz}^2}. \qquad (7.7)$$

The maximum shearing stress, equal to half the difference between principal stresses, is

$$\tau_{max} = \frac{\sigma_1 - \sigma_2}{2} = \pm\sqrt{\left(\frac{\sigma_x}{2}\right)^2 + \tau_{xz}^2}. \qquad (7.8)$$

In the design of a shaft to carry a given loading, allowable working stresses in tension (or compression) and in shear will usually be prescribed. It then becomes necessary to find out from eqs. (7.7) and (7.8) which of these stresses governs the design and then to select the required diameter d of the shaft accordingly. For a brittle material like cast-iron, the maximum normal stress σ_1 should be used, while for a ductile material like structural steel, the maximum shear stress $\tau_{\max}$ is most commonly used.

To find the required diameter d of the shaft, it is necessary to substitute expressions (a) and (b) into eqs. (7.7) and (7.8). Doing this, we obtain

$$\sigma_1 = \frac{16}{\pi d^3} (M + \sqrt{M^2 + T^2}) \tag{7.7a}$$

and

$$\tau_{\max} = \frac{16}{\pi d^3} \sqrt{M^2 + T^2}. \tag{7.8a}$$

Then to select the diameter d on the basis of an allowable working stress σ_w in tension, we use eq. (7.7a) which gives

$$d = \sqrt[3]{\frac{16}{\pi \sigma_w} (M + \sqrt{M^2 + T^2})}. \tag{7.9}$$

Similarly, to select the required diameter d on the basis of an allowable working stress τ_w in shear, we use eq. (7.8a) and find

$$d = \sqrt[3]{\frac{16}{\pi \tau_w} \sqrt{M^2 + T^2}}. \tag{7.10}$$

If allowable stresses in both tension and shear are prescribed, we calculate d from both eqs. (7.9) and (7.10) and select the larger value.

Usually the bending of a shaft will be produced by transverse forces rather than by bending couples as shown in Fig. 7.8. In such case, we have to consider also shearing stresses associated with non-uniform bending. As observed in Art. 7.2, these shearing stresses usually vanish at those points where the normal stresses due to bending are a maximum so that they do not influence the maximum principal stresses and can be dis-

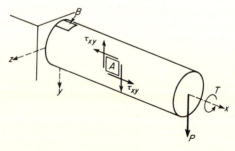

Fig. 7.9

regarded. However, when bending by transverse forces is combined with torsion, some consideration of the shear stresses associated with bending may be necessary. Consider, for example, the shaft loaded as shown in Fig. 7.9. In such case, a surface element A on the front side of the shaft will sustain shearing stress

$$\tau'_{xy} = \frac{16T}{\pi d^3} \tag{c}$$

due to the torque T and also shearing stress

$$\tau''_{xy} = \frac{4P}{3A} \tag{d}$$

due to the transverse load P. Thus the maximum shear stress $\tau_{xy} = \tau'_{xy} + \tau''_{xy}$. Even though the normal stress σ_x for this element vanishes, it might be in a worse condition than an element B at the top surface.

EXAMPLE 1. A steel shaft supported in bearings at A and B and carrying pulleys at C and D, is to transmit 100 hp at 500 rpm from the drive pulley D to the offtake pulley C as shown in Fig. 7.10. The following numerical data are given: $P_1 = 2P_2$, $Q_1 = 2Q_2$, $R_d = 6$ in., $R_c = 8$ in., $l = 4$ ft, $a = 1$ ft, and the working stress in shear is $\tau_w = 6000$ psi. Calculate the required diameter d of the shaft.

SOLUTION. From eq. (4.7), p. 73, the torque required to transmit 100 hp at 500 rpm is

$$T = \frac{63,000 \times 100}{500} = 12,600 \text{ in.-lb.}$$

Then since $T = (P_1 - P_2) R_d$ we find, with the given data,

$$P_1 = 2P_2 = 4200 \text{ lb.}$$

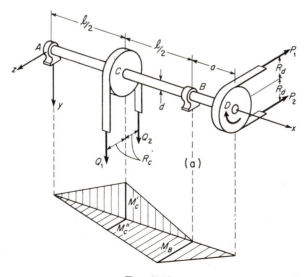

FIG. 7.10

Similarly, $T = (Q_1 - Q_2) R_c$, from which

$$Q_1 = 2Q_2 = 3150 \text{ lb.}$$

With these values of the loads determined, we turn our attention to the bending moments in the shaft caused by the transverse loads. It will be noted that the shaft undergoes bending both in the horizontal xz-plane and the vertical xy-plane. The corresponding bending moment diagrams are shown in Fig. 7.10b. A study of these diagrams shows that the maximum bending moment in the shaft will occur either at cross-section B or at cross-section C. The bending moment at B is

$$M_B = (P_1 + P_2) a = 6300 \times 12 = 75,600 \text{ in.-lb.}$$

At C, the bending moment in the vertical plane is

$$M'_c = \left(\frac{Q_1 + Q_2}{2}\right)\frac{l}{2} = \frac{4725}{2} \times 24 = 56,700 \text{ in.-lb,}$$

while that in the horizontal plane is

$$M''_c = \left(\frac{P_1 + P_2}{4}\right)\frac{l}{2} = \frac{6300}{4} \times 24 = 37,800 \text{ in.-lb.}$$

These two bending moments in orthogonal axial planes of the shaft may be added vectorially and we obtain

$$M_c = \sqrt{(M'_c)^2 + (M''_c)^2} = \sqrt{46.44(10)^8} = 68,200 \text{ in.-lb.}$$

This resultant bending moment at C being slightly smaller than that at B and the torque T being the same at both these cross-sections, we conclude that an element on the front side of the shaft at B is the critical element. Substituting $M = 75,600$ in.-lb, $T = 12,600$ in.-lb, and $\tau_w = 6000$ psi into eq. (7.10), we obtain

$$d = \sqrt[3]{\frac{16}{6000\pi}} \sqrt{(75,600)^2 + (12,600)^2} = \sqrt[3]{\frac{16 \times 76,700}{6000\pi}} = 4.02 \text{ in.}$$

EXAMPLE 2. A torsion pendulum consists of a solid right circular disk suspended by a thin steel shaft of circular cross-section as shown in Fig. 7.11a. The disk has weight $W = 100$ lb; the shaft has length $l = 20$ in. and diameter $d = \frac{1}{8}$ in. For the shaft, allowable stresses in tension and shear, respectively, are $\sigma_w = 16,000$ psi and $\tau_w = 8000$ psi. What is the maximum angle of twist ϕ that the shaft may have during torsional oscillations of the disk without exceeding either given working stress?

SOLUTION: In this case, the shaft will be subjected to combined tension and torsion. The uniform axial stress is

$$\sigma_x = \frac{W}{A} = \frac{100}{.0123} = 8130 \text{ psi.} \tag{e}$$

The shear stress due to torsion will be

$$\tau_{xy} = Gr\theta, \tag{f}$$

where θ is the angle of twist per unit length of the shaft and r is its radius.

To find the permissible value of θ, we will use Mohr's circle as shown in Fig. 7.11b. In this diagram, we first lay off, to a suitable scale, the known normal stresses $\sigma_x = 8130$ psi and $\sigma_y = 0$. This locates the center C of the circle as shown. Next, we mark the maximum allowable tensile stress $\sigma_1 = 16,000$ psi as shown by point A on the σ-axis. This done, we see that if we draw the circle through point A, its radius, representing $\tau_{\max}$, will be

$$16,000 - 4065 = 11,935 \text{ psi.}$$

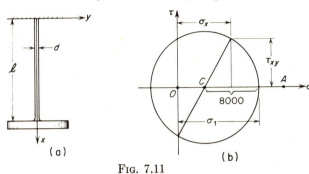

Fig. 7.11

Since this is greater than the allowable shear stress of 8000 psi, we conclude that the given working stress in shear governs. Accordingly, we draw the circle with radius $\tau_{\max} = 8000$ psi as shown.

We may now scale τ_{xy} from the diagram, whereby we obtain $\tau_{xy} = 6900$ psi. Using this value in eq. (f), we find

$$\theta = \frac{\tau_{xy}}{Gr} = \frac{6900}{12(10)^6(1/16)} = 0.0092.$$

The corresponding angle of twist for the shaft is

$$\phi = \theta l = 0.0092 \times 20 = 0.184 \text{ rad} = 10° 33'.$$

Note now that the maximum tensile stress in the shaft is $\sigma_1 = 12,065$ psi.

PROBLEM SET 7.3

1. A steel shaft supported in bearings A and B at its ends carries a pulley 2 ft in diameter at C as shown in Fig. A. Power is applied by a torque T at A and taken off through a belt overrunning the pulley, the tensions in the two branches of the belt being 300 lb on the taut side and 100 lb on the slack side. Find the required diameter d for the shaft if the working stresses are $\sigma_w = 12,000$ psi and $\tau_w = 6000$ psi. *Ans.* $d = 2$ in.

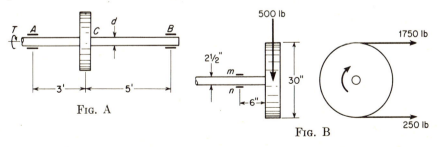

Fig. A   Fig. B

2. A 2.5-in.-diameter shaft supported in bearings carries a 30-in.-diameter pulley weighing 500 lb at an overhanging end of the shaft as shown in Fig. B. Calculate the principal tensile stress at the section mn if the horizontal belt tensions are as shown. *Ans.* $\sigma_1 = 12{,}400$ psi.

3. Referring to Fig. C, determine the required diameter d of the shaft to transmit 50 hp at 300 rpm from pulley C to pulley D. The power is applied at C by a horizontal belt and taken off at D through a vertical belt. Each pulley is 30 in. in diameter, and the ratio of belt tension on the taut side to that on the slack side is $P_1/P_2 = 3$. Use $\tau_w = 6000$ psi. *Ans.* $d = 2.36$ in.

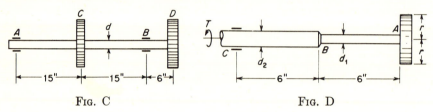

FIG. C   FIG. D

4. A stepped shaft of circular cross-section is supported in a bearing at C, Fig. D, and another at D (not shown) and carries a gear wheel at an overhanging end A. The gear wheel has a pitch radius $r = 4$ in. and in driving another gear (not shown) is subjected to a vertical tangential force P at its rim. Find the ratio d_1/d_2 of shaft diameters such that maximum shear stresses will be equal at sections B and C of the shaft. *Ans.* $d_1/d_2 = 0.83$.

5. A 3-ft length of $\frac{3}{4}$-in.-diameter steel pipe extends horizontally from a concrete wall. A workman fits a pipe wrench to the end of the pipe and pushes downward with a force P on the end of the handle, which is 15 in. long. If the section modulus for the pipe cross-section is $Z = 0.0705$ in.^{3} and the allowable stress in shear is $\tau_w = 20{,}000$ psi, what is the safe value of the force P? *Ans.* $P = 72$ lb.

6. A hollow steel pipe is to be used as a standard to support a highway road sign as shown in Fig. E. The maximum wind pressure on the face of the board is assumed to be 50 lb/ft.^{2} The standard is unsupported laterally and its outside-to-inside diameter ratio is 1.12. The allowable working stress in shear is given as $\tau_w = 8400$ psi. Calculate the required outside diameter d of the pipe. *Ans.* $d = 5.20$ in.

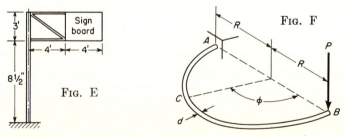

FIG. E   FIG. F

7. A shaft of diameter d, bent in the form of a semicircle AB of radius R, is built-in at A and loaded at B by a force P acting perpendicular to the plane of the ring as shown in Fig. F. Thus any cross-section C of the ring is subjected to both bending and torsion. Assuming that d is small compared with R so that the theory of bending of straight bars may be used, find the value of ϕ for which the principal stress σ_1 will be a maximum. *Ans.* $\phi = 120°$.

8. Solve the preceding problem if the load P at B is replaced by a twisting couple T, the plane of which coincides with the vertical plane through AB. What is the magnitude of this principal stress? *Ans.* $\phi = 90°$; $\sigma_1 = 32T/\pi d^3$.

9. Referring to the torsion pendulum shown in Fig. 7.11a, we have for the shaft: $l = 20$ in., $d = \frac{1}{8}$ in., $G = 12(10)^6$ psi. In such case, what is the maximum allowable weight W of the disk in order that the pendulum may have an amplitude $\phi = 12°$ during torsional oscillation without exceeding the allowable working stress $\tau_w = 8000$ psi in shear? *Ans.* $W = 36.7$ lb.

7.4 Analysis of Strain

Very often the strains induced in an element of material subjected to plane stress are of practical interest. To completely define the state of strain in the xy-plane of such an element, it is necessary to specify the linear strains ϵ_x and ϵ_y in two perpendicular directions and the change in angle between these two directions (the shearing strain γ_{xy}). In Fig. 7.12a, let OX and OY represent two such perpendicular lines of unit length coinciding with two edges of the element before deformation. During deformation, point X moves to X' and point Y moves to Y' relative to the origin O, X' and Y' remaining in the plane XOY. Then the strain in the neighborhood of O is completely defined by the two extensions $\epsilon_x = XX'$ and $\epsilon_y = YY'$ and by the change γ_{xy} in the original right angle XOY

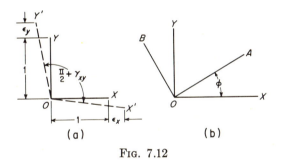

(a)   (b)

Fig. 7.12

as it deforms into the final angle $X'OY'$. Extensional strains are considered positive and compressive strains, negative. The shear strain γ_{xy} will be considered *positive* when the right angle XOY is *increased*. With these sign conventions, positive strains will be seen to correspond to positive stresses, as previously defined.

We now consider the following question: When a state of strain in one plane is defined by given values of ϵ_x, ϵ_y, and γ_{xy}, in orthogonal directions OX and OY, what are the corresponding strains associated with the mutually perpendicular directions OA and OB which make the angle ϕ with OX and OY, respectively, as shown in Fig. 7.12b? To answer this question, consider a rectangular element $OXAY$ whose diagonal OA makes the angle ϕ with OX as shown in Fig. 7.13. Let dx, dy, denote the lengths of its sides and ds the length of its diagonal. As a result of the given strains ϵ_x, ϵ_y, γ_{xy}, OX elongates by the amount $\epsilon_x dx$, OY elongates by the amount $\epsilon_y dy$, and the original right angle YOX increases by the amount γ_{xy}. For clarity, these three deformations of the element are shown separately in Figs. 7.13a, b, c. The corresponding changes in the length of the diagonal OA are $+\epsilon_x dx \cos\phi$, $+\epsilon_y dy \sin\phi$, and $-\gamma_{xy} dx \sin\phi$, as shown in the figures. The net change in OA is the algebraic sum of these three quantities and the corresponding strain in the direction OA is obtained by

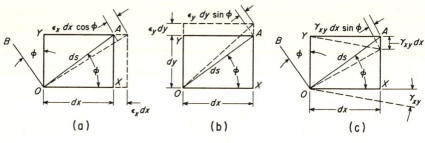

FIG. 7.13

dividing this sum by the length ds. Thus with the observation that $dx/ds = \cos\phi$ and $dy/ds = \sin\phi$, we obtain

$$\epsilon_a = \epsilon_x \cos^2\phi + \epsilon_y \sin^2\phi - \gamma_{xy} \sin\phi \cos\phi. \tag{a}$$

The strain ϵ_b in the direction of OB may be obtained simply by replacing ϕ in expression (a) by $90° + \phi$, giving

$$\epsilon_b = \epsilon_x \sin^2\phi + \epsilon_y \cos^2\phi + \gamma_{xy} \sin\phi \cos\phi. \tag{b}$$

By making the trigonometric substitutions $\sin^2\phi = \frac{1}{2}(1 - \cos 2\phi)$, $\cos^2\phi = \frac{1}{2}(1 + \cos 2\phi)$, and $\sin\phi \cos\phi = \frac{1}{2}\sin 2\phi$ in expressions (a) and (b), they may be written in the form

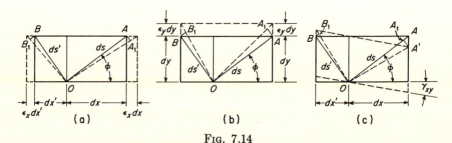

FIG. 7.14

$$\left.\begin{array}{l}\epsilon_a = \tfrac{1}{2}(\epsilon_x + \epsilon_y) + \tfrac{1}{2}(\epsilon_x - \epsilon_y)\cos 2\phi - \tfrac{1}{2}\gamma_{xy}\sin 2\phi, \\ \epsilon_b = \tfrac{1}{2}(\epsilon_x + \epsilon_y) + \tfrac{1}{2}(\epsilon_x - \epsilon_y)\cos 2\phi + \tfrac{1}{2}\gamma_{xy}\sin 2\phi.\end{array}\right\} \quad (7.11)$$

A general expression for the shearing strain γ_{ab} may be obtained in a similar manner. Consider, in Fig. 7.14, two adjacent rectangular elements so proportioned that their diagonals OA and OB are mutually perpendicular Then from Fig. 7.14a representing the influence of ϵ_x alone, we see that the change in the original right angle BOA is as follows:

$$\angle AOA_1 + \angle BOB_1 = \frac{AA_1}{OA} + \frac{BB_1}{OB} = \frac{\epsilon_x\, dx \sin\phi}{ds} + \frac{\epsilon_x\, dx' \cos\phi}{ds'}$$

$$= \epsilon_x \cos\phi \sin\phi + \epsilon_x \sin\phi \cos\phi = \epsilon_x \sin 2\phi.$$

Similarly, from Fig. 7.14b, the change in the right angle BOA due to the strain ϵ_y alone is

$$- \angle AOA_1 - \angle BOB_1 = -\frac{AA_1}{OA} - \frac{BB_1}{OB} = -\frac{\epsilon_y\, dy \cos\phi}{ds} - \frac{\epsilon_y\, dy \sin\phi}{ds'}$$

$$= -\epsilon_y \sin\phi \cos\phi - \epsilon_y \cos\phi \sin\phi = -\epsilon_y \sin 2\phi.$$

Finally, referring to Fig. 7.14c, the change in the right angle BOA caused by the shearing strain γ_{xy} alone is

$$\angle A'OA_1 - \angle BOB_1 = \frac{A'A_1}{OA} - \frac{BB_1}{OB} = \frac{\gamma_{xy}\, dx \cos\phi}{ds} - \frac{\gamma_{xy}\, dx' \sin\phi}{ds'}$$

$$= \gamma_{xy} \cos^2\phi - \gamma_{xy} \sin^2\phi = \gamma_{xy} \cos 2\phi.$$

The algebraic sum of the above angle changes gives the net change in the angle BOA, i.e., the desired shearing strain γ_{ab} as follows:

$$\gamma_{ab} = (\epsilon_x - \epsilon_y) \sin 2\phi + \gamma_{xy} \cos 2\phi,$$

which can arbitrarily be written in the form

$$\tfrac{1}{2}\gamma_{ab} = \tfrac{1}{2}(\epsilon_x - \epsilon_y) \sin 2\phi + \tfrac{1}{2}\gamma_{xy} \cos 2\phi. \quad (7.12)$$

Comparing the first of eqs. (7.11) and eq. (7.12) with eqs. (7.1) of Art. 7.1, we see that there is a complete analogy between a state of plane strain and one of plane stress. The linear strains ϵ_x, ϵ_y, and ϵ_a in eqs. (7.11) and (7.12) correspond, respectively, to the stresses σ_x, σ_y, and σ_n in eqs. (7.1), while the *half shear strains* $\gamma_{xy}/2$ and $\gamma_{ab}/2$ correspond to the shear stresses τ_{xy} and τ, respectively. This suggests that a Mohr's circle for strains may be constructed in the same manner as was previously done for stresses.

In Fig. 7.15a, let ϵ_x and ϵ_y represent given linear strains in the x and y directions and let γ_{xy} represent the corresponding shearing strain, i.e., the amount that the original right angle XOY increases due to deformation. This shearing strain is represented in Fig. 7.15a by a clockwise rotation of

OX through the angle $\gamma_{xy}/2$ and by a counterclockwise rotation of OY through the angle $\gamma_{xy}/2$. *Clockwise* rotation of a line is to be considered as *positive* and *counterclockwise* rotation as *negative*.

Now to construct a Mohr's circle for this state of plane strain, we begin with the coordinate plane ϵ, $\gamma/2$, as shown in Fig. 7.15b. Laying out the extensional strain ϵ_x as a positive abscissa and the clockwise rotation

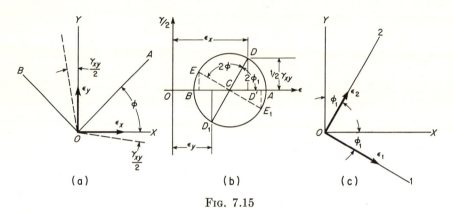

(a)   (b)   (c)

FIG. 7.15

$\gamma_{xy}/2$ of the line OX as a positive ordinate, we obtain point D on the circle corresponding to the x direction in the plane of the plate. Similarly, laying out the extensional strain ϵ_y as a positive abscissa and the counterclockwise rotation $\gamma_{xy}/2$ of the line OY as a negative ordinate, we obtain point D_1 on the circle corresponding to the y direction in the plane of the plate. Drawing the diameter DD_1, the center C is located and the circle can be drawn as shown. This done, the strains ϵ_a, ϵ_b, and $\gamma_{ab}/2$ associated with the orthogonal lines OA and OB which make any angle ϕ with the lines OX and OY, respectively, can be found from the circle as follows: Since ϕ is a counterclockwise angle from X to A, we lay out from D on the circle the counterclockwise angle 2ϕ to locate point E representing the direction OA in Fig. 7.15a, and draw the diameter EE_1 as shown. The coordinates of points E and E_1 represents the strains ϵ_a, ϵ_b, and $\gamma_{ab}/2$ associated with the directions OA and OB in Fig. 7.15a. The positive coordinates of point E indicate that the line OA undergoes extension and rotates clockwise during deformation, while the coordinates of E_1 (positive abscissa and negative ordinate) indicate that OB also undergoes extension but that it rotates counterclockwise. This means that after deformation the angle AOB is slightly greater than a right angle.

Of particular interest in the case of a state of plane strain defined by given values of ϵ_x, ϵ_y, and γ_{xy}, is the direction of *principal strains*. That is, the directions of two mutually perpendicular lines $O1$ aad $O2$ which remain at right angles after deformation and along which the linear strains ϵ_1 and ϵ_2 are, respectively, a maximum and a minimum. From Mohr's circle, Fig. 7.15b, we see that these principal directions are represented by points A and B, since these points have zero ordinates indicating no shearing strain. The clockwise angle $DCA = 2\phi_1$ in Mohr's circle indicates that the direction of the principal strain ϵ_1 makes a clockwise angle ϕ_1 with the direction OX. Similarly the clockwise angle $D_1CB = 2\phi_1$ indicates that the direction of principal strain ϵ_2 makes the same clockwise angle ϕ_1 with the direction OY. Thus the principal axes of strain $O1$ and $O2$ will be oriented as shown in Fig. 7.15c. The magnitudes of ϵ_1 and ϵ_2 are seen to be $\epsilon_1 = OA$ and $\epsilon_2 = OB$ on Mohr's circle. They are also seen to be the maximum and minimum linear strains in the plane of the plate.

Analytic expressions for the principal strains ϵ_1 and ϵ_2 in terms of given values of ϵ_x, ϵ_y, and γ_{xy} can easily be found from the geometry of Mohr's circle (Fig. 7.15b) as follows :

$$\left. \begin{aligned} \epsilon_1 &= OA = OC + CA = OC + CD, \\ \epsilon_2 &= OB = OC - BC = OC - CD. \end{aligned} \right\} \tag{c}$$

Then noting that $OC = \frac{1}{2}(\epsilon_x + \epsilon_y)$ while $CD = \sqrt{(CD')^2 + (DD')^2}$, where $CD' = \frac{1}{2}(\epsilon_x - \epsilon_y)$ and $DD' = \frac{1}{2}\gamma_{xy}$, eqs. (c) become

$$\left. \begin{aligned} \epsilon_1 &= \frac{1}{2}(\epsilon_x + \epsilon_y) + \sqrt{\left(\frac{\epsilon_x - \epsilon_y}{2}\right)^2 + \left(\frac{\gamma_{xy}}{2}\right)^2}, \\ \epsilon_2 &= \frac{1}{2}(\epsilon_x + \epsilon_y) - \sqrt{\left(\frac{\epsilon_x - \epsilon_y}{2}\right)^2 + \left(\frac{\gamma_{xy}}{2}\right)^2}. \end{aligned} \right\} \tag{7.13}$$

The clockwise angle ϕ_1 defining the direction of these principal strains with reference to OX is given by the equation

$$\tan 2\phi_1 = -\frac{DD'}{CD'} = \frac{-\gamma_{xy}/2}{\frac{1}{2}(\epsilon_x - \epsilon_y)} = \frac{-\gamma_{xy}}{\epsilon_x - \epsilon_y}. \tag{7.14}$$

obtained from the trigonometry of the right triangle CDD' in Fig. 7.15b. Equations (7.13) and (7.14) are seen to be analogous to eqs. (7.4) and (7.2) of Art. 7.1 for the magnitudes and directions of principal stresses. They can be used for the calculation of magnitude and direction of principal strain without the need of constructing Mohr's circle, if preferred.

PROBLEM SET 7.4

1. A state of plane strain in a steel plate is defined by the following data: $\epsilon_x = +0.00050$, $\epsilon_y = +0.00014$, $\gamma_{xy} = +0.00036$. Construct a Mohr's circle and find the magnitudes and directions of principal strains. *Ans.* $\epsilon_1 = 575 \ (10)^{-6}$; $\epsilon_2 = 65 \ (10)^{-6}$; $\phi_1 = -22°30'$.

2. Solve Problem 1 if $\epsilon_x = -0.00014$, $\epsilon_y = -0.00050$, $\gamma_{xy} = +0.00036$. *Ans.*
$\epsilon_1 = -65 \ (10)^{-6}$; $\epsilon_2 = -575 \ (10)^{-6}$; $\phi_1 = -22°30'$

3. Using eqs. (7.13) and (7.14), find the magnitudes and directions of principal strains for the state of plane strain defined by $\epsilon_x = +0.00050$, $\epsilon_y = +0.00030$, $\gamma_{xy} = -0.00105$. *Ans.* $\epsilon_1 = 935 \ (10)^{-6}$; $\epsilon_2 = -135 \ (10)^{-6}$; $\phi_1 = +39°36'$.

7.5 The Strain Rosette

In the case of a material subjected to plane stress, it is often desirable to obtain the stresses by direct measurement. Since stresses cannot be measured directly, this requires the measurement of strains or deformations which take place in the material during loading. Such measurements are usually made with very sensitive strain gages attached to the surface of the body before it is loaded so that they can record the amount of strain that takes place during loading. The question naturally arises as to how to arrange these strain gages so as to get sufficient strain data to be able to compute therefrom the corresponding stresses. An obvious answer to this question would be to measure the linear strains ϵ_x and ϵ_y in any two perpendicular directions at a chosen point on the surface of the body and the change in angle γ_{xy} between these two directions. Then the principal strains could be found as explained in the preceding article, and having these, the principal stresses would be calculated from eqs. (3.6) of Art. (3.2). However, accurate measurement of the shearing strain γ_{xy} is found to be very difficult. It is easier and more accurate to measure, in the neighborhood of a chosen point on the surface of the body, three linear strains ϵ_a, ϵ_b, ϵ_c, in three different directions and then deduce from these measurements the magnitudes and directions of the principal strains ϵ_1 and ϵ_2. Such a group of strain gages is called a *strain rosette*. One standard arrangement of the strain rosette will now be discussed in detail.*

The 45° Strain Rosette. If linear strains ϵ_a, ϵ_b, and ϵ_c are measured along lines oa, ob, and oc, 45° apart as shown in Fig. 7.16a, Mohr's circle can be constructed without ambiguity and the principal strains ϵ_1 and ϵ_2 can be found therefrom. This Mohr's circle will be constructed in the following manner: In the ϵ, $\gamma/2$-plane (Fig. 7.16b) draw three verticals aa, bb, and cc, having respectively, the abscissas ϵ_a, ϵ_b, ϵ_c, as shown. (If any of the measured strains were negative, the corresponding verticals would lie to the left of the origin O.) Locate the center C of the circle midway between the two verticals aa and cc as shown. If the intermediate vertical bb lies to the left of C, as assumed here, lay off on the vertical aa the positive ordinate $D'D = CF'$ and on the vertical cc, lay off the negative ordinate $D'_1D_1 = CF'$,

*For additional information on the strain rosette, see K. J. Bossart and G. A. Brewer, "A Graphical Method of Rosette Analysis," *Proc. Soc. Exp. Stress Analysis* Vol. IV, No. I, p. 1.

F' being the point where bb intersects the ϵ-axis. The line DD_1, so obtained, is the diameter of the required Mohr's circle which can now be drawn. This circle will cut the intermediate vertical bb at a point F such that CF will be perpendicular to DD_1, since by construction the triangles CDD' and FCF' have been made congruent. Thus points D, F, and D_1, having, respectively, abscissa ϵ_a, ϵ_b, and ϵ_c, and being 90° apart in counterclockwise order in Fig. 7.16b, represent completely the strains associated with the directions oa, ob, and oc which are 45° apart in Fig. 7.16a. The principal strains are $\epsilon_1 = OA$ and $\epsilon_2 = OB$. The direction of the principal strain ϵ_1 makes the angle $\phi_1 = \frac{1}{2} \angle DCA$ with the direction oa as shown in Fig. 7.16c.

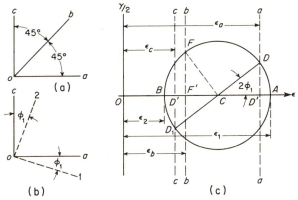

FIG. 7.16

Knowing the principal strains ϵ_1 and ϵ_2, the principal stresses σ_1 and σ_2 can be calculated from eqs. (3.6); namely

$$\sigma_1 = \frac{(\epsilon_1 + \mu\epsilon_2)E}{1 - \mu^2}, \quad \sigma_2 = \frac{(\epsilon_2 + \mu\epsilon_1)E}{1 - \mu^2}. \tag{a}$$

This assumes, of course, that the material has not been stressed beyond its elastic limit.

EXAMPLE 1. Data taken from a 45°-strain rosette (see Fig. 7.17a) reads as follows: $\epsilon_a = 750$ microinches/in., $\epsilon_b = -110$ microinches/in., and $\epsilon_c = 210$ microinches/in. Find the magnitudes and directions of principal strains and the corresponding principal stresses. Assume $E = 30(10)^6$ psi and $\mu = 0.30$.

SOLUTION. Beginning with the ϵ, $\gamma/2$-plane, Fig. 7.17b, construct the verticals aa, bb, cc, with abscissas $\epsilon_a = 750$ units, $\epsilon_b = -110$ units, and $\epsilon_c = 210$ units, as shown. Locate center C on the ϵ-axis midway between the verticals aa and cc. Then since the vertical bb lies to the left of C, lay off the positive ordinate $D'D = CF'$ along the vertical aa which locates point D. Similarly, lay off the negative ordinate $D'_1D_1 = CF'$ along the vertical cc to locate point D_1. With C as a center and DD_1 as a diameter, draw Mohr's circle as shown. This circle intersects the vertical bb at F making CF perpendicular to DD_1 as it should be. Scaling OA and

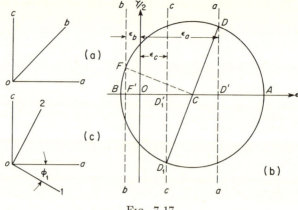

FIG. 7.17

OB from the diagram, we find that the principal strains are $\epsilon_1 = 1130 \ (10)^{-6}$, $\epsilon_2 = -170 \ (10)^{-6}$. The angle DCA is measured with a protractor and found to be $65°30'$. Hence $\phi_1 = 32°45'$, clockwise from oa to ol as shown in Fig. 7.17c. Substituting the scaled values of ϵ_1 and ϵ_2 into eqs. (a) above, the corresponding principal stresses are found to be $\sigma_1 = 35,700$ psi and $\sigma_2 = 5900$ psi. It will be noted that although the strain ϵ_2 is negative, the corresponding stress σ_2 is positive.

PROBLEM SET 7.5

1. Referred to the directions shown in Fig. 7.16a, a 45°-strain rosette gives the following data: $\epsilon_a = 670$ microinches/in., $\epsilon_b = 330$ microinches/in., $\epsilon_c = 150$ microinches/in. Construct a Mohr's circle for this state of plane strain and find the principal stresses σ_1 and σ_2 and the angle ϕ_1 defining their directions with reference to the direction oa. Use $E = 30 \ (10)^6$ psi and $\mu = 0.30$. *Ans.* $\sigma_1 = 23,800$ psi; $\sigma_2 = 11,300$ psi; $\phi_1 = -8°33'$.

2. Solve Problem 1 if the measured strains are $\epsilon_a = 232$, $\epsilon_b = 123$, $\epsilon_c = -80$, all values being microinches per inch. *Ans.* $\sigma_1 = 7030$ psi; $\sigma_2 = -500$ psi; $\phi_1 = +8°23'$.

3. Solve Problem 1 if the measured strains are $\epsilon_a = 815$; $\epsilon_b = -72$, $\epsilon_c = 165$, all values being microinches per inch. *Ans.* $\sigma_1 = 36,000$ psi; $\sigma_2 = 6000$ psi; $\phi_1 = -30°00'$.

4. Sometimes, instead of a 45°-strain rosette, a 60°-strain rosette is used, i.e., three linear strains ϵ_a, ϵ_b, and ϵ_c are measured along three sides of an equilateral triangle. Prove that in such case, the center C of Mohr's circle for strain will always have the abscissa equal to $(\epsilon_a + \epsilon_b + \epsilon_c)/3$.

5. When a 60°-strain rosette is used and the center of Mohr's circle is obtained according to the preceding problem, prove that the correct radius of the circle is that one which makes equal to zero the algebraic sum of the three $\gamma/2$-ordinates associated with the three linear strains ϵ_a, ϵ_b, ϵ_c. This condition makes it fairly easy to construct the correct circle by trial and error.

8

Deflection of Beams

8.1 Differential Equation of the Elastic Line

In discussing pure bending of a prismatic bar in Art. 5.3, it was shown that the curvature of the neutral surface was given by the equation

$$\frac{1}{\rho} = \frac{M}{EI}. \tag{a}$$

Thus, for pure bending, where M is constant along the length of the bar, its axis bends in a circular arc. It is customary to call this curved axis of the bar the *elastic line* or the *deflection curve*.

In the case of a beam bent by transverse loads acting in a plane of symmetry, the bending moment M varies along the length of the beam, and we represent this variation by a bending moment diagram. For such non-uniform bending it is usually assumed that eq. (a) holds at each cross-section. Thus the curvature $1/\rho$ is seen to vary along the beam as the bending moment varies and the elastic line becomes a rather complicated curve, the shape of which is defined by eq. (a). To express the shape of this curve in rectangular coordinates, we consider any portion of a bent beam as shown in Fig. 8.1. Through any point O on this elastic line, we take coordinate axes x and y as shown, the x-axis coinciding with the original straight axis of the beam, positive to the right, and the y-axis positive downwards. At the ends of any element ds of this elastic line, we construct normals which intersect at C, thus defining the radius of curvature ρ of the element. Denoting

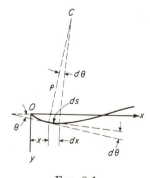

Fig. 8.1

the angle between these two normals by $d\theta$, which is also the angle between the tangents to the elastic line at the ends of the element, we have the relationship $ds = \rho\,d\theta$, from which

$$\frac{d\theta}{ds} = \frac{1}{\rho}. \tag{b}$$

For a beam which is bent within the elastic range of its material, the elastic line will usually be a very flat curve. Limiting the discussion to such conditions, and referring to Fig. 8.1, we now introduce the approximations $ds \approx dx$ and $\theta \approx dy/dx$, so that eq. (b) becomes

$$\left|\frac{d^2y}{dx^2}\right| = \frac{1}{\rho},$$ (c)

indicating that we consider only the absolute value of d^2y/dx^2.

Combining expressions (a) and (c), we have

$$\frac{d^2y}{dx^2} = \pm \frac{M}{EI}.$$ (d)

The sign in this expression must now be chosen such that it will be consistant with the choice of coordinate axes in Fig, 8.1 and the definition of positive bending moment as that which produces curvature concave upwards (see p. 96). For the coordinate axes as shown in Fig. 8.1, we see that when the curvature is concave upwards, the slope dy/dx is algebraically decreasing with x and hence d^2y/dx^2 is negative. Likewise, when the curvature is concave downwards (negative bending moment), the slope dy/dx is algebraically increasing with x and d^2y/dx^2 is positive. Thus d^2y/dx^2 is always opposite in sign to M and we take expression (d) in the form

$$\frac{d^2y}{dx^2} = -\frac{M}{EI}.$$ (8.1)

This is the *differential equation* of the elastic line for a beam subjected to bending in a plane of symmetry. Its solution $y = f(x)$ defines the shape of the elastic line or the *deflection curve* as it is frequently called.

Equation (8.1) carries with it two important limitations resulting from the assumptions made in its derivation. First, the moment-curvature relationship (a), derived in Art. 5.3, assumes that stress is proportional to strain, i.e., that Hooke's law applies. Thus the equation is valid only for beams that are not stressed beyond the elastic limit of their materials. Second, since expression (c) assumes that the curvature is always small, the equation is limited to the treatment of small deflections. Most beams encountered in engineering practice will be well within the validity of these two limitations. Furthermore, eq. (8.1), derived originally for pure bending does not account for any deflection resulting from shear deformation of the material. The additional deflections due to such deformation will be discussed in Art. 8.4. It will be shown there that for beams of ordinary proportions, the deflections due to shear deformation are usually small and can be neglected.

If we twice differentiate eq. (8.1) with respect to x, we obtain, with reference to eqs. (5.1) and (5.2)

$$EI \frac{d^3y}{dx^3} = -\frac{dM}{dx} = -V,$$ (8.2)

and

$$EI \frac{d^4y}{dx^4} = -\frac{dV}{dx} = w,$$ (8.3)

where V is the shear force at any cross-section and w is the intensity of distributed load. Equation (8.3) will be found particularly useful in those cases in which the load has a non-uniform distribution.

The application of eqs. (8.1) and (8.3) to find the deflection curves for various transverse loadings of prismatic beams will now be shown by several examples.

EXAMPLE 1. A simply supported prismatic beam AB carries a uniformly distributed load of intensite w over its span l as shown in Fig. 8.2. Develop the equation of the elastic line and find the maximum deflection δ at the middle of the span.

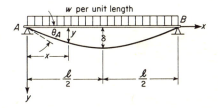

FIG. 8.2

SOLUTION. Taking coordinate axes x and y as shown, we have for the bending moment at any point x

$$M_x = \frac{wl}{2}x - \frac{wx^2}{2},$$

and eq. (8.1) becomes

$$EI \frac{d^2y}{dx^2} = -\frac{wlx}{2} + \frac{wx^2}{2}.$$ (e)

Multiplying both sides by dx and integrating, we obtain

$$EI \frac{dy}{dx} = -\frac{wlx^2}{4} + \frac{wx^3}{6} + C_1,$$ (f)

where C_1 is an integration constant. To evaluate this constant, we note from symmetry that when $x = l/2$, $dy/dx = 0$. From this condition, we find

$$C_1 = \frac{wl^3}{24}.$$

and eq. (f) becomes

$$EI \frac{dy}{dx} = -\frac{wlx^2}{4} + \frac{wx^3}{6} + \frac{wl^3}{24}.$$ (g)

Again multiplying both sides by dx and integrating,

$$EIy = -\frac{wlx^3}{12} + \frac{wx^4}{24} + \frac{wl^3x}{24} + C_2.$$ (h)

The integration constant C_2 is found from the condition that $y = 0$ when $x = 0$.

Thus $C_2 = 0$ and the required equation for the elastic line becomes

$$y = \frac{wx}{24EI}(l^3 - 2lx^2 + x^3). \tag{i}$$

To find the maximum deflection at mid-span, we set $x = l/2$ in eq. (i) and obtain

$$\delta = \frac{5wl^4}{384EI}.$$

The maximum slope θ_A at the left end of the beam can be found by setting $x = 0$ in eq. (g), which gives

$$\left(\frac{dy}{dx}\right)_{x=0} = \theta_A = \frac{wl^3}{24EI}.$$

EXAMPLE 2. A simply supported beam AB carries a triangularly distributed load as shown in Fig. 8.3. Find the equation of the deflection curve referred to the coordinate axes x and y as shown. Determine also the maximum deflection δ.

SOLUTION. In this case, we begin directly with eq. (8.3) and write

$$EI\frac{d^4y}{dx^4} = \frac{wx}{l}.$$

Separating variables and integrating twice, we obtain

$$EI\frac{d^2y}{dx^2} = \frac{wx^3}{6l} + C_1x + C_2. \tag{j}$$

Again, separating variables and performing two more integrations, we obtain

$$EIy = \frac{wx^5}{120l} + C_1\frac{x^3}{6} + C_2\frac{x^2}{2} + C_3x + C_4. \tag{k}$$

To find the four constants of integration, we now note that the bending moment, represented by eq. (j), and the deflection, represented by eq. (k), both vanish when $x = 0$ and when $x = l$. From these four boundary conditions, we find

$$C_1 = -\frac{wl}{6}, \quad C_2 = 0, \quad C_3 = \frac{7wl^3}{360}, \quad C_4 = 0.$$

Substituting these values back into eq. (k) and rearranging terms,

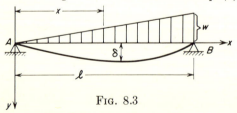

FIG. 8.3

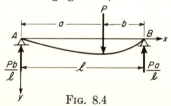

FIG. 8.4

we obtain

$$y = \frac{wx}{360lEI}(7l^4 - 10l^2x^2 + 3x^4).$$

To find the maximum deflection δ, we first set $dy/dx = 0$ and find $x = 0.519l$. Then using this value of x in the expression for y, the maximum deflection becomes

$$\delta = y_{max} = 0.00652\frac{wl^4}{EI}.$$

Setting $x = l$ in eq. (k), we obtain, for the deflection at B,

$$\delta = \frac{Pl^3}{3EI}.$$

EXAMPLE 3. A simply supported prismatic beam AB carries a concentrated load P as shown in Fig. 8.4. Locate the point of maximum deflection on the elastic line and find the value of this deflection.

SOLUTION. Choosing coordinate axes x and y as shown, we have for $0 < x < a$,

$$M_x = \frac{Pb}{l}x,$$

while for $a < x < l$

$$M_x = \frac{Pb}{l}x - P(x - a).$$

Substituting these expressions for bending moment into eq. (8.1), we obtain for the two portions of the deflection curve, the following two differential equations

$0 < x < a$	$a < x < l$
$EI\dfrac{d^2y}{dx^2} = -\dfrac{Pbx}{l}$	$EI\dfrac{d^2y}{dx^2} = -\dfrac{Pbx}{l} + P(x - a)$

Successive integrations of these equations give

$$EI\frac{dy}{dx} = -\frac{Pbx^2}{2l} + C_1 \quad \text{(l)} \qquad EI\frac{dy}{dx} = -\frac{Pbx^2}{2l} + \frac{P(x - a)^2}{2} + D_1 \quad \text{(n)}$$

$$EIy = -\frac{Pbx^3}{6l} + C_1x + C_2 \quad \text{(m)} \qquad EIy = -\frac{Pbx^3}{6l} + \frac{P(x - a)^3}{6} + D_1x + D_2$$

$$\text{(o)}$$

where C_1, C_2, D_1, D_2, are constants of integration. To find these four constants, we have the following conditions:

1. At $x = 0$, $y = 0$.   2. At $x = l$, $y = 0$.

3. At $x = a$, $\dfrac{dy}{dx} = \dfrac{dy}{dx}$ at $x = a$.

4. At $x = a$, $y = y$ at $x = a$.

Using condition (1) in eq. (m), we find $C_2 = 0$. Using condition (3) in eqs. (l) and (n), we find $C_1 = D_1$. Using condition (4) in eqs. (m) and (o) and noting that $C_1 = D_1$ while $C_2 = 0$, we find that $D_2 = 0$. Finally, using condition (2) in eq. (o) and noting that $a = l - b$, we find

$$C_1 = \frac{Pb}{6l}(l^2 - b^2) = D_1.$$

Using the constants as determined, eqs. (m) and (o) defining the two portions of the elastic line of the beam become

$$EIy = \frac{Pbx}{6l}(l^2 - b^2 - x^2) \quad \text{(p)} \quad \bigg| \quad EIy = \frac{Pb}{6l}\left[\frac{l}{b}(x - a)^3 + (l^2 - b^2)x - x^3\right]. \quad \text{(q)}$$

For $a > b$, the maximum deflection will occur in the left portion of the span, to which eq. (p) applies. Setting the derivative of this expression equal to zero gives

$$x = \sqrt{\frac{l^2 - b^2}{3}},$$

which defines the abscissa of the point having a horizontal tangent and hence the point of maximum deflection. Substituting this value of x into eq. (p), we find

$$y_{\max} = \frac{Pb}{9\sqrt{3lEI}} \sqrt{(l^2 - b^2)^3}. \tag{r}$$

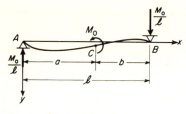

Fig. 8.5

EXAMPLE 4. A simply supported prismatic beam AB is acted upon by a couple M_0 applied at an intermediate point C as shown in Fig. 8.5. Derive the general equation for the portion AC of the elastic line. Find also the deflection of point C and the slope at A.

SOLUTION. The reaction $R_A = M_0/l$ and the bending moments in the two portions of the span become

$0 < x < a$	$a < x < l$
$M_x = \dfrac{M_0 x}{l}$	$M_x = \dfrac{M_0 x}{l} - M_0.$

Substituting these expressions into eq. (8.1), we obtain

$$EI \frac{d^2 y}{dx^2} = -\frac{M_0 x}{l} \qquad\qquad EI \frac{d^2 y}{dx^2} = -\frac{M_0 x}{l} + M_0.$$

Two integrations of each of these equations produce

$$EI \frac{dy}{dx} = -\frac{M_0 x^2}{2l} + C_1 \quad \text{(s)} \qquad EI \frac{dy}{dx} = -\frac{M_0 x^2}{2l} + M_0 x + D_1 \text{ (u)}$$

$$EIy = -\frac{M_0 x^3}{6l} + C_1 x + C_2 \quad \text{(t)} \qquad EIy = -\frac{M_0 x^3}{6l} + \frac{M_0 x^2}{2} + D_1 x + D_2. \text{ (v)}$$

The integration constants are determined from the conditions

(1) $y = 0$ at $x = 0$ $\qquad\qquad$ (2) $y = 0$ at $x = l$

3) at $x = a$, $\dfrac{dy}{dx} = \dfrac{dy}{dx}$, at $x = a$

4) at $x = a$, $y = y$, at $x = a$.

Substituting these conditions into the above equations, we find

$$C_1 = M_0 a - \frac{M_0 l}{3} - \frac{M_0 a^2}{2l},$$ $$D_1 = -\frac{Ml}{3} - \frac{M_0 a^2}{2l},$$

$$C_2 = 0$$ $$D_2 = \frac{M_0 a^2}{2}.$$

With these values, eqs. (s) and (t) for the portion AC of the elastic line become

$$EI \frac{dy}{dx} = \frac{M_0}{6l}[6al - 3(a^2 + x^2) - 2l^2], \qquad (w)$$

$$EIy = \frac{M_0 x}{6l}[6al - 3a^2 - x^2 - 2l^2]. \qquad (x)$$

Setting $x = a$ in eq. (x), we find for the deflection of point C

$$\delta_c = \frac{M_0 a}{3lEI}[3al - 2a^2 - l^2]. \qquad (y)$$

Setting $x = 0$ in eq. (w), we find for the slope at A

$$\theta_A = \frac{M_0}{6EIl}[6al - 3a^2 - 2l^2]. \qquad (z)$$

PROBLEM SET 8.1

1. With reference to the coordinate axes x and y as shown in Fig. A, derive the equation defining the deflection curve of a uniformly loaded cantilever beam. From this, evaluate the deflection δ at the free end. *Ans.* $\delta = wl^4/8EI$.

2. Repeat the solution of the previous problem for the case of a distributed load the intensity of which increases uniformly from zero at the free end to w at the built-in end. *Ans.* $\delta = wl^4/30EI$.

3. Repeat the solution of Problem 1 for the case of a clockwise couple of moment M_0 applied at the free end of the beam. *Ans.* $\delta = M_0 l^2/2EI$.

4. Repeat the solution of Problem 1 for the case of a concentrated vertical load P at the free end of the beam. *Ans.* $\delta = Pl^3/3EI$.

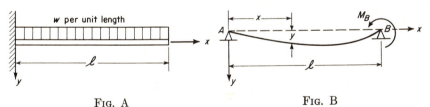

Fig. A   Fig. B

5. A simply supported beam is acted upon by a counterclockwise couple of moment M_B at the end B as shown in Fig. B. Derive the equation of the deflection curve and find the maximum deflection. *Ans.* $y_{max} = \dfrac{M_B l^2}{9\sqrt{3}\ EI}$ at $x = l/\sqrt{3}$.

6. Derive the equation of the elastic line for the cantilever beam in Fig. A if the distributed load varies parabolically from zero intensity at the built-in end to intensity w at the free end. What is the maximum deflection δ at the free end in this case? *Hint.* Begin with eq. (8.3) instead of (8.1). *Ans.* $\delta = 13wl^4/180EI$.

7. A simply supported beam carries a distributed load the intensity of which varies sinesoidally as shown in Fig. C, w_0 being the intensity at mid-span. Find the equation of the elastic line and determine therefrom the maximum deflection δ at the middle of the beam. *Ans.* $\delta = w_0 l^4/\pi^4 EI$.

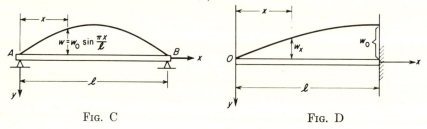

Fig. C   Fig. D

8. A cantilever beam carries a sinesoidally distributed load as shown in Fig. D. The intensity of load at any section defined by x is $w_x = w_0 \sin(\pi x/2l)$, w_0 being the intensity at the built-in end. Derive the equation of the elastic line and find the deflection at the free end. *Ans.*

$$\delta = \frac{2w_0 l^4}{\pi EI}\left(\frac{1}{3} - \frac{8}{\pi^3}\right).$$

9. For the simply supported beam loaded as shown in Fig. 8.3, find the slope of the elastic line at the end B, i.e., the value of dy/dx at $x = l$. *Ans.*

$$\left(\frac{dy}{dx}\right)_{x=l} = -\frac{wl^3}{45EI}.$$

10. A cantilever beam of length l carries a concentrated load at its free end. The cross-sectional moment of inertia of the beam varies linearly from I_0 at the built-in end to $I_0/2$ at the free end. Find the deflection δ under the load P. *Ans.* $\delta = 0.386 \, Pl^3/EI_0$.

8.2   The Moment-Area Method

We shall now discuss a semigraphical method of dealing with the problem of deflection of beams subjected to bending. Combining eqs. (a) and (b) of the preceding article, we obtain

$$\frac{d\theta}{ds} = \frac{M}{EI}, \tag{a}$$

where $d\theta$ is the angle subtended by an arc element ds and M is the bending moment to which the element is subjected. Then, as before, for flat deflection curves, we may take $ds \approx dx$ (see Fig. 8.1) and write

$$d\theta \approx \frac{M dx}{EI}. \tag{b}$$

This relationship can now be given a very simple graphical interpretation with reference to the elastic line of the beam and its bending moment diagram. In Fig. 8.6, AB is any portion of the elastic line and A_1B_1 is the corresponding bending moment diagram. AO and BO are tangents at A and B, intersecting at O, and θ is the angle between these tangents, assumed small. The vertical distance $B'B$ is the deflection of point B away from the tangent at A, while $A'A$ is the deflection of A away from the tangent at B. All these quantities are understood to be very small.

Now let $ds \approx dx$ be any element of the elastic line at the distance x from B and note that the angle between tangents at its ends is $d\theta$. Then from eq. (b), we conclude that this angle is equal to the area $M\,dx$ of the shaded strip of the bending moment diagram in Fig. 8.6 divided by EI. Integrating eq. (b) between A and B, we obtain for the total angle θ between tangents at A and B

$$\theta = \int_A^B \frac{M\,dx}{EI}. \tag{8.4}$$

Since this integral represents the total area of the bending moment diagram A_1B_1 divided by EI, we have the following conclusion:

THEOREM I. *The angle θ between tangents at any two points A and B on the elastic line is equal to the total area of the corresponding portion of the bending moment diagram, divided by EI.*

Let us now consider the deflection of point B relative to the tangent at A, i.e., the vertical distance $B'B$ in Fig. 8.6. Keeping in mind that the angles between these tangents are very small, we note from the figure that bending

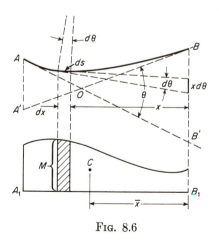

FIG. 8.6

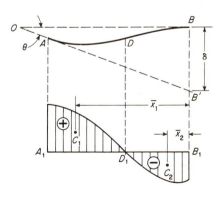

FIG. 8.7

of the element ds contributes to this deflection by the amount $x\,d\theta$. Then by summation, the total distance $B'B$ becomes

$$\delta = \int_A^B x\,d\theta = \int_A^B \frac{Mx\,dx}{EI}. \tag{8.5}$$

Since $M\,dx$ is the area of the shaded strip of the bending moment diagram and x is its distance from B, we conclude that the right-hand side of eq. (8.5) represents the *statical moment* with respect to B of the total bending moment area between A and B, divided by EI. Thus, we have the conclusion:

THEOREM II. *The deflection of B away from the tangent at A is equal to the statical moment, with respect to B, of the bending moment area between A and B, divided by EI.*

This statical moment will be obtained simply as the product of the total area of the bending moment diagram between A and B multiplied by the distance $\bar{x}$ to its centroid C (Fig. 8.6).

If there is an inflection point in the elastic line between A and B as shown in Fig. 8.7, caution must be exercised in using the above theorems. In such a case, the bending moment diagram divides itself into a positive portion A_1D_1 and a negative portion D_1B_1, with centroids C_1 and C_2, respectively. Then to find the angle θ between tangents at A and B, we have from eq. (8.4)

$$\theta = \int_A^D \frac{M\,dx}{EI} - \int_D^B \frac{M\,dx}{EI}. \tag{c}$$

In short, to obtain the angle θ between tangents at A and B, we must take the *net area* of the bending moment diagram between A and B, divided by

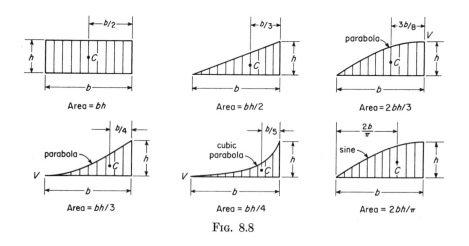

FIG. 8.8

EI. Similarly, for the deflection of B away from the tangent at A, eq. (8.5) becomes

$$\delta = \int_A^D \frac{Mx\,dx}{EI} - \int_D^B \frac{Mx\,dx}{EI}.$$ (d)

That is, the required deflection is obtained as the difference between the statical moments with respect to B of the positive and negative portions of the bending moment diagram. This can also be expressed as follows:

$$\delta = \left[\frac{\text{area}}{EI}\right]_A^D \cdot \bar{x}_1 - \left[\frac{\text{area}}{EI}\right]_D^B \cdot \bar{x}_2,$$ (e)

where $\bar{x}_1$ and $\bar{x}_2$ are the centroidal distances shown in Fig. 8.7.

Since most of the bending moment areas with which we have to deal are simple rectangles, triangles, and parabolic segments, their areas and the location of their centroids are easily found. A few of these are summarized in Fig. 8.8 for ready reference. In these diagrams, C denotes the centroid of the shaded area and V, the vertex of the bounding curve.

Applications of the moment-area method will now be illustrated by several examples.

EXAMPLE 1. Determine the deflection δ and the slope θ at the free end A of the cantilever beam AB loaded as shown in Fig. 8.9a.

SOLUTION. The bending moment diagram is shown in Fig. 8.9b. Since the tangent to the elastic line at B coincides with the undeflected axis of the beam, the required deflection δ will be the deflection of A away from the tangent at B. Thus, using Theorem II, we have

$$\delta = \frac{Pl^2}{2EI} \times \frac{2l}{3} = \frac{Pl^3}{3EI}.$$ (f)

Likewise, the slope at A is the angle between tangents at A and at B and from Theorem I, we have

$$\theta = \frac{Pl^2}{2EI}.$$ (g)

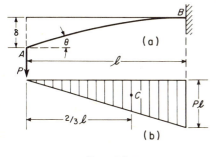

FIG. 8.9

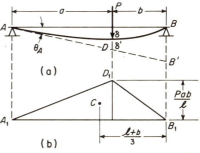

FIG. 8.10

EXAMPLE 2. A simply supported beam AB carries a concentrated load P at point D as shown in Fig. 8.10a. Find the deflection δ of point D from the cord line AB and the angle θ_A between this cord line and the tangent at A.

SOLUTION. The bending moment diagram is shown in Fig. 8.10b. The area of this diagram is $Pab/2$ and the distance of its centroid C from B is $\frac{1}{3}(l + b)$ as shown. Taking the statical moment of this area with respect to point B, we obtain the deflection $B'B$ of B away from the tangent at A. Thus

$$B'B = \frac{Pab}{2EI} \times \frac{(l + b)}{3}.$$

Then noting from the figure that $\theta_A = B'B \div l$, we have

$$\theta_A = \frac{Pab}{6lEI}(l + b). \tag{h}$$

We see also from the figure that the required deflection of point D from the chord line AB is

$$\delta = a\theta_A - \delta', \tag{i}$$

where δ' is the deflection of D away from the tangent at A. This deflection δ' can be found by using Theorem II for the portion A_1D_1 of the bending moment diagram. The area of this is $Pa^2b/2l$ and its centroid is at the distance $a/3$ to the left of D. Thus

$$\delta' = \frac{Pa^2b}{2lEI} \times \frac{a}{3}. \tag{j}$$

Substitution of expressions (h) and (j) into eq. (i) gives

$$\delta = \frac{Pa^2b}{6lEI}(l + b) - \frac{Pa^3b}{6lEI} = \frac{Pa^2b^2}{3lEI}. \tag{k}$$

EXAMPLE 3. Locate the point of maximum deflection on the simple beam discussed in the preceding example and evaluate this deflection by the moment-area method.

SOLUTION. The maximum deflection occurs at point E (Fig. 8.11a) where the tangent to the elastic line is horizontal, i.e., parallel to the chord line AB. Let x denote the distance of this point from A. Then the angle between tangents at A and E must be equal to θ_A as already found in eq. (h) of the preceding example. Thus the area of that portion of the bending moment diagram between A and E (Fig. 8.11b) must be such that

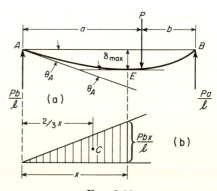

FIG. 8.11

$$\theta_A = \frac{Pab}{6lEI}(l + b) = \frac{Pbx^2}{2lEI}$$

from which

$$x = \sqrt{\frac{a(l + b)}{3}} = \sqrt{\frac{l^2 - b^2}{3}}. \qquad (1)$$

Then since δ_{max} is the deflection of A away from the tangent at E, we have

$$\delta_{max} = \frac{Pbx^2}{2lEI} \cdot \frac{2x}{3} = \frac{Pb}{9\sqrt{3lEI}}\sqrt{(l^2 - b^2)^3}. \qquad (m)$$

This is seen to agree with eq. (r) on p. 202.

EXAMPLE 4. A prismatic cantilever beam AB carries a uniformly distributed load over the portion b of its length as shown in Fig. 8.12a. Find the deflection δ of the free end A.

SOLUTION. The bending moment diagram is shown in Fig. 8.12b. Its area $wb^3/6$ and the position of its centroid C are found by reference to Fig. 8.8. From Theorem II the deflection δ will be obtained by dividing by EI the statical moment of this area with respect to point A_1. Thus

$$\delta = \frac{wb^3}{6EI}(a + \tfrac{3}{4}b). \qquad (n)$$

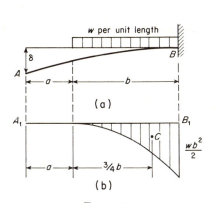

(a)

(b)

FIG. 8.12

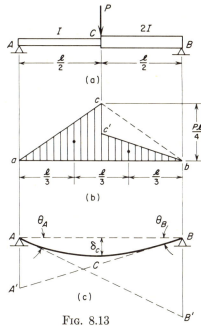

(a)

(b)

(c)

FIG. 8.13

EXAMPLE 5. One of the advantages of the moment-area method is that it permits a fairly simple treatment of the deflection of beams that have abrupt changes in cross-section. In Fig. 8.13a, for example, a simply supported beam AB carrying a load P at the middle has cross-sectional moment of inertia I over the left half of the span and $2I$ over the right half. Using the moment-area method, find the angles of rotation θ_A and θ_B of the end tangents and the deflection δ_C under the load P.

SOLUTION. The bending moment diagram for the beam is shown by the triangle acb in Fig. 8.13b. To convert this into an M/I-diagram, we simply divide by two all ordinates for the right-hand half of the beam, thus obtaining the shaded part of the diagram as shown.

We now consider the deflection curve as shown in Fig. 8.13c. To obtain the deflection of point A away from the tangent at B, we have, by using the second moment-area theorem,

$$\delta_{A/B} = AA' = \frac{Pl}{4EI} \cdot \frac{l}{4} \cdot \frac{l}{3} + \frac{Pl}{8EI} \cdot \frac{l}{4} \cdot \frac{2l}{3} = \frac{Pl^3}{24EI}.$$

Then, from the geometry of the figure,

$$\theta_B = \frac{AA'}{l} = \frac{Pl^2}{24EI}.$$

In the same way, we find

$$\delta_{B/A} = BB' = \frac{Pl}{8EI} \cdot \frac{l}{4} \cdot \frac{l}{3} + \frac{Pl}{4EI} \cdot \frac{l}{4} \cdot \frac{2l}{3} = \frac{5Pl^3}{96EI}$$

and

$$\theta_A = \frac{BB'}{l} = \frac{5Pl^2}{96EI}.$$

We note that the ratio $\theta_A/\theta_B = \frac{5}{4}$.

To find the deflection δ_C at the middle of the beam, we first look for the deflection of point C away from the tangent at B. By the second moment-area theorem, this is

$$\delta_{C/B} = \frac{Pl}{8EI} \cdot \frac{l}{4} \cdot \frac{l}{6} = \frac{Pl^3}{192EI}.$$

Then from the geometry of Fig. 8.13c, we see that

$$\delta_C = \frac{AA'}{2} - \delta_{C/B} = \frac{Pl^3}{48EI} - \frac{Pl^3}{192EI} = \frac{Pl^3}{64EI}.$$

PROBLEM SET 8.2

1. Using the moment-area method, find the deflection δ_A at the free end A of the cantilever beam loaded as shown in Fig. A. The flexural rigidity EI of the beam is uniform throughout its length. *Ans.* $\delta_A = Pa^2(3l - a)/6EI$.

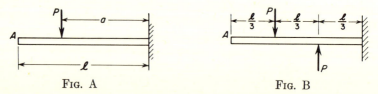

FIG. A   FIG. B

2. Solve the preceding problem for the case in which a concentrated couple of moment M replaces the force P in Fig. A. *Ans.* $\delta_A = Ma(2l - a)/2EI$.

3. A cantilever beam of length l carries two forces P at its third-points as shown in Fig. B. Using the moment-area method, find the deflection δ_A at the free end A. The flexural rigidity of the beam is uniform throughout its length. *Ans.* $\delta_A = 10Pl^3/81EI$.

4. A simply supported beam AB of uniform flexural rigidity EI carries at its third-points two concentrated couples of moments M and $2M$ as shown in Fig. C. Using the moment-area method, find the angles of rotation θ_A and θ_B of the end tangents at A and B, respectively. *Ans.* $\theta_A = Ml/6EI$; $\theta_B = 0$.

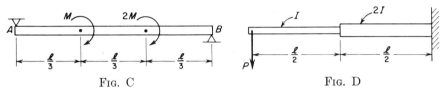

Fig. C   Fig. D

5. A stepped shaft supported as a cantilever beam carries a concentrated load P at its free end as shown in Fig. D. The cross-sectional moments of inertia of the two parts of the shaft are I and $2I$ as shown. Using the moment-area method, find the deflection δ_A at the free end A. *Ans.* $\delta_A = 3Pl^3/16EI$.

6. A simply supported beam AB loaded at the middle as shown in Fig. E has cross-sectional moments of inertia I and $2I$ as indicated. Using the moment-area method, find the deflection δ at the middle of the beam. *Ans.* $\delta = 35Pl^3/2592EI$.

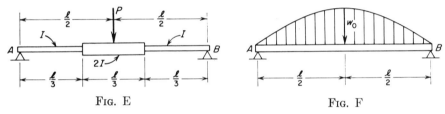

Fig. E   Fig. F

7. A simply supported beam AB carries a distributed load the intensity of which varies according to a sine law as shown in Fig. F. Using the data given in the last case of Fig. 8.8 and the moment-area method, find the deflection δ at the middle of the beam. *Ans.* $\delta = w_0l^4/\pi^4EI$.

8. A cantilever beam of length l and uniform flexural rigidity EI is subjected to continuously distributed externally applied moment of intensity m in.-lb per inch of length of the beam. Using the area-moment method, show that the deflection of the free end of the beam is $\delta = ml^3/3EI$, and explain why this is the same deflection as that obtained for the case of a concentrated force $P = m$ applied at the end of the beam.

9. Referring to the beam shown in Fig. 8.5, find the angle of rotation θ_A by using the moment-area method. Show that the result agrees with that given by eq. (z) on page 203.

8.3 Deflections by Superposition

The differential equation of the elastic line of a beam bent by transverse loads (eq. 8.1) was derived in Art. 8.1. Since the equation is linear in y and its derivatives, it follows that its solutions for various loading conditions may be superimposed. This means, for example, that if we have $y_1(x)$ defining the deflection curve of a cantilever beam carrying a concentrated load P at its free end and $y_2(x)$ defining the deflection curve under uniformly

Table 8.1   BEAM DEFLECTION FORMULAS

BEAM TYPE	SLOPE AT FREE END	DEFLECTION AT ANY SECTION IN TERMS OF x: y IS POSITIVE DOWNWARD	MAXIMUM DEFLECTION
1. Cantilever Beam — Concentrated load P at the free end			
	$\theta = \dfrac{Pl^2}{2EI}$	$y = \dfrac{Px^2}{6EI}(3l - x)$	$\delta_{\max} = \dfrac{Pl^3}{3EI}$
2. Cantilever Beam — Concentrated load P at any point			
	$\theta = \dfrac{Pa^2}{2EI}$	$y = \dfrac{Px^2}{6EI}(3a - x)$ for $0 < x < a$ $y = \dfrac{Pa^2}{6EI}(3x - a)$ for $a < x < l$	$\delta_{\max} = \dfrac{Pa^2}{6EI}(3l - a)$
3. Cantilever Beam — Uniformly distributed load of w lb per unit length			
	$\theta = \dfrac{wl^3}{6EI}$	$y = \dfrac{wx^2}{24EI}(x^2 + 6l^2 - 4lx)$	$\delta_{\max} = \dfrac{wl^4}{8EI}$
4. Cantilever Beam — Uniformly varying load; maximum intensity w lb per unit length			
	$\theta = \dfrac{wl^3}{24EI}$	$y = \dfrac{wx^2}{120lEI}(10l^3 - 10l^2x + 5lx^2 - x^3)$	$\delta_{\max} = \dfrac{wl^4}{30EI}$
5. Cantilever Beam — Couple M applied at the free end			
	$\theta = \dfrac{Ml}{EI}$	$y = \dfrac{Mx^2}{2EI}$	$\delta_{\max} = \dfrac{Ml^2}{2EI}$
6. Beam Freely Supported at Ends — Concentrated load P at the center			
	$\theta_1 = \theta_2 = \dfrac{Pl^2}{16EI}$	$y = \dfrac{Px}{12EI}\left(\dfrac{3l^2}{4} - x^2\right)$ for $0 < x < \dfrac{l}{2}$	$\delta_{\max} = \dfrac{Pl^3}{48EI}$

Beam Type	Slope at Ends	Deflection at any section in terms of x: y is positive downward	Maximum and center deflection
7. Beam Freely Supported at Ends — Concentrated load at any point			
	Left End. $$\theta_1 = \frac{Pb(l^2 - b^2)}{6lEI}$$ Right End. $$\theta_2 = \frac{Pab(2l - b)}{6lEI}$$	$$y = \frac{Pbx}{6lEI}(l^2 - x^2 - b^2)[0 < x < a]$$ $$y = \frac{Pb}{6lEI}\left[\frac{l}{b}(x-a)^3 + (l^2-b^2)x - x^3\right][a < x < l]$$	$$\delta_{max} = \frac{Pb(l^2 - b^2)^{3/2}}{9\sqrt{3}\,lEI}$$ $$at\ x = \sqrt{\frac{l^2 - b^2}{3}}$$ At center, if $a > b$ $$\delta = \frac{Pb}{48EI}(3l^2 - 4b^2)$$
8. Beam Freely Supported at Ends — Uniformly distributed load of w lb per unit length			
	$$\theta_1 = \theta_2 = \frac{wl^3}{24EI}$$	$$y = \frac{wx}{24EI}(l^3 - 2lx^2 + x^3)$$	$$\delta_{max} = \frac{5wl^4}{384EI}$$
9. Beam Freely Supported at Ends — Couple M at the right end			
	$$\theta_1 = \frac{Ml}{6EI}$$ $$\theta_2 = \frac{Ml}{3EI}$$	$$y = \frac{Mlx}{6EI}\left(1 - \frac{x^2}{l^2}\right)$$	$$\delta_{max} = \frac{Ml^2}{9\sqrt{3}\,EI}$$ $$at\ x = l/\sqrt{3}$$ At center $$\delta = \frac{Ml^2}{16EI}$$
10. Beam Freely Supported at Ends — Uniformly varying load: max. intensity w			
	$$\theta_1 = \frac{7wl^3}{360EI}$$ $$\theta_2 = \frac{wl^3}{45EI}$$	$$y = \frac{wx}{360lEI}(7l^4 - 10l^2x^2 + 3x^4)$$	$$\delta_{max} = .00652\frac{wl^4}{EI}$$ $$at\ x = 0.519l$$ At center $$\delta = .00651\frac{wl^4}{EI}$$

distributed load, then the equation of the deflection curve for the simultaneous action of both loadings is simply

$$y(x) = y_1(x) + y_2(x). \tag{a}$$

Specifically, the deflection at the free end under the load P is $\delta_1 = Pl^3/3EI$. Under the uniform load of intensity w, the deflection at the free end is $\delta_2 = wl^4/8EI$. Hence, due to the combined loading, the deflection at the free end is simply

$$\delta = \frac{Pl^3}{3EI} + \frac{wl^4}{8EI}. \tag{b}$$

For ready reference, a table of deflections, slopes, and complete elastic line equations for ten primary cases of loading of cantilever and simply supported beams is presented in Table 8.1. These various results have already been found either in the examples or problems of the two preceding articles. Using this table and the method of superposition, it is possible to obtain various required deflections and slopes for beams and loadings comprising various combinations of these primary cases. The idea of superposition can be employed in a variety of ways and some of these will be illustrated by the examples which follow.

EXAMPLE 1. A uniform cantilever beam AB is loaded as shown in Fig. 8.14. Find the deflection δ of the free end B.

SOLUTION. Using Case 2 of Table 8.1, we have

$$\delta = \frac{P_1 a^2}{6EI} (3l - a) + \frac{P_2 b^2}{6EI} (3l - b).$$

As a particular case, if $P_1 = P_2 = P$, $a = l/3$, and $b = 2l/3$, this reduces to

$$\delta = \frac{2Pl^3}{9EI}.$$

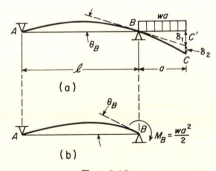

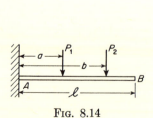

FIG. 8.14   FIG. 8.15

EXAMPLE 2. A simply supported prismatic beam with span l and overhang a is loaded as shown in Fig. 8.15a. Find the deflection δ_c at the end C of the overhang.

SOLUTION. Owing to the distributed load on the overhang, the portion AB of the beam is in the condition of a simple beam subjected to a couple of moment $M_B = wa^2/2$ at the end B as shown in Fig. 8.15b. Hence using Case 9 of Table 8.1, we have

$$\theta_B = \frac{M_B l}{3EI} = \frac{wa^2 l}{6EI}.$$

Due to this rotation of the tangent at B, the overhanging portion of the beam, considered as absolutely rigid, would take the position BC' in Fig. 8.15a, and we see that C' would have a deflection

$$\delta_1 = a\theta_B = \frac{wa^3 l}{6EI}.$$

In addition to this, the overhanging portion BC does bend due to the distributed loading and we have, with reference to Case 3 of Table 8.1,

$$\delta_2 = \frac{wa^4}{8EI},$$

as the deflection $C'C$ of point C away from the tangent at B. Thus the total deflection of point C becomes

$$\delta = \delta_1 + \delta_2 = \frac{wa^3 l}{6EI} + \frac{wa^4}{8EI}$$

EXAMPLE 3. A prismatic cantilever beam of length l carries a uniform load of intensity w between $x = a$ and $x = l$ as shown in Fig. 8.16a. Find the deflection δ at the free end of the beam.

SOLUTION. Consider the beam acted upon by just one element $w\,dx$ of the distributed load as shown in Fig. 8.16b. Then from Case 2 of Table 8.1, we have

$$d\delta = \frac{wx^2\,dx}{6EI}\,(3l - x).$$

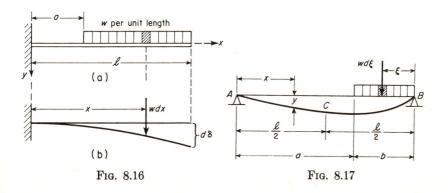

FIG. 8.16   FIG. 8.17

Now since each element of the distributed load between $x = a$ and $x = l$ produces a like increment of deflection at the free end, we obtain by summation

$$\delta = \int_a^l \frac{w\,x^2\,dx}{6EI}\,(3l - x) = \frac{w}{24EI}\,(3l^4 - 4a^3l + a^4).$$

If the load extends over the full length of the beam, $a = 0$ and this reduces to

$$\delta = \frac{wl^4}{8EI},$$

which checks with Case 3 of Table 8.1, as it should.

EXAMPLE 4. A simple beam AB of length l carries a uniformly distributed load of intensity w over a portion b of the span as shown in Fig. 8.17. Assuming $a > b$, find the deflection δ_c at the mid-point C of the span.

SOLUTION. We first consider just one element $wd\xi$ of the load at the distance ξ from B. Then using Case 7 of Table 8.1,

$$y = \int_0^b \frac{wd\xi \cdot \xi x}{6lEI}\,(l^2 - x^2 - \xi^2) = \frac{wb^2x}{24lEI}\,(2l^2 - 2x^2 - b^2).$$

Setting $x = l/2$ in this expression, we obtain

$$\delta_c = \frac{wb^2}{96EI}\,(3l^2 - 2b^2).$$

For the particular case where $b = l/2$, this becomes

$$\delta_c = \frac{1}{2}\left(\frac{5wl^4}{384EI}\right),$$

as is to be expected from a consideration of Case 8 in Table 8.1.

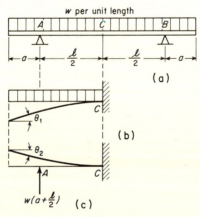

FIG. 8.18

EXAMPLE 5. A simply supported beam having a span l and overhangs of length a at each end carries a uniformly distributed load of intensity w over its entire length as shown in Fig. 8.18a. Find the ratio a/l in order that the tangents to the elastic line at each end of the beam will be horizontal.

SOLUTION. Since both the beam and the loading are symmetrical about point C, the tangent to the elastic line at C remains horizontal. Hence we may consider half of the beam as a cantilever built in at C (Fig. 8.18b, c). Then from Case 3 of Table 8.1, the slope θ_1 at the free end due to the uniform loading alone (Fig. 8.18b) is

$$\theta_1 = \frac{w\left(a + \dfrac{l}{2}\right)^3}{6EI}.$$

Likewise, due to the concentrated force $R_A = w(a + l/2)$ acting alone, the slope θ_2 at the free end (Case 2 of Table 8.1) becomes

$$\theta_2 = \frac{w\left(a + \dfrac{l}{2}\right)\left(\dfrac{l}{2}\right)^2}{2EI}.$$

These slopes are of opposite sign; hence setting $\theta_1 = \theta_2$ we realize the required condition of the problem. This yields the quadratic equation

$$a^2 + al - \frac{l^2}{2} = 0,$$

from which

$$a = -\frac{l}{2} \pm \frac{\sqrt{3}}{2}l.$$

Taking only the positive root as physically meaningful, we obtain $a/l = 0.366$. The standard meter stick in Paris is supported in this manner to insure that its two end cross-sections will remain parallel.

PROBLEM SET 8.3

1. The beam in Fig. A has uniform flexural rigidity EI throughout its full length. Using the method of superposition, find the deflection δ_D under the load P. Neglect weight of beam. *Ans.* $\delta_D = Pa^2(l + a)/3EI$.

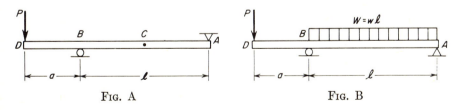

FIG. A   FIG. B

2. A simply supported beam with overhang is loaded as shown in Fig. B. If $a = l/2$, find the ratio P/W to make the deflection δ_D under the load P equal to zero. *Ans.* $P/W = \frac{1}{6}$.

3. Referring to the beam shown in Fig. A and using the method of superposition, calculate the upward deflection δ_C of the mid-point C of the span AB. *Ans.* $\delta_C = Pal^2/16EI$.

4. Referring again to the beam shown in Fig. A, assume that the load P is moved from D to C and then calculate the upward deflection δ_D of the free end of the overhang. *Ans.* $\delta_D = Pal^2/16EI$.

5. A simply supported beam of uniform flexural rigidity EI and span l carries two symmetrically placed loads P at its third-points. Using the method of superposition, find the deflection δ at the middle of the beam. *Ans.* $\delta = 23Pl^3/648EI$.

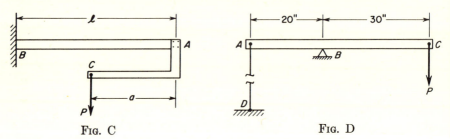

FIG. C   FIG. D

6. A cantilever beam AB of length l and uniform flexural rigidity EI has a bracket AC attached to its free end as shown in Fig. C, and a vertical load P is applied to the free end C of the bracket. Find the ratio a/l required in order that the deflection of point A will be zero. *Ans.* $a/l = \frac{2}{3}$.

7. In Fig. D, a beam AC having uniform flexural rigidity $EI = 30(10)^6$ lb/in.^{2} is supported by a fulcrum at B and a vertical tie-rod at A. The tie-rod AD is a steel wire $\frac{1}{8}$ in. in diameter and 10 ft long. The beam is loaded at C with a vertical force $P = 100$ lb. Find the deflection δ_C of point C. *Ans.* $\delta_C = 0.123$ in.

8. A simply supported beam with overhang carries a uniformly distributed load of intensity w as shown in Fig. E. If it is desired to have zero deflection at C, what is the required distance a from the end A of the beam to the beginning of the distributed load? *Ans.* $a = 10.8$ ft.

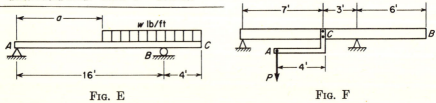

FIG. E   FIG. F

9. The beam shown in Fig. F has rigidly attached to it at point C an arm AC to the free end A of which a vertical load $P = 10$ kips is applied. Compute the vertical deflection of the end B of the beam if $E = 30(10)^6$ psi and $I = 339.2$ in.^{4} *Ans.* $\delta_B = 0.0925$ in., up.

10. Using the result for $\delta_{\max}$ in case 2 of Table 8.1 and the method of superposition, verify the result for $\delta_{\max}$ in case 4 of Table 8.1.

11. A simply supported beam having uniform flexural rigidity EI and span $l = 2a$ carries a triangularly distributed load over the right-hand half of the span. The intensity of this load varies linearly from zero at mid-span to w_0 at the right-hand support. Using the method of superposition, find the deflection δ at the middle of the beam. *Ans.* $\delta = 3w_0a^4/80EI$.

12. Two simply supported beams of uniform cross-section are geometrically similar in every respect, each dimension of one being n times the corresponding dimension of the other, and they are made of the same material. Compare the mid-point deflections due to their own distributed weights. *Ans.* $\delta_2/\delta_1 = n^2$.

13. A steel cantilever beam 10 ft long has an 8WF17 section and carries a concentrated vertical load $P = 3000$ lb at its free end. To prevent overstressing, the inboard half of the beam is reinforced by two $5\frac{1}{4}$ in. by $\frac{1}{4}$ in. steel cover plates top and bottom extending from the wall to the middle of the beam. Using the method of superposition, calculate the deflection at the free end of the beam. *Ans.* $\delta = 0.626$ in.

8.4 Strain Energy of Bending

Consider in Fig. 8.19a, a prismatic beam subjected to *pure bending* within the elastic limit of the material. For such loading, the bending moment M is constant along the length l of the beam and the elastic line is a circular arc of curvature M/EI(see p. 197) and the angle θ subtended by this arc is

$$\theta = \frac{Ml}{EI}. \tag{a}$$

This represents a linear relationship between M and θ as shown in Fig. 8.19b. Hence we conclude that as the applied couples at the ends of the beam are gradually increased from zero to any value M, the work which they do is represented by the shaded area OAB of the diagram in Fig. 8.19b. This work, equal to the strain energy stored in the beam, has the magnitude

$$U = \frac{M\theta}{2}, \tag{b}$$

which is analogous to the strain energy of torsion in a bar subjected to twist [see eq. (d), p. 83].

Using expression (a), the strain energy (b) may be written in either of the following two forms:

$$U = \frac{M^2 l}{2EI} \quad \text{or} \quad U = \frac{EI\theta^2}{2l}. \tag{8.6}$$

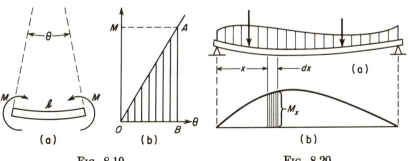

FIG. 8.19   FIG. 8.20

If we know the bending moment M, we use the first form to calculate the strain energy. If we know the curvature, defined by θ/l, we use the second form to calculate the strain energy.

It is sometimes useful to express the strain energy U in a bar subjected to pure bending in terms of the maximum fiber stress $\sigma_{max} = Mc/I$, as given by eq. (5.5a), p. 115. In the particular case of a beam of rectangular cross-section of width b and depth h, eq. (5.5a) gives $M = \frac{1}{6}bh^2\,\sigma_{max}$ and the first of eqs. (8.6) becomes

$$U = \frac{1}{3}(bhl)\frac{\sigma_{max}^2}{2E}. \tag{c}$$

This shows that for the same maximum fiber stress, a bar in pure bending can store only one-third as much strain energy as if it were in simple tension.

In discussing the strain energy in a beam subjected to non-uniform bending (Fig. 8.20a), we shall neglect strain energy of shear and consider only strain energy of bending.* Considering any element of the beam of length dx, the bending moment M_x will be essentially constant over the length of the element. Hence eqs. (8.6) apply to the element provided we replace M by M_x, l by dx, and θ by $d\theta = dx/\rho \approx (d^2y/dx^2)dx$. Then summing up such expressions for all elements of the beam, we obtain

$$U = \int_o^l \frac{M_x^2 dx}{2EI} \quad \text{or} \quad U = \frac{EI}{2}\int_o^l \left(\frac{d^2y}{dx^2}\right)^2 dx. \tag{8.7}$$

To illustrate the use of the first of eqs. (8.7), consider, for example, a cantilever beam loaded as shown in Fig. 8.21a. The bending moment at any cross-section defined by x is $M_x = -Px$ and the first of eqs. (8.7) becomes

$$U = \int_o^l \frac{P^2x^2 dx}{2EI} = \frac{P^2l^3}{6EI}. \tag{d}$$

If the beam has a rectangular cross-section of width b and depth h, the maximum bending stress at the built-in end is

$$\sigma_{max} = \frac{Mc}{I} = \frac{6Pl}{bh^2}.$$

Expressed in terms of this maximum stress, eq. (d) becomes

$$U = \frac{1}{9}(bhl)\frac{\sigma_{max}^2}{2E}. \tag{e}$$

*Strain energy due to shear will be small in comparison with that due to bending except for short thick beams. This question is discussed further in the following article.

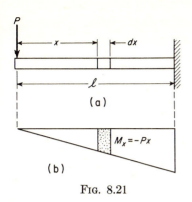

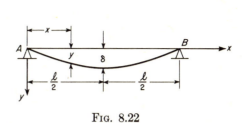

FIG. 8.21   FIG. 8.22

Comparing this with expression (c) above, we observe that for the same maximum bending stress, the cantilever beam carries only one-third as much strain energy as the beam subjected to uniform bending.

To illustrate the use of the second of eqs. (8.7), consider the simply supported beam AB in Fig. 8.22. This beam is so loaded that its deflection curve has the form of a half sine wave defined by the equation

$$y = \delta \sin \frac{\pi x}{l}. \tag{f}$$

The mid-point deflection δ being specified, it is desired to find the total strain energy stored in the beam. Differentiating expression (f) twice with respect to x, we find

$$\frac{d^2 y}{dx^2} = -\frac{\delta \pi^2}{l^2} \sin \frac{\pi x}{l}.$$

Substitution of this into the second of eqs. (8.7) gives

$$U = \frac{EI}{2} \int_o^l \frac{\delta^2 \pi^4}{l^4} \sin^2 \frac{\pi x}{l} \, dx = \frac{EI \pi^4 \delta^2}{4l^3}. \tag{g}$$

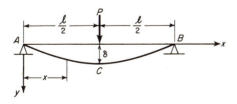

FIG. 8.23

EXAMPLE 1. Calculate the deflection δ at the middle of a simply supported prismatic beam AB, under the action of a load P as shown in Fig. 8.23.

SOLUTION. The bending moment at any cross-section defined by x is

$$M_x = \frac{Px}{2}.$$

Then, using the first form of eq. (8.7), we obtain for the strain energy of bending in the beam

$$U = 2\int_0^{l/2} \frac{P^2x^2dx}{8EI} = \frac{P^2l^3}{96EI}. \tag{h}$$

Equating this strain energy to the work done by the applied load as it is gradually increased from 0 to P during deflection:

$$\frac{P\delta}{2} = \frac{P^2l^3}{96EI},$$

from which
$$\delta = \frac{Pl^3}{48EI}. \tag{i}$$

EXAMPLE 2. A simply supported beam as shown in Fig. 8.24 is struck at its mid-point C by a ball of weight W freely falling from a height h above the beam. Neglecting the weight of the beam and assuming that it behaves elastically, find the total deflection δ that will be produced at point C.

SOLUTION. The gravity force W of the ball falling through the total distance $h + \delta$ does work equal to

$$W(h + \delta). \tag{j}$$

Now let P denote the force exerted by the ball on the beam in the extreme position of maximum deflection δ. Then for this configuration the strain energy in the deflected beam is

$$\frac{P\delta}{2}. \tag{k}$$

Equating this strain energy (k) to the work (j), we obtain

$$\frac{P\delta}{2} = W(h + \delta),$$

from which
$$P = \frac{2W}{\delta}(h + \delta).$$

Substituting this expression for P in eq. (i) above, we obtain

$$\delta = \frac{2Wl^3}{48EI}\frac{h + \delta}{\delta}. \tag{l}$$

Noting that
$$\delta_{st} = \frac{Wl^3}{48EI}$$

represents the static deflection of the beam under the load W, eq. (l) can be written in the form

$$\delta^2 - 2\delta_{st}\delta - 2h\delta_{st} = 0. \tag{m}$$

Taking only the positive root of this quadratic equation, we obtain, for the maximum deflection,

$$\delta = \delta_{\text{st}} + \sqrt{\delta_{\text{st}}^2 + 2h\delta_{\text{st}}}. \tag{n}$$

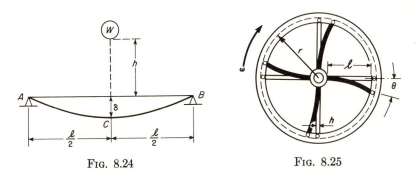

FIG. 8.24   FIG. 8.25

It is seen from this that the dynamic deflection δ is always larger than the static deflection δ_{st}. Even when $h = 0$, i.e., when the weight W has no free fall but is suddenly applied to the beam, eq. (n) gives $\delta = 2\delta_{\text{st}}$. That is, the suddenly applied load produces twice as much deflection as when gradually applied.

As another extreme, we may take $h \gg \delta_{\text{st}}$ in which case the quantity δ_{st}^2 under the radical may be neglected and $\delta \approx \delta_{\text{st}} + \sqrt{2h\delta_{\text{st}}}$.

EXAMPLE 3. The rim of a flywheel of weight W and mean center-line radius r is attached to a hub by four spokes as shown in Fig. 8.25 Each spoke has a rectangular cross-section of dimensions $b \times h$ and a length l. The spokes are built in at the hub and pinned at the rim. While the wheel rotates with constant angular velocity ω, the hub is suddenly locked. What maximum bending stress will be induced in each spoke where it joins the hub? Neglect the weight of the spokes.

SOLUTION. During free rotation, the flywheel rim has kinetic energy

$$T = \frac{I\omega^2}{2} = \frac{W}{g}\frac{r^2\omega^2}{2}. \tag{o}$$

When the hub suddenly locks, the rim continues to rotate through some angle θ before coming to rest and the spokes bend enough to absorb the kinetic energy of the rim in the form of strain energy of bending. Since each spoke is a cantilever beam of rectangular cross-section loaded at its pinned end, eq. (e) applies and the strain energy of bending in four spokes is

$$U = \frac{4}{9}(bhl)\frac{\sigma_{\max}^2}{2E}. \tag{p}$$

Equating kinetic energy (o) to strain energy (p),

$$\frac{4}{9}(bhl)\frac{\sigma_{\max}^2}{2E} = \frac{W}{g}\frac{r^2\omega^2}{2},$$

from which $$\sigma_{\max} = \frac{3}{2}r\omega\sqrt{\frac{W}{g}\frac{E}{(bhl)}}. \tag{q}$$

1. A simply supported steel beam carries a vertical concentrated load $W = 200$ lb at mid-span. Calculate the amount of strain energy stored in the beam if the span $l = 8$ ft and the cross-section is a rectangle of width $b = 2$ in. and depth $h = 1$ in. *Ans.* $U = 73.7$ in.-lb.

2. Calculate the strain energy in the beam of the preceding problem if the load $W = 200$ lb is distributed uniformly over the full span instead of concentrated at the middle. *Ans.* $U = 29.5$ in.-lb.

3. Two cantilever beams of the same material are similar in every respect except size, one having all of its linear dimensions just n times those of the other. What is the ratio of their strain energies when they are subjected to their own distributed weights? *Ans.* $U_2/U_1 = n^5$.

4. A thin, high-strength-steel hacksaw blade, of length l between the holes at its ends, is bent into a complete circle and the two overlapping ends fastened together with a bolt. How much strain energy is sorted in the blade if its flexural rigidity is EI and the bending stresses remain below the elastic limit? *Ans.* $U = 2\pi^2 EI/l$.

5. A solid circular rotor of weight $W = 386$ lb and radius $r = 10$ in. is supported by a bearing at the end of a simply supported beam as shown in Fig. A. Initially, the rotor turns at a constant angular speed of 30 rpm. If the bearing suddenly freezes so that the rotor stops almost instantly, what dynamical reaction will be produced at A? Bearing and support at B are independent, and the beam has a rectangular cross-section 1 in. wide by 4 in. deep. *Ans.* $R_A = 797$ lb, down.

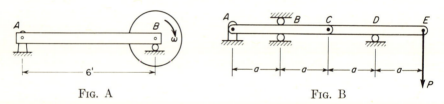

FIG. A   FIG. B

6. The compound beam shown in Fig. B carries a concentrated vertical force P at E. If the flexural rigidity EI is the same for both portions of the beam, calculate the total amount of strain energy stored. Using this strain energy, find the deflection δ_E under the load P. *Ans.* $\delta_E = 4Pa^3/3EI$.

7. A simply supported wooden beam of rectangular cross-section and span $l = 9$ ft is struck at mid-span by a weight $W = 40$ lb falling from a height $h = 12$ in. Determine the required cross-sectional area A if the maximum bending stress is not to exceed 1000 psi and $E = 1.5 \ (10)^6$ psi. *Ans.* $A = 120$ sq in.

8. It is decided to give the beam in the preceding problem a cross-section 10×12 in. When the 12-in. dimension is vertical, the falling weight produces a maximum deflection δ_1. When the 10-in. dimension is vertical, the maximum deflection is δ_2. What is the ratio $\delta_1:\delta_2$? *Ans.* $\delta_1:\delta_2 \approx 0.834$.

8.5 Deflections Due to Shearing Strain

In the preceding discussions of beam deflections, only the deformation due to bending stresses was considered. In all cases of non-uniform bending where there is a shear force at each cross-section, there will be some additional deflection of the beam due to deformation associated with the shearing stresses. Due to the fact that these shear stresses are not uniformly distributed from top to bottom of the beam, cross-sections become warped as shown in Fig. 8.26, which shows curvature of the beam axis produced by shear deformation alone. Small rectangular elements on the neutral axis become rhombuses, the vertical edges of which remain vertical during deformation. Thus the slope of the deflection curve due to shear at any cross-section is simply equal to the shear strain γ at the neutral axis. Denoting then by y_1 the deflection due to shear, we write the following expression for the slope of the elastic line

$$\frac{dy_1}{dx} = \frac{\tau_{\max}}{G} = \frac{kV_x}{AG}, \tag{a}$$

where V_x/A is the average shear stress in the cross-section and k is a numerical factor by which this average shear stress must be multiplied to obtain the maximum shear stress at the neutral axis. For a rectangular cross-section, $k = 3/2$ (see eq. 5.9, p. 128) and for a circular cross-section, $k = 4/3$ (see eq. (g), p. 132). Owing to the fact that the flanges of an I-beam carry very little vertical shear, the factor k may have a value of 2 or 3 in this case depending on the proportions of the I-section.

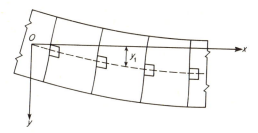

FIG. 8.26

Assuming some continuous distribution of transverse load of intensity w_x on the beam, the shear force V_x will be a continuous function of x which may be differentiated with respect to x. Then the curvature of the elastic line becomes

$$\frac{d^2y_1}{dx^2} = \frac{k}{AG}\frac{dV_x}{dx} = -\frac{kw}{AG}. \tag{8.8}$$

Analogous to eq. (8.1), this represents the differential equation of the elastic line of the beam for shear deflection alone. The deflections calculated from eq. (8.8) can be added directly to those calculated from eq. (8.1) to obtain the total deflections of a beam subjected to non-uniform bending. Usually, the deflections due to shear will be found small compared with those due to bending and the former may be disregarded. However, in the case of short deep beams, the shear deflections may become significant.

To illustrate, let us consider the case of a simple beam uniformly loaded as shown in Fig. 8.27. In this case, the shear force at any section defined by x is

$$V_x = \frac{wl}{2} - wx,$$

and eq. (8.8) becomes

$$\frac{d^2y_1}{dx^2} = -\frac{kw}{AG}. \tag{b}$$

Integrating once

$$\frac{dy_1}{dx} = -\frac{kwx}{AG} + C_1 \tag{c}$$

and again

$$y_1 = -\frac{kwx^2}{2AG} + C_1x + C_2. \tag{d}$$

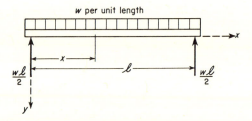

w per unit length

FIG. 8.27

Using the end conditions: $y_1 = 0$ at $x = 0$, $y_1 = 0$ at $x = l$, we find $C_2 = 0$ and $C_1 = kwl/2AG$. Then eq. (d) becomes

$$y_1 = \frac{kwlx}{2AG}\left(1 - \frac{x}{l}\right). \tag{e}$$

Taking $x = l/2$, this gives for the maximum deflection at the middle of the span

$$\delta_1 = \frac{k}{8}\frac{wl^2}{AG}. \tag{f}$$

Introducing the notation $I = Ar^2$, where r is the radius of gyration of the cross-section, eq. (f) can be written in the form

$$\delta_1 = \frac{5wl^4}{384EI}\left[9.6k\left(\frac{r}{l}\right)^2\frac{E}{G}\right]. \tag{g}$$

Taking, as a particular case, a steel beam of rectangular cross-section, we have $k = 3/2$, $E/G = 2.5$, and $r^2 = h^2/12$ for which eq. (g) becomes

$$\delta_1 = \frac{5wl^4}{384EI}\left(3\,\frac{h^2}{l^2}\right). \tag{g'}$$

This shows that for a depth-span ratio $h/l = 1/10$, the maximum deflection due to shear is approximately 3 per cent of that due to bending. As the depth-span ratio increases, the deflection due to shear becomes more important.

In the case of an I-beam, Fig. 8.28, eq. (5.7′), p. 129, gives

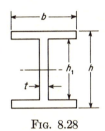

$$\tau_{\max} = \frac{kV}{A} = \frac{V}{It}\left[\frac{b}{2}\left(\frac{h^2}{4} - \frac{h_1^2}{4}\right) + \frac{th_1^2}{8}\right],$$

from which

$$k = \frac{A}{It}\left[\frac{bh^2}{8} - \frac{h_1^2}{8}(b - t)\right]. \tag{h}$$

FIG. 8.28

Then for a 24WF120 section (see Table B.2 of Appendix B) we find $A = 35.29$ in.^{2}, $I = 3635$ in.^{4}, $r = 10.15$ in., $t = 0.556$ in., $b = 12.09$ in., $h = 24.31$ in., $h_1 = 22.45$ in., and eq. (h) gives $k = 2.91$. With this value of k, and with $E/G = 2.5$, eq. (g) becomes

$$\delta_1 = \frac{5wl^4}{384EI}\left(12.2\,\frac{h^2}{l^2}\right). \tag{g''}$$

This shows that for a span $l = 12$ ft, the maximum deflection due to shear would be approximately 34 per cent of that due to bending, which is by no means negligible.

In the case of a simply supported beam that carries a concentrated load P at mid-span (Fig. 8.29), the shear force $V = \pm P/2$ is constant in magnitude but changes sign at the middle of the beam. When V is constant, we see from eq. (a) that the slope of the deflection curve due to shear deformation is also constant. Thus it may be concluded in this case that the deflection curve due to shear is represented by two straight lines AC and CB as shown in Fig. 8.29c, where the constant slope of the line AC is

$$\frac{dy_1}{dx} = \frac{kV}{AG} = \frac{kP}{2AG}.$$

Hence the maximum shear deflection at C is simply

$$\delta_1 = \frac{l}{2}\frac{dy_1}{dx} = \frac{kPl}{4AG}. \tag{i}$$

Again introducing the notation $I = Ar^2$, this may be written in the form

$$\delta_1 = \frac{Pl^3}{48EI}\left[12k\,\frac{r^2}{l^2}\frac{E}{G}\right].$$ (j)

Taking again, as a particular case, a steel beam of rectangular cross-section, we have $k = 3/2$, $E/G = 2.5$, and $r^2 = h^2/12$, and eq. (j) becomes

$$\delta_1 = \frac{Pl^3}{48EI}\left(3.75\,\frac{h^2}{l^2}\right).$$ (j')

It has been assumed throughout the foregoing discussion that all cross-sections of the beam are free to warp as shown in Fig. 8.26. In the case of a uniformly loaded beam, this condition is approximately satisfied. The shear force at the middle of such a beam is zero and there will be no warping of this cross-section. The warping increases gradually with the shear force as we proceed along the beam to the left or to the right of the middle. The condition of symmetry of deformation with reference to the middle section is therefore satisfied .

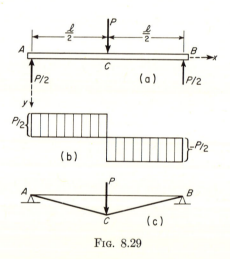

Fig. 8.29

Consider now the case of the simple beam with a concentrated load at the middle, Fig. 8.29. From the condition of symmetry, the middle cross-section must remain a plane section. At the same time, adjacent sections to the left or to the right of the middle section carry a shear force equal to $P/2$ and warping of these sections should take place. However, from the condition of continuity of deformation, there can be no abrupt change from a plane middle section to warped adjacent sections. There must be a gradual increase in warping as we proceed along the beam in either direction from the middle. Thus only at some finite distance from the middle can the warping attain the value compatible with the shear force on the section. From this discussion, it must be concluded that in the neighborhood of the middle section the normal stress distribution cannot be that predicted by the elementary theory of bending (see p. 113). Warping will be partially prevented and the additional deflection due to shear will be somewhat less than that predicted by eq. (j) above. A more detailed investigation of this question shows that for a simply supported steel beam of rectangular cross-section with a concentrated load at the middle, the maximum deflection due to both bending and shear is

$$\delta = \frac{Pl^3}{48EI}\left[1 + 2.95\frac{h^2}{l^2} + 0.02\frac{h}{l}\right]. \tag{k}$$

Additional deflection of a transversely loaded beam due to shear deformation may also be calculated by using the strain energy of shear. To illustrate, let us consider again the case of the simply supported beam of rectangular cross-section loaded at the middle as shown in Fig. 8.29a. In such case, the shear force at any cross-section between A and C is simply $V = P/2$, and from eq. (5.8), p. 128, the shear stress on any element situated at the distance y from the neutral surface is

$$\tau = \frac{P}{4I}\left(\frac{h^2}{4} - y^2\right). \tag{l}$$

Now using the first of eqs. (4.12), p. 82, the strain energy of shear in this one element of volume $bdydx$ is

$$dU = \frac{P^2}{32GI^2}\left(\frac{h^2}{4} - y^2\right)^2 bdydx, \tag{m}$$

and the total strain energy of shear in the entire beam becomes

$$U = 2\int_0^{l/2}\int_{-\frac{h}{2}}^{+\frac{h}{2}} \frac{P^2}{32GI^2}\left(\frac{h^4}{4} - y^2\right)^2 bdydx = \frac{P^2lh^2}{80GI}. \tag{n}$$

Equating this strain energy to the work $P\delta_1/2$ done by the applied load as it gradually increases from zero to the final value P through the deflection δ_1 of its point of application C, we obtain

$$\delta_1 = \frac{Plh^2}{40GI} \tag{o}$$

which can arbitrarily be written in the form

$$\delta_1 = \frac{Pl^3}{48EI}\left(1.2\frac{h^2}{l^2}\frac{E}{G}\right). \tag{p}$$

Taking the particular case of a steel beam for which $E/G = 2.5$, this becomes

$$\delta_1 = \frac{Pl^3}{48EI}\left(3.00\frac{h^2}{l^2}\right). \tag{p'}$$

Comparing eqs. (p') and (j'), we see that there is a discrepancy between the two results. Eq. (p'), obtained by considering the strain energy of shear, gives a smaller deflection than eq. (j'), obtained by assuming that the deflection curve (Fig. 8.29c) was two straight lines having slopes equal to the shear strain at the neutral surface of the beam. To explain this discrepancy, we recall that in arriving at eq. (j'), we assumed that all cross-sections were free to warp. However, as already observed, this assumption violates the condition of symmetry at the middle of the beam. Since eq. (p'), obtained on the basis of strain energy of shear, does not involve this assumption, it must be considered as the better result. In fact this can be seen to be the case by comparing both eqs. (j') and (p') with the more rigorous result represented by eq. (k).

PROBLEM SET 8.5

1. Considering only shear deformation and using eq. (8.8), develop the equation of the deflection curve for the uniformly loaded cantilever beam shown in Fig. A, Plot this deflection curve and note how it differs from that due to bending deformation. Find also the deflection δ_1 at the free end of the beam. The intensity w of the load, the cross-sectional area A of the beam, the shear modulus G of the material, and the shape factor k are all assumed to be given. *Ans.* $\delta_1 = kwl^2/2AG$.

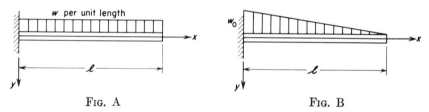

FIG. A   FIG. B

2. Repeat the solution of the preceding problem for the case where the intensity of the distributed load varies linearly from w_0 at the built-in end $(x = 0)$ to zero at the free end $(x = l)$, as shown in Fig. B. *Ans.* $\delta_1 = kw_0l^2/6AG$.

3. If the cantilever beam in Fig. A has a rectangular cross-section 2 in. wide by 6 in. deep and is made of steel for which the ratio $E/G = 2.5$, find the ratio δ_1/δ of shear deflection to bending deflection at the free end. Assume $l = 18$ in. *Ans.* $\delta_1/\delta = 0.139$.

9

Statically Indeterminate Beams

9.1   Method of Superposition

Consider in Fig. 9.1 a beam AB supported by a *hinge* at A and *rollers* at B and C, and subjected to applied loads P_1 and P_2 acting in a plane of symmetry of the beam.   Due to the action of these loads, reactions will be induced at the points of support and a free-body diagram of the beam will be as shown in Fig. 9.1b.   The reactions Y_B and Y_C at B and C will be vertical because the rollers can offer no horizontal restraint.   The reaction at A, which can have any direction in the plane of the figure, is represented by its horizontal and vertical components X_A and Y_A.

Considering now the static equilibrium of this free body, we have the general case of a system of forces in one plane.   For such a system there are three equations of equilibrium, namely:

$$\Sigma X_i = 0 \qquad \Sigma Y_i = 0 \qquad \Sigma M_i = 0. \tag{a}$$

Since these three equations are insufficient to determine four unknowns, the beam is said to be *statically indeterminate*.   There is no way, from the standpoint of the statics of a rigid body, to determine the reactions at the supports.

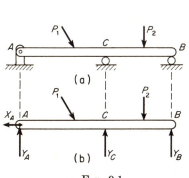

Fig. 9.1

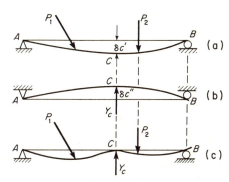

Fig. 9.2

231

If the roller at C is taken away, we are left with a simple beam AB supported at its ends. Since such supports are sufficient for complete constraint of the beam in one plane, we say that the intermediate support at C is *redundant*. If the roller at B instead of that at C is removed, we have a simple beam AC with overhang, which again is completely constrained in one plane. Thus either the support at B or the one at C may be considered as redundant. In general, any supports of a beam in excess of those both necessary and sufficient for its complete constraint in the plane of loading are said to be *redundant supports*.

Let us now choose the roller at C as the redundant support. Then removing this support, we consider the simple beam AB as our *primary system*. In Fig. 9.2, we study the deflection curves of this beam under two separate conditions of loading. In Fig. 9.2a, only the applied forces P_1 and P_2 are acting and δ'_c represents the downward deflection of point C when the support there is removed. In Fig. 9.2b, only a vertical force Y_C is acting at point C, and δ''_c represents the upward deflection of this point. Using the methods of the preceding chapter, the deflections δ'_c and δ''_c can be calculated without difficulty. Superimposing the two states of loading in Fig. 9.2a and b, we conclude that the net deflection of point C is $\delta_c = \delta'_c - \delta''_c$. Now since the support at C allows no deflection, the true value of δ_c is zero and we conclude that

$$\delta'_c - \delta''_c = 0. \tag{b}$$

When the deflections δ'_c and δ''_c are expressed in terms of P_1, P_2, and Y_C, condition (b) defines the true value of Y_C. As soon as Y_C is known, the

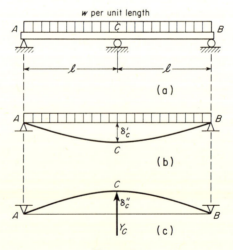

Fig. 9.3

remaining reactions X_A, Y_A and Y_B may be found from eqs. (a).

To illustrate the procedure more specifically, we consider the beam shown in Fig. 9.3. Choosing the support C as redundant, we obtain as our primary system a simple beam AB of span $2l$. Under the action of the uniform load, the downward deflection of point C, from Case 8 of Table 8.1, p. 213, will be

$$\delta'_c = \frac{5w(2l)^4}{384EI} = \frac{5wl^4}{24EI}.$$

Likewise from Case 6 of Table 8.1, the upward deflection of point C due to a vertical force Y_C will be

$$\delta''_c = \frac{Y_C(2l)^3}{48EI} = \frac{Y_c l^3}{6EI}.$$

Thus condition (b) above gives

$$\frac{5wl^4}{24EI} - \frac{Y_c l^3}{6EI} = 0,$$

from which

$$Y_C = \frac{5}{4}\,wl.$$

With this value of Y_C, eqs. (a) give $Y_A = Y_B = \frac{3}{8}wl$ and the reactions are completely determined.

The above example may be looked at in a somewhat different way by noting that the tangent to the elastic line at C remains horizontal because of symmetry conditions. Thus we may regard each half of the beam as a cantilever built-in at C and supported at its free end. Considering the portion CB, we have then a so-called "*propped cantilever*" as shown in Fig. 9.4. When this beam deflects under the applied loading, rotation of the tangent to the elastic line at C is prevented and a counterclockwise moment M_C is developed at C as shown. Removing this constraining moment as the redundant reaction, we obtain a simple beam CB as our primary system. Then under the action of the distributed load, the tangent at C rotates through the angle θ'_c as shown in Fig. 9.4b, while under the action of the redundant moment M_c at C it rotates in the opposite direction through an angle θ''_c, Fig. 9.4c. The true value of M_c is found by making $\theta'_c - \theta''_c = 0$, since this realizes the condition of a built-in end. From Case 8 of Table 8.1, we find

$$\theta'_c = \frac{wl^3}{24EI}.$$

From Case 9, we find

$$\theta''_c = \frac{M_c l}{3EI}.$$

Equating these values in accordance with the condition that $\theta'_c - \theta''_c = 0$, we obtain

$$M_c = \frac{wl^2}{8}. \tag{c}$$

The positive value indicates that the assumed direction of M_c in Fig. 9.4a was correct; according to our sign convention for bending moment, it represents a negative bending moment at the built-in end C of the beam. Returning now to Fig. 9.4a and equating to zero the algebraic sum of moments of all forces around point C, we obtain

$$Y_B l - \frac{wl^2}{2} + M_c = 0.$$

Substituting expression (c) for M_c, this gives

$$Y_B = \frac{3}{8}\,wl,$$

as found before.

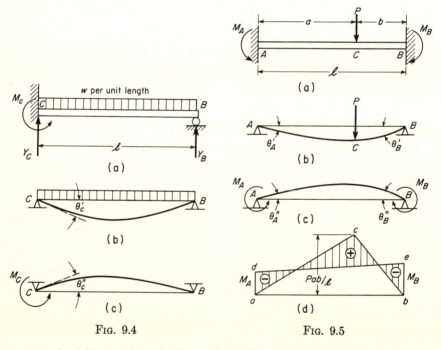

Fig. 9.4   Fig. 9.5

As a final illustration of the method of superposition in dealing with statically indeterminate beams, consider the beam in Fig. 9.5 having both ends built-in and carrying a concentrated load P. Removing the resistance against rotation at each end of the beam as a redundant constraint, we obtain, as our primary system, a simply supported beam AB. In this case there are two redundant reactions, M_A and M_B, and we say that the beam with built-in ends is *twice* statically indeterminate. To find M_A and M_B, consider first the primary system under the action of the applied load P, Fig. 9.5b. Then from Case 7 of Table 8.1, we have, for the angles of rotation of the end tangents

$$\theta'_A = \frac{Pb(l^2 - b^2)}{6lEI} \quad \text{and} \quad \theta'_B = \frac{Pab(2l - b)}{6lEI}.$$

Considering now the primary system under the action of end moments M_A and M_B as shown in Fig. 9.5c, and using Case 9 of Table 8.1, we have

$$\theta''_A = \frac{M_A l}{3EI} + \frac{M_B l}{6EI} \quad \text{and} \quad \theta''_B = \frac{M_B l}{3EI} + \frac{M_A l}{6EI}.$$

Finally, since the net angle of rotation at each end of the beam must be zero, we have $\theta'_A - \theta''_A = 0$ and $\theta'_B - \theta''_B = 0$. Thus

$$\frac{Pb(l^2 - b^2)}{6lEI} - \frac{M_A l}{3EI} - \frac{M_B l}{6EI} = 0,$$

$$\frac{Pab(2l - b)}{6lEI} - \frac{M_B l}{3EI} - \frac{M_A l}{6EI} = 0.$$

These two equations yield for M_A and M_B the following values:

$$M_A = \frac{Pab^2}{l^2} \quad \text{and} \quad M_B = \frac{Pa^2 b}{l^2}. \tag{d}$$

Having the end moments M_A and M_B, the bending moment diagram for the beam is easily constructed by superimposing the bending moment diagrams for the loadings shown in Figs. 9.5b and c. This superposition is shown in Fig. 9.5d, where the triangle abc represents the bending moment diagram for the load P and the trapezoid $abde$, that for the end moments. By plotting the negative ordinates due to M_A and M_B on the same side of the base line ab as the positive ordinates due to P, the overlapping portions automatically cancel and the net bending moment is given by the ordinates of the shaded portions of the diagram.

Assuming $b < a$, the numerically largest negative bending moment will be M_B as given by the last of eqs. (d). This has its greatest value when $b = l/3$ for which $|M_B|_{max} = 4Pl/27$. The positive bending moment at the point of application of the load P, by reference to Fig. 9.5d is

$$M_C = \frac{Pab}{l} - \frac{M_A + M_B}{2} = \frac{Pab}{2l}.$$

This has its greatest value when $a = b = l/2$, so that $|M_C|_{max} = Pl/8$, which is seen to be slightly less than $|M_B|_{max}$.

EXAMPLE 1. Two wood beams cut from the same timber are arranged as shown in Fig. 9.6a, the free end of the cantilever DC being supported at the middle of the simple beam AB. Both beams are horizontal and at right angles to one another. Find the vertical deflection δ_c at the point of contact due to a vertical load P applied to the end of the cantilever as shown.

SOLUTION. Let X represent the magnitude of the force with which the two beams interact as the point of contact. Then the cantilever has a net load $P - X$ at its end, Fig. 9.6b, and the deflection of this point, from Case 1 of Table 8.1, becomes

$$\delta'_c = \frac{(P - X)l^3}{3EI}.$$

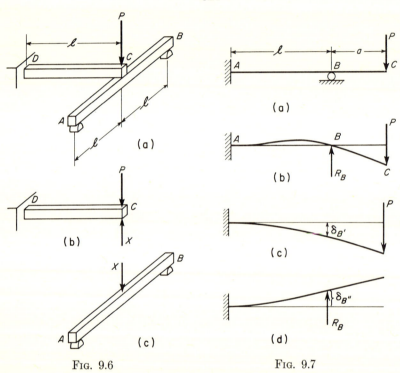

FIG. 9.6   FIG. 9.7

At the same time the mid-point of the beam AB is subjected to a downward force X and deflects by the amount

$$\delta''_c = \frac{X(2l)^3}{48EI},$$

by Case 6 of Table 8.1. Since these two deflections must be equal, we obtain

$$\frac{P}{3} - \frac{X}{3} = \frac{X}{6},$$

from which

$$X = \frac{2}{3}P.$$

Substituting this back into either of the deflection expressions gives

$$\delta'_c = \delta''_c = \delta_c = \frac{Pl^3}{9EI}.$$

This assumes, of course, that there was contact but no pressure between the two beams at C prior to the application of the load P.

EXAMPLE 2. A cantilever beam ABC is built-in at A, propped at B, and carries a load P at the free end as shown in Fig. 9.7. Find the magnitude of the redundant reaction at B.

SOLUTION. Replacing the support at B by the reactive force R_B, we obtain as our primary system, the cantilever beam AC loaded at B and C as shown in Fig. 9.7b. From Case 1 of Table 8.1, the downward deflection of point B due to the load P is

$$\delta'_B = \frac{Pl^2}{6EI} [3(l+a) - l].$$

Likewise, from Case 1, the upward deflection of point B due to the force R_B is

$$\delta''_B = \frac{R_B l^3}{3EI}.$$

Since the net deflection $\delta'_B - \delta''_B = 0$, we obtain

$$\frac{Pl^2}{6EI}(2l + 3a) - \frac{R_B l^3}{3EI} = 0,$$

from which

$$R_B = P\left(1 + \frac{3a}{2l}\right).$$

Using this value for R_B, we find for the positive bending moment at the built-in end of the beam

$$M_A = R_B l - P(a+l) = \frac{Pa}{2}.$$

PROBLEM SET 9.1

1. Find the redundant reaction R_B for the beam AB uniformly loaded as shown in Fig. A by taking the cantilever built-in at A and free at B as the primary system. Ans. $R_B = \frac{3}{8}wl$.

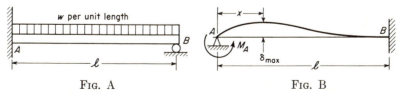

w per unit length

FIG. A   FIG. B

2. A propped cantilever beam is subjected to the action of a couple of moment M_A at the end A as shown in Fig. B. What reactive moment M_B will be induced at the built-in end B? Ans. $M_B = M_A/2$.

3. Referring again to the propped cantilever beam in Fig. B, find the maximum deflection produced by the applied couple M_A. Ans. $\delta_{max} = M_A l^2/27EI$ at $x = l/3$.

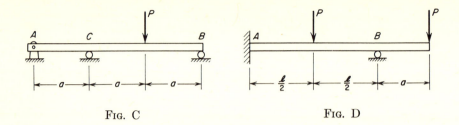

FIG. C   FIG. D

4. Find the redundant reaction R_A at the support A of the statically indeterminate beam supported and loaded as shown in Fig. C. *Ans.* $R_A = P/4$.

5. Find the redundant moment M_A at the built-in end A of the statically indeterminate beam supported and loaded as shown in Fig. D. *Ans.* $M_A = P(8a - 3l)/16$.

6. Find the redundant reaction R_A at the support A of the statically indeterminate beam supported and loaded as shown in Fig. E. The beam is continuous at C and its flexural rigidity is constant throughout. *Ans.* $R_A = 3Pa/2l$.

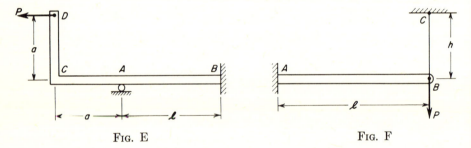

FIG. E   FIG. F

7. A cantilever steel beam AB is built-in at A and supported at B by a vertical steel tie-rod BC as shown in Fig. F. Before the load P is applied at B, the tie-rod is just taut but without initial tension. Find the tension S in the tie-rod after the load P is applied at B. The beam has flexural rigidity EI and the tie-rod has tensile rigidity AE. *Ans.*

$$S = \frac{P}{1 + (3hI/Al^3)}.$$

8. Referring again to Fig. F, assume that the load P at B is removed and that the system is free from stress at temperature $T = 70°$ F. Then if the temperature drops to $0°$ F, what tension S will exist in the vertical tie-rod? The beam has a rectangular cross-section 3 in. wide by 1 in. deep and the tie-rod has a circular cross-section of diameter $d = \frac{1}{4}$ in. The coefficient of thermal expansion for the tie-rod is $\alpha = 7.0 (10)^{-6}$ in./in./$°$F; $E = 30 (10)^6$ psi for both beam and tie-rod, and $l = h = 10$ in. *Ans.* $S = 96$ lb.

9. Referring to the system shown in Fig. 9.6a, assume that the concentrated force P at C is replaced by a uniformly distributed load of intensity w between D and C. In such case, find the force X transmitted from one beam to the other at C. Both beams have the same flexural rigidity EI. *Ans.* $X = wl/4$.

10. Find the reaction R_A at the support A of the continuous prismatic beam supported and loaded as shown in Fig. G. *Ans.* $R_A = P/4$, up.

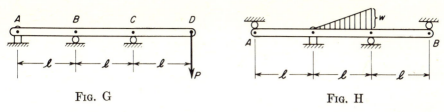

FIG. G   FIG. H

11. Find the reactions R_A and R_B at the two ends of the continuous prismatic beam supported and loaded as shown in Fig. H. *Ans.* $R_A = wl/45$; $R_B = wl/36$.

12. The continuous frame ABC shown in Fig. I is built-in at A, redundantly supported by a roller at C, and subjected to the action of a horizontal force P at B. Find the reaction R_C at C, neglecting axial extension of the vertical member AB. The flexural rigidity EI is constant throughout. *Ans.* $R_C = 3P/8$.

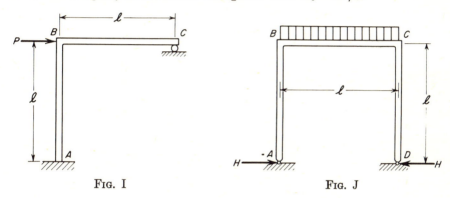

FIG. I   FIG. J

13. Solve the preceding problem if the concentrated force P at B in Fig. I is replaced by a uniformly distributed load of intensity w along the vertical member between A and B. *Ans.* $R_C = wl/8$.

14. The continuous frame $ABCD$ shown in Fig. J is pin-supported at the lower end of each leg and carries a uniformly distributed load of intensity w on the horizontal member BC. Each member has length l and flexural rigidity EI. Find the horizontal thrust H at each support. *Ans.* $H = wl/20$.

15. Referring to Fig. 9.6a, assume that the cantilever beam extends beyond C with an overhang of length a and that the load P is applied to the free end of this overhang as in Fig. 9.7a. In such case, find the force X transmitted to the lower beam AB at its mid-point C. Both beams have the same flexural rigidity EI. *Ans.* $X = P(2l + 3a)/3l$.

9.2 Theorem of Three Moments

In the case of a uniform *continuous beam* on many supports (Fig. 9.8a), one support is usually considered as an immovable hinge while all the others are treated as rollers. In such an arrangement, each intermediate support represents just one redundant constraint and hence the continuous beam is as many times statically indeterminate as there are intermediate supports.

Thus the beam in Fig. 9.8a is five times statically indeterminate. In such case, we might choose the five intermediate reactions R_1, R_2, ..., R_5 as the redundants and then set the deflections δ_1, δ_2, ..., δ_5 of their points of application equal to zero. These five equations together with the three equations of statics for the entire beam would be sufficient to determine the eight unknown reaction components. This, however, turns out to be a cumbersome and involved set of equations. Instead, it is much simpler to cut the beam at each intermediate support and introduce the bending moments M_1, M_2, ..., M_5, over the supports as the redundants. In this way, the primary system becomes six simple beams, each carrying its own externally applied loads together with two redundant moments at its two ends. By dealing with these simple beams two at a time, the complexity of the problem will be greatly reduced.

In Fig. 9.8b, we consider any two adjacent spans with supports $n - 1$, n, and $n + 1$. Let l_n and l_{n+1} denote these two span lengths and let M_{n-1}, M_n, and M_{n+1} denote the three redundant bending moments at these three supports. Whether these bending moments are positive or negative will depend upon the conditions of external loading. We will assume that they are positive as shown in the figure. Then if calculation produces for any one of them a negative value, this will automatically indicate a negative bending moment.

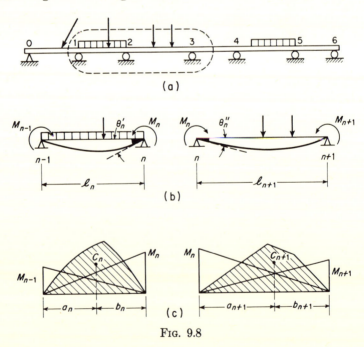

Fig. 9.8

The loadings shown in Fig. 9.8b produce some bending of the beam in spans l_n and l_{n+1} and we indicate rotation of the end tangents at the support n by θ'_n for the left span and by θ''_n for the right span. These angles of rotation are considered to be positive when they agree in direction with the end moments M_n. Therefore, as a condition of continuity of slope of the uncut elastic line over the support n, we have

$$\theta'_n = -\theta''_n. \tag{a}$$

This is the key to the solution of the problem.

To express θ'_n and θ''_n in terms of the applied loads and end moments for each span separately, we use the moment-area method discussed in Art. 8.2. The bending moment diagrams for the end moments are the triangles shown in Fig. 9.8c. The bending moment diagrams for the applied loads are represented by the shaded areas in the same figure. We denote these areas by A_n and A_{n+1}, respectively, and their centroids by C_n and C_{n+1}, the positions of which are defined by a_n, b_n, and a_{n+1}, b_{n+1}, as shown. Now considering the left span l_n and using superposition, we find

$$\theta'_n = \frac{M_n l_n}{3EI} + \frac{M_{n-1} l_n}{6EI} + \frac{A_n a_n}{l_n EI}. \tag{b}$$

Considering the right span l_{n+1}, we have

$$\theta''_n = \frac{M_n l_{n+1}}{3EI} + \frac{M_{n+1} l_{n+1}}{6EI} + \frac{A_{n+1} b_{n+1}}{l_{n+1} EI}. \tag{c}$$

Substituting (b) and (c) into eq. (a), we get

$$M_{n-1} l_n + 2M_n(l_n + l_{n+1}) + M_{n+1} l_{n+1} = -\frac{6A_n a_n}{l_n} - \frac{6A_{n+1} b_{n+1}}{l_{n+1}}. \tag{9.1}$$

This is called the *three-moment equation;* it can be written once for each intermediate support of the continuous beam. Then from this system of simultaneous equations the bending moments at the intermediate supports can be found.

Throughout the foregoing discussion, it has been assumed that the two extreme ends of the continuous beam were simply supported. If one or both of these ends should be built-in, the number of redundancies will exceed the number of intermediate supports. In such case, for each built-in end, an additional equation expressing the condition that no rotation can take place at such a support will be available. Suppose, for example, that the extreme left end of the beam is built-in. Then referring to Fig. 9.8b and taking $n = 1$, we have

$$\theta_0 = \frac{M_0 l_1}{3EI} + \frac{M_1 l_1}{6EI} + \frac{A_1 b_1}{l_1 EI}, \tag{d}$$

where θ_0 is the angle of rotation of the tangent at the left support. Setting this equal to zero, we obtain

$$M_0 = -\frac{M_1}{2} - \frac{3A_1b_1}{l_1{}^2}.\tag{9.2}$$

Having found the bending moments at all of the supports of a continuous beam, there is no difficulty about finding the reactions. Taking again any two adjacent spans, as shown in Fig. 9.8b, let R'_n be the simple beam reaction at n due to loads in the span l_n and R''_n, the reaction at n due to loads in the span l_{n+1}. In addition, there will be reactions produced by the end moments M_{n-1}, M_n, and M_{n+1}. Taking the directions of these moments as shown in Fig. 9.8b, the total additional reaction at the support n will be

$$\frac{M_{n-1} - M_n}{l_n} + \frac{-M_n + M_{n+1}}{l_{n+1}}.$$

the first part coming from the span l_n and the second part from the span l_{n+1}. Adding to this, the reactions R'_n and R''_n, due to applied loads, we obtain for the total reaction at any intermediate support

$$R_n = R'_n + R''_n + \frac{M_{n-1} - M_n}{l_n} + \frac{-M_n + M_{n+1}}{l_{n+1}}.\tag{9.3}$$

Having the reactions and the bending moments at all supports, the shear force and bending moment diagram for the continuous beam may be constructed. Applications of eqs. (9.1), (9.2) and (9.3) to several specific cases of continuous beams will be illustrated in the following examples.

EXAMPLE 1. Construct bending moment and shearing force diagrams for the continuous three-span beam under uniform load of intensity w as shown in Fig. 9.9a. The three spans are equal and the beam is of constant cross-section.

SOLUTION. For a simple beam under uniform load, the bending moment diagram is a parabola with maximum ordinate $wl^2/8$. The area of the parabolic segment is

$$A = \frac{2}{3}l\,\frac{wl^2}{8} = \frac{wl^3}{12},$$

and its centroid is at mid-span, so that $a = b = l/2$. Considering now the first two spans on the left and noting that $M_0 = 0$, eq. (9.1) becomes

$$0 + 2M_1\,(2l) + M_2l = -\frac{wl^3}{4} - \frac{wl^3}{4}.\tag{e}$$

From conditions of symmetry, it is evident that $M_1 = M_2$, and eq. (e) gives $M_1 = -wl^2/10 = M_2$. The complete bending moment diagram as shown in Fig. 9.9b can now be drawn.

From eq. (9.3) the reaction at support 1 becomes

$$R_1 = \frac{wl}{2} + \frac{wl}{2} + \frac{wl}{10} = \frac{11wl}{10}.$$

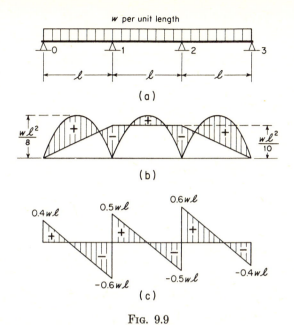

FIG. 9.9

From symmetry, we conclude that $R_2 = R_1$ while $R_0 = R_3$. Then since $R_0 + R_1 + R_2 + R_3 = 3wl$, $R_0 = R_3 = 4wl/10$. Having the reactions, the shear diagram is constructed as shown in Fig. 9.9c.

EXAMPLE 2. Construct bending moment and shearing force diagrams for the continuous three-span beam loaded as shown in Fig. 9.10a.

SOLUTION. For a simple beam with a concentrated load P at the middle, the bending moment diagram is a triangle with maximum ordinate $Pl/4$. The area of this triangle is

$$A = \frac{Pl}{4} \cdot \frac{l}{2} = \frac{Pl^2}{8},$$

and its centroid is at mid-span so that $a = b = l/2$.

Writing eq. (9.1), once for the first two spans on the left and again for the last two spans on the right, we have

$$\left.\begin{array}{l} 0 + 2M_1(2l) + M_2l = 0, \\[2mm] M_1l + 2M_2(2l) + 0 = 0 - \dfrac{3Pl^2}{8}. \end{array}\right\} \tag{f}$$

These simultaneous equations are easily solved for the bending moments M_1 and M_2 and we find

$$M_1 = +\frac{Pl}{40}, \quad M_2 = -\frac{Pl}{10}.$$

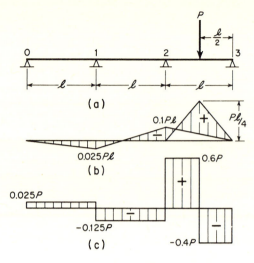

FIG. 9.10

The corresponding bending moment diagram for the entire beam is shown in Fig. 9.10b. The maximum bending moment occurs under the load P and has the value

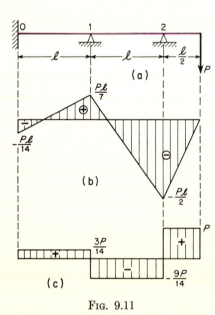

FIG. 9.11

$$M_{max} = \frac{Pl}{4} - \frac{Pl}{20} = \frac{Pl}{5}.$$

Using eq. (9.3), the reactions R_1 and R_2 at the intermediate supports are found as follows

$$R_1 = 0 + 0 - \frac{P}{40} - \frac{P}{40} - \frac{P}{10} = -\frac{3P}{20},$$

$$R_2 = 0 + \frac{P}{2} + \frac{P}{40} + \frac{P}{10} + \frac{P}{10} = +\frac{29}{40}P.$$

Then with the entire beam as a free body, we find $R_0 = +P/40$ and $R_3 = +16P/40$. The shearing force diagram may now be constructed as shown in Fig. 9.10c.

EXAMPLE 3. Construct bending moment and shearing force diagrams for the continuous beam shown in Fig. 9.11a, which overhangs the right-hand support and carries a load P at the free end.

SOLUTION. Writing eq. (9.1) for the two adjacent spans 01 and 12, we have

$$M_0 l + 2M_1(2l) + M_2 l = 0. \tag{g}$$

Then since the left hand support is built-in and there are no external loads on the first span, eq. (9.2) gives $M_0 = -M_1/2$. It must be noted also that $M_2 = -Pl/2$ since the beam freely overhangs the last support on the right. Substituting these values into eq. (g), we find $M_1 = +Pl/7$. Then $M_0 = -Pl/14$ and $M_2 = -Pl/2$. The corresponding bending moment diagram is shown in Fig. 9.11b. From statics, the reactions are now found to be $R_0 = +3P/14$, $R_1 = -12P/14$, $R_2 = +23P/14$, and the shear diagram is as shown in Fig. 9.11c.

PROBLEM SET 9.2

1. Find the bending moments M_1 and M_2 at the supports 1 and 2 of the three-span continuous beam loaded as shown in Fig. A. *Ans.* $M_1 = -22Pl/405$, $M_2 = -32Pl/405$.

2. A uniform three-span continuous beam with overhanging ends carries a uniformly distributed load as shown in Fig. B. Find the ratio a/l in order to make the bending moments at the three supports all equal. *Ans.* $a/l = 1/\sqrt{6} = 0.408$.

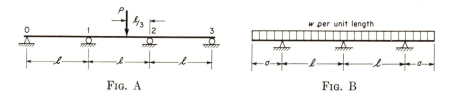

FIG. A   FIG. B

3. Referring again to the beam in Fig. B, find the ratio a/l in order to make the reactions at the three supports equal. *Ans.* $a/l = 0.44$.

4. A two-span continuous beam with overhang at the right hand end is loaded as shown in Fig. C. Calculate the reactions at the three points of support. *Ans.* $R_0 = 833$ lb, up; $R_1 = 2750$ lb, down; $R_2 = 917$ lb, up.

5. A two-span continuous beam built-in at the left end is loaded as shown in Fig. D. Calculate the bending moments at the three supports. *Ans.* $M_0 = +5140$ ft.-lb; $M_1 = -10,280$ ft.-lb; $M_2 = 0$.

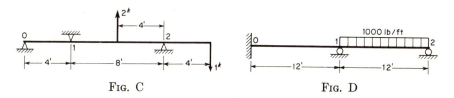

FIG. C   FIG. D

6. The right-hand span of a continuous beam carries a uniform load of intensity w over half its length as shown in Fig. E. Find the bending moments at the two intermediate supports. *Ans.* $M_1 = 7wl^2/960$; $M_2 = -28wl^2/960$.

7. Construct bending moment and shear force diagrams for the continuous beam shown in Fig. F. *Ans.* $M_1 = M_2 = +4080$ ft-lb; $R_0 = R_3 = 10,680$ lb; $R_1 = R_2 = -3480$ lb.

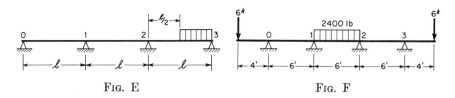

FIG. E   FIG. F

8. Calculate the bending moments and reactions at each support of the continuous beam in Fig. G. *Ans.* $M_1 = -1.54$ kip-ft; $M_2 = -3.74$ kip-ft; $M_3 = -1.65$ kip-ft; $R_0 = -0.386$ kips; $R_1 = +2.69$ kips; $R_2 = +6.22$ kips; $R_3 = +3.75$ kips; $R_4 = -0.275$ kips.

9. Calculate the bending moment M_0 and the reaction R_0 at the built-in end of the continuous beam shown in Fig. H. *Ans.* $M_0 = -5260$ ft.-lb; $R_0 = 3080$ lb.

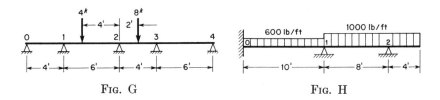

FIG. G   FIG. H

9.3   The Theorem of Castigliano

In preceding chapters, general expressions for the strain energy stored in an elastic bar subjected to tension, torsion, or bending have been developed. This concept of elastic strain energy can be very useful in the study of deflections of various points of a structure under load. Consider, for example, a prismatic bar under simple tension as shown in Fig. 2.14a, p. 40. The strain energy in this case is, from eq. (2.4),

$$U = \frac{P^2 l}{2AE}.$$
(a)

By taking the derivative of this expression with respect to the applied load P, we obtain

$$\frac{dU}{dP} = \frac{Pl}{AE} = \delta.$$

Thus the derivative of the strain energy with respect to the applied load P gives the deflection of its point of application in the direction of the load.

Again, in the case of a shaft of circular cross-section subjected to torsion as shown in Fig. 4.12, p. 82, the strain energy is, from eq. (4.13′),

$$U = \frac{T^2 l}{2GJ}.$$
(b)

The derivative of this expression with respect to the applied torque T becomes

$$\frac{dU}{dT} = \frac{Tl}{GJ} = \phi,$$

which is the angle of twist of one end of the shaft with respect to the other. If we interpret the pair of statically balanced torques at the ends of the shaft as a "generalized force" and the angle of twist between these two ends as the "corresponding displacement," we conclude again that the "displacement" is given by the first derivative of the strain energy with respect to the "force."

Finally, for a cantilever beam bent by a transverse load P at the free end, Fig. 8.21a, p. 221, the strain energy of bending is

$$U = \frac{P^2 l^3}{6EI}. \qquad (c)$$

The derivative of this expression with respect to P is

$$\frac{dU}{dP} = \frac{Pl^3}{3EI} = \delta,$$

which is seen to be the deflection of the end of the beam in the direction of the applied load P.

Each of the foregoing cases is simply an example of a general theorem regarding strain energy which is called the *theorem of Castigliano*. We shall now proceed with a general derivation of this theorem. In Fig. 9.12, let AB be any elastic body or structure completely constrained in space and subjected to applied forces P_1, P_2, P_3, ... If the material follows Hooke's law and the deformations are small, the displacements of the points of loading will usually be linear functions of the loads, i.e., the principle of superposition

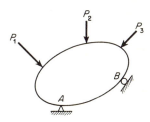

Fig. 9.12

will hold.* In such cases, the strain energy stored in the loaded system will be equal to the work done by the applied forces and independent of the order in which they are applied. If, for example, the forces are applied simultaneously and gradually increased in the same proportion, then the work done will be

$$U = \frac{1}{2}(P_1 \delta_1 + P_2 \delta_2 + P_3 \delta_3 + \cdots), \qquad (9.4)$$

where δ_1, δ_2, δ_3, ... are the deflections of points 1, 2, 3, ... in the directions

*For exceptional cases, see Timoshenko and Young, *Theory of Structures*, 2nd. ed., p. 221.

of the corresponding forces P_1, P_2, P_3, ... It must be clearly understood here that δ_1, δ_2, δ_3, ... may not be the total deflections of the points 1, 2, 3, ... but, in each case, only that component of the total deflection in the direction of the corresponding force. From our assumption that the displacements δ_1, δ_2, ... are linear functions of the forces P_1, P_2, ... , we conclude that if we substitute for δ_1, δ_2, ... , their expressions in terms of P_1, P_2, ... , eq. (9.4) becomes a homogeneous quadratic function of the forces.

Assume now that after the loads P_i have all been applied and the strain energy in the system is U, as given by eq. (9.4), we increase any load P_n by the amount dP_n. This will cause a slight change in the deformation of the body and the stored strain energy will be increased slightly. This increase may be expressed as the rate of change of U with respect to P_n times the change made in P_n. In this way, the new amount of strain energy becomes

$$U + \frac{\partial U}{\partial P_n} \cdot dP_n. \tag{d}$$

Since the final strain energy does not depend upon the order in which the forces are applied, we will now assume that dP_n is applied first and afterwards the forces P_1, P_2, P_3, ... When dP_n is applied first, it produces only infinitesimal deflections and the corresponding strain energy, of second order, can be ignored. Applying now the forces P_1, P_2, P_3, ... , it must be seen that their effect on the structure will be unaltered by the presence of dP_n and they will store the same strain energy U as before. However, during the application of P_1, P_2, P_3, ... , the force dP_n, already applied earlier, will ride through the displacement δ_n produced by P_1, P_2, P_3, ... In so doing, it produces additional work of the amount $dP_n \cdot \delta_n$ and the final strain energy in this case will be

$$U + dP_n \cdot \delta_n. \tag{e}$$

Equating this expression for final strain energy to expression (d) obtained before, we have

$$U + dP_n \delta_n = U + \frac{\partial U}{\partial P_n} dP_n,$$

from which

$$\delta_n = \frac{\partial U}{\partial P_n}. \tag{9.5}$$

This states the theorem of Castigliano in general form. For any elastic system which obeys the law of superposition, *the partial derivative of the strain energy, represented as a quadratic function of the forces, with respect to any one of these forces, gives the corresponding component of displacement of the point of application of that force.* The terms "force" and "displacement" must be taken in the generalized sense. If the "force" is a couple, the corresponding displacement will be an angle of rotation in the same sense as the couple. If it is a true force, the corresponding displacement will be a linear deflection in the direction of that force. It must also be appreciated that we speak only of statically independent forces, to any one of which we can give an arbitrary increment δP_n without affecting any of the other forces. This means that reactions determined by statics cannot be considered as independent forces.

EXAMPLE 1. A simply supported beam with overhang is loaded as shown in Fig. 9.13a. Using the theorem of Castigliano, find the vertical deflection of point C.

SOLUTION. The bending moment diagram for the beam is shown in Fig. 9.13b. Between A and B, a general expression for bending moment at any distance x to the right of A is

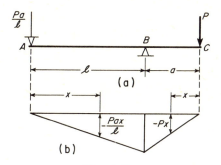

FIG. 9.13

$$M_x = -\frac{Pax}{l}. \tag{f}$$

Between B and C, the bending moment at any distance x to the left of C is

$$M_x = -Px. \tag{g}$$

Using the first of eqs. (8.7), p. 220, the strain energy of bending in the beam becomes

$$U = \int_0^l \frac{P^2a^2x^2}{2EIl^2}\,dx + \int_0^a \frac{P^2x^2dx}{2EI} = \frac{P^2a^2}{6EI}\,(l+a). \tag{h}$$

Then from eq. (9.5)

$$\delta = \frac{\partial U}{\partial P} = \frac{Pa^2}{3EI}\,(l+a).$$

EXAMPLE 2. The axis of a cantilever ring, built-in at B and loaded at the free end A, forms a horizontal quarter circular arc of radius R, Fig. 9.14. Find the vertical deflection δ of the free end A, assuming the ring to have a circular cross-section the diameter of which is small compared with the radius R of its center line.

SOLUTION. Such a curved cantilever is subjected to torsion as well as bending. At any cross-section D, defined by the angle θ, we see that the bending moment is

$$M_\theta = PR \sin \theta, \tag{i}$$

while the twisting moment on the same cross-section is

$$T_\theta = PR (1 - \cos \theta). \tag{j}$$

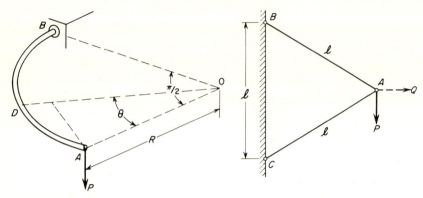

FIG. 9.14   FIG. 9.15

The total strain energy due to the combined bending and torsion of the ring is

$$U = \int_0^{\pi/2} \frac{M_\theta^2 \, Rd\theta}{2EI} + \int_0^{\pi/2} \frac{T_\theta^2 \, Rd\theta}{2GJ}.$$

Then by eq. (9.5)

$$\delta_v = \frac{\partial U}{\partial P} = \int_0^{\pi/2} \frac{M_\theta\left(\dfrac{\partial M_\theta}{\partial P}\right)Rd\theta}{EI} + \int_0^{\pi/2} \frac{T_\theta\left(\dfrac{\partial T_\theta}{\partial P}\right)Rd\theta}{GJ}, \tag{k}$$

where, from eqs. (i) and (j),

$$\frac{\partial M_\theta}{\partial P} = R \sin \theta, \tag{i'}$$

$$\frac{\partial T_\theta}{\partial P} = R (1 - \cos \theta). \tag{j'}$$

Substituting (i), (j), (i'), (j'), into eq. (k) and performing the indicated integrations, we find

$$\delta_v = \frac{\pi PR^3}{4EI} + \frac{(3\pi - 8) \, PR^3}{4GJ}.$$

EXAMPLE 3. A simple truss composed of two bars each of length l carries a vertical load P at joint A as shown in Fig. 9.15. Find the horizontal and vertical components of the total deflection δ of point A. The bars are of the same material, AB having a cross-sectional area A and AC, a cross-sectional area $A_1 = 2A$.

SOLUTION. From statics, we see that the tensile force in the bar AB and the compressive force in the bar AC are each equal to P. Hence the strain energy in the system is

$$U = \frac{P^2l}{2AE} + \frac{P^2l}{2A_1E} = \frac{3P^2l}{4AE}.$$

Then, for the vertical component of the deflection of A, we have

$$\delta_v = \frac{\partial U}{\partial P} = \frac{3Pl}{2AE}.$$

To find the horizontal component of the deflection of joint A, we introduce at A a fictitious horizontal force Q with respect to which we can differentiate the strain energy expression. With this force acting in addition to P, the strain energy becomes

$$U = \frac{(Q/\sqrt{3} + P)^2l}{2AE} + \frac{(Q/\sqrt{3} - P)^2l}{2(2A)E}.$$

Then from eq. (9.5)

$$\delta_h = \frac{\partial U}{\partial Q} = \frac{(Q + P/\sqrt{3})l}{2AE}.$$

For the case in which we are interested, $Q = 0$ and the horizontal component of deflection becomes

$$\delta_h = \frac{Pl}{2\sqrt{3}\,AE}.$$

PROBLEM SET 9.3

1. Using the Castigliano theorem, calculate the vertical deflection δ at the middle of a simply supported beam (Fig. A) which carries a uniformly distributed load of intensity w over the full span. The flexural rigidity EI of the beam is constant and only strain energy of bending is to be considered. *Hint:* Introduce a fictional load $Q = 0$ at the middle of the beam. *Ans.* $\delta = 5wl^4/384EI$.

2. Using the Castigliano theorem, calculate the angle of rotation θ of either end-tangent for the simply supported beam AB loaded as shown in Fig. A. Assume constant EI and consider only strain energy of bending. *Hint:* Introduce a fictional couple of moment $M = 0$ at one end of the beam. *Ans.* $\theta = wl^3/24EI$.

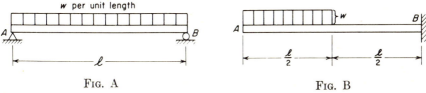

w per unit length

FIG. A   FIG. B

3. Using the Castigliano theorem, calculate the vertical deflection δ at the free end of the cantilever beam of length l loaded as shown in Fig. B. *Ans.* $\delta = 41wl^4/384EI$.

4. A thin cantilever ring AB having a circular axis of mean radius R is built-in at B and carries a vertical load P at its free end A as shown in Fig. C. Using the Castigliano theorem and considering only strain energy of bending, find the vertical deflection of point A. *Ans.* $\delta_v = \pi PR^3/4EI$.

5. Referring again to the cantilever ring in Fig. C and using the same procedure as in the preceding problem, find the horizontal deflection of point A. *Ans.* $\delta_h = PR^3/2EI$.

6. Referring again to the ring in Fig. C, find the angle of rotation θ of the tangent at A due to the applied load P. *Ans.* $\theta = PR^2/EI$.

7. A simple truss ABC with pinned joints is loaded as shown in Fig. D. Both bars are made of steel and have the same cross-sectional area A. Using the Castigliano theorem, find the vertical deflection of joint A. *Ans.* $\delta_v = 7.62Pl/AE$.

8. Referring again to the truss in Fig. D, find the horizontal displacement of the joint A caused by the action of the vertical load P. *Ans.* $\delta_h = -1.73Pl/AE$.

9. A thin, circular steel ring of mean radius $R = 3$ in. has a $\frac{1}{4}$ in. by $\frac{1}{4}$ in. square cross-section and is cut through at one point on its circumference. The ring is then spread open and a $\frac{1}{4}$-in.-thick block is inserted in the gap. What maximum bending stress will be produced in the ring? *Ans.* $\sigma_{max} = 22,100$ psi.

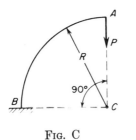

FIG. C

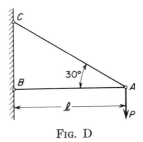

FIG. D

9.4 Applications of Castigliano's Theorem to Statically Indeterminate Problems

The theorem of Castigliano is very useful in the treatment of statically indeterminate problems. Consider, for example, the case of a continuous beam having several redundant supports. Denoting by X, Y, Z, ... the statically indeterminate reactions, the strain energy in the beam will be a function of these forces and the displacements of their points of application will be obtained as dU/dX, dU/dY, dU/dZ, ... However, these displacements are all known to be zero in accordance with the conditions of constraint. Hence, we have

$$\frac{\partial U}{\partial X} = 0, \quad \frac{\partial U}{\partial Y} = 0, \quad \frac{\partial U}{\partial Z} = 0, \ldots \qquad (9.6\text{a})$$

There will always be as many of these equations as there are redundant reactions so that the reactions can be found.

If the bending moments over the supports are taken as the redundant quantities in the case of a continuous beam, then the strain energy can be expressed as a function of these bending moments M_1, M_2, M_3, ... In such case the partial derivatives $\partial U/\partial M_1$, $\partial U/\partial M_2$, $\partial U/\partial M_3$, ... will represent the relative rotations between tangents to the elastic line on the two sides of each support. However, from continuity of the elastic line over each support, we know that these relative rotations are all zero. Hence, again

$$\frac{\partial U}{\partial M_1} = 0, \quad \frac{\partial U}{\partial M_2} = 0, \quad \frac{\partial U}{\partial M_3} = 0, \dots . \qquad (9.6b)$$

In general, to find the redundant forces in a statically indeterminate system, we remove the redundant constraints and replace them by the corresponding forces. Then expressing the strain energy of the system in terms of the forces and applying the Castigliano theorem, we obtain eqs. (9.6) from which the redundant forces can be calculated.

To illustrate, let us reconsider the case of a propped cantilever beam under uniform load, Fig. 9.16. This is a system having one statically indeterminate reaction. Choosing the vertical reaction X at B as the redundant force, the bending moment at any point along the beam will be

$$M_x = Xx - \frac{wx^2}{2}. \qquad (a)$$

and the strain energy is

$$U = \int_0^l \frac{M_x^2 dx}{2EI},$$

which is a function of X, since M_x is a function of X. Thus, differentiating under the integral sign

$$\frac{\partial U}{\partial X} = \int_0^l \frac{M_x \dfrac{\partial M_x}{\partial X} dx}{EI} = 0, \qquad (b)$$

wherein, from eq. (a),

$$\frac{\partial M_x}{\partial X} = x. \qquad (c)$$

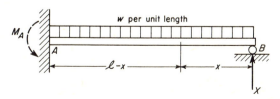

Fig. 9.16

Substituting expressions (a) and (c) into eq. (b),

$$\frac{1}{EI} \int_0^l \left(Xx - \frac{wx^2}{2} \right) (x) \, dx = 0.$$

Performing the indicated integration, we obtain

$$\frac{Xl^3}{3} - \frac{wl^4}{8} = 0,$$

from which $X = \frac{3}{8}wl$, as obtained previously on p. 233.

Choosing the restraining couple M_A at the built-in end of the beam as the redundant reaction, the bending moment at any cross-section becomes

$$M_x = \frac{wlx}{2} - \frac{M_A x}{l} - \frac{wx^2}{2}. \tag{d}$$

Then since the angle of rotation of the tangent at A is known to be zero, the Castigliano theorem gives

$$\frac{\partial U}{\partial M_A} = \int_0^l \frac{M_x \dfrac{\partial M_x}{\partial M_A} \, dx}{EI} = 0, \tag{e}$$

wherein, from eq. (d),

$$\frac{\partial M_x}{\partial M_A} = -\frac{x}{l}. \tag{f}$$

Substituting expressions (d) and (f) into eq. (e),

$$\frac{1}{EI} \int_0^l \left(\frac{wlx}{2} - \frac{M_A x}{l} - \frac{wx^2}{2} \right) \left(-\frac{x}{l} \right) dx = 0.$$

Performing the indicated integration, we find $M_A = wl^2/8$, as previously obtained on p. 234.

EXAMPLE 1. Three bars each of length l and pinned at their ends are arranged in a vertical plane as shown in Fig. 9.17. The vertical bar has a cross-sectional area A and each inclined bar has cross-sectional area A_1. A vertical load P acts at joint C and it is desired to find the ratio A_1/A of cross-sectional areas to make the tension in DC numerically equal to the compressive forces in AC and BC.

SOLUTION. Let X represent the tensile force in DC, chosen as the redundant bar. Then the compressive force in each inclined bar is $(P - X)/\sqrt{2}$. Thus the strain energy of the system becomes

$$U = \frac{X^2 l}{2AE} + \frac{(P - X)^2 l}{2A_1 E}$$

In this case, the end D of the vertical bar must have a displacement equal to zero.

Hence, from the Castigliano theorem,

$$\frac{dU}{dX} = \frac{Xl}{AE} - \frac{(P - X)l}{A_1 E} = 0,$$

from which

$$X = \frac{P}{1 + \dfrac{A_1}{A}} \cdot$$ (g)

The statement of the problem requires that

$$X = \frac{P - X}{\sqrt{2}}.$$ (h)

Eliminating X between eqs. (g) and (h), we find $A_1/A = \sqrt{2}$.

EXAMPLE 2. Two wood beams of identical cross-section are supported at their ends and cross at their mid-points as shown in Fig. 9.18. When unloaded, they are just in contact at C. What interactive force X will exist between the two beams at C when a vertical load P is applied to the upper beam as shown?

SOLUTION. The net downward load on the beam AB is $P - X$; that on the beam DE is X. The total strain energy in a simple beam loaded at the middle by a force Q is, from eq. (h), p. 222,

$$U = \frac{Q^2 l^3}{96EI}.$$

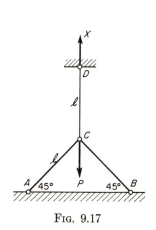

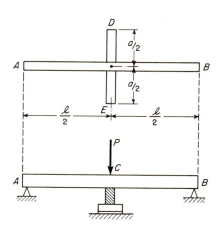

FIG. 9.17   FIG. 9.18

Thus the strain energy in the two beams becomes

$$U = \frac{(P - X)^2 l^3}{96EI} + \frac{X^2 a^3}{96EI}.$$ (i)

Regarding the pair of interactive forces X as a generalized force and observing that the corresponding displacement is the relative displacement between the midpoints of the two beams, which is zero because they remain in contact at C, we have by the Castigliano theorem

$$\frac{dU}{dX} = -\frac{(P - X)l^3}{48EI} + \frac{Xa^3}{48EI} = 0,$$

from which

$$X = \frac{Pl^3}{l^3 + a^3}.$$

EXAMPLE 3. To reduce deflection, a simply supported wood beam AB, loaded at the middle, is trussed by steel cables AD and BD and a post CD arranged as shown in Fig. 9.19. Neglecting axial shortening of both the beam and the post, find the compressive force X induced in the post. The beam has flexural rigidity EI and the cables have cross-sectional area A_1 and modulus of elasticity E_1.

SOLUTION. The net downward load on the middle of the beam is $Q = P - X$ and the tension in each cable is $S_1 = X/2 \sin \alpha$. Hence, neglecting strain energy of compression in the beam and post, the strain energy of the system is

$$U = \frac{Q^2 l^3}{96EI} + 2\left(\frac{S_1^2 l_1}{2A_1 E_1}\right).$$

Substituting for Q and S_1 and noting that $l_1 = \frac{1}{2}l \sec \alpha$, this becomes

$$U = \frac{(P - X)^2 l^3}{96EI} + \frac{X^2 l \sec \alpha}{8A_1 E_1 \sin^2 \alpha}.$$

Now shortening of the post is the generalized displacement corresponding to the compressive forces X and we have specified that this is to be neglected, i.e., that it is to be taken as zero. Hence, the Castigliano theorem gives

$$\frac{dU}{dX} = -\frac{(P - X)l^3}{48EI} + \frac{Xl \sec \alpha}{4A_1 E_1 \sin^2 \alpha} = 0,$$

from which

$$X = \frac{P}{1 + \dfrac{12 \sec \alpha}{l^2 \sin^2 \alpha} \dfrac{EI}{A_1 E_1}}.$$

Taking $\alpha = 20°$ and $EI/A_1 E_1 l^2 = 0.01$, this gives $X = 0.478P$. In such case the deflection of the beam will be reduced by approximately 50 per cent.

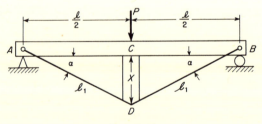

FIG. 9.19

PROBLEMSET 9.4

1. A prismatic beam AB with built-in ends and span l carries a single concentrated load P at the middle. Using the theorem of Castigliano, find the bending moments M_A and M_B at the ends of the beam. *Ans.* $M_A = M_B = -Pl/8$.

2. Repeat the solution of the preceding problem for the case where the beam carries a uniformly distributed load of intensity w over the full span. *Ans.* $M_A = M_B = -wl^2/12$.

3. A continuous prismatic beam having two equal spans l carries a uniformly distributed load of intensity w over one span only, as shown in Fig. A. Find the reaction X at the middle support by using the theorem of Castigliano. *Ans.* $X = 5wl/8$.

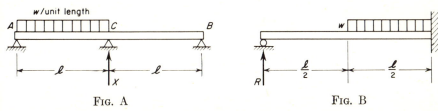

FIG. A   FIG. B

4. Using the theorem of Castigliano, find the bending moment M_C at the middle support of the two-span continuous beam loaded as shown in Fig. A. *Ans.* $M_C = wl^2/16$.

5. A propped-cantilever beam of uniform cross-section carries a uniformly distributed load of intensity w over the right-hand half of the span as shown in Fig. B. Using the theorem of Castigliano, find the magnitude of the reaction R at the propped end of the beam. *Ans.* $R = 7wl/128$.

6. A thin semicircular arch ring of mean radius R is pin-supported on unyielding foundations at its ends A and B and carries a vertical load P at the crown C, as shown in Fig. C. Using the theorem of Castigliano, find the horizontal components H of the reactions at A and B. *Ans.* $H = P/\pi$.

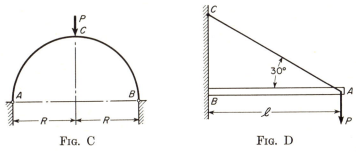

FIG. C   FIG. D

7. A thin circular ring of uniform cross-section and mean radius R is subjected to the action of two equal and opposite forces P pulling along a diameter AB. Using the theorem of Castigliano, find the bending moments M induced in the ring at the ends of the diameter which is perpendicular to AB. *Ans.* $M = 0.182PR$.

8. In Example 3, the strain energy of compression in the beam and post were neglected in calculating the compressive force X in the post. Calculate this compressive force X more exactly by taking account of the additional strain energy of compression. Use the data given in Example 3 and assume, in addition, that the post is cut from the same piece of timber as the beam with $AE = 4A_1E_1$. *Ans.* $X = 0.427P$.

9. A wood cantilever beam 5 ft long and having a rectangular cross-section 4 in. wide by 6 in. deep is built-in at B and supported at A by an inclined guy wire, as shown in Fig. D. The beam has modulus of elasticity $E = 2 (10)^6$ psi, and the guy wire has modulus of elasticity $E_1 = 30 (10)^6$ psi and cross-sectional area $A_1 = 0.0533$ sq in. Find the tensile force S induced in the guy wire due to the vertical load P applied at A. *Ans.* $S = 1.48P$.

9.5   Limit Analysis of Statically Indeterminate Beams

In all the foregoing discussions of statically indeterminate beams, it has been assumed that the beam behaves elastically throughout. It is also of interest to investigate the behavior of such beams under loads which induce plastic bending of the beam, as discussed in Art. 6.1. Consider, for example, a propped cantilever steel beam subjected to a load P as shown in Fig. 9.20a. For any value of this load less than the elastic limit value P_E, the bending moment diagram will be as shown in Fig. 9.20b. From this diagram, it is seen that points B and D are points of largest bending moment. If P is increased beyond the value P_E, a plastic hinge will begin to form either at D or at B, depending upon whether the elastic

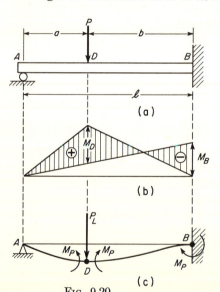

FIG. 9.20

moment M_D or M_B is greater. This, of course, will be determined by the position of the load P on the beam, i.e., by the ratio b/a. As a matter of fact, it can be shown that the elastic moment M_D will be numerically greater than the elastic moment M_B if $b/a < \sqrt{2}$. Assuming this to be the case, a plastic hinge will form first at

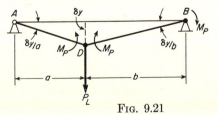

FIG. 9.21

point D. After this plastic hinge is fully developed at point D, the beam behaves as a statically determinate system represented by a compound beam ADB acted upon at D by the load P_L and an internal moment M_P.

As the load P is further increased, the negative bending moment M_B continues to grow until it also attains the limiting value M_P. When this condition is reached, the beam will not resist any further increase in load and becomes a *mechanism* with hinges at A, D, and B, which will collapse freely without further increase in load. The value of the load required to produce this condition represents the *limit load* P_L for the beam. This condition is represented in Fig. 9.20c.

To calculate the limit load, it is not necessary to trace the behavior from beginning to end as outlined above. We simply assume plastic hinges at B and at D, each offering constant bending moment M_P, and then solve for the value of P_L required to maintain equilibrium. Referring to Fig. 9.20c, the reaction at A in the limit condition will be

$$R_A = \frac{P_L b}{l} - \frac{M_P}{l}.$$

Then the bending moment at D becomes

$$M_D = R_A a = \frac{P_L ab}{l} - \frac{M_P a}{l} = M_P,$$

from which

$$P_L = \frac{M_P(l + a)}{a(l - a)}, \tag{a}$$

since $b = l - a$. Taking the safe working load $P_w = P_L/n$, we attain a factor of safety n against complete collapse.

It will be noted from expression (a) that the limit load P_L is a function of the ratio a/l defining the position of the load on the span. To find the most critical position, i.e., the ratio a/l to make P_L a minimum, we set $dP_L/da = 0$. This gives

$$a^2 + 2la - l^2 = 0,$$

from which

$$\frac{a}{l} = \sqrt{2} - 1 = 0.414.$$

Substituting this value of a/l into eq. (a), we obtain $(P_L)_{\min} = 5.84\ M_P/l$.

Another method for calculating the limit load P_L for the mechanism in Fig. 9.20c is to use the principle of *virtual work*. We take the system in the limit condition, i.e., as a mechanism with plastic hinges at B and D (black circles) and a simple hinge at A (white circle) as shown in Fig. 9.21. The load P_L and the plastic moments at B and D we consider as *active forces* on this movable system. Then a virtual displacement of the system can be defined by a small vertical displacement δy of the hinge D. At the same time the portion AD will rotate by the angle $\delta y/a$ about the immovable hinge A and the portion BD will rotate by the angle $\delta y/b$ about the immovable hinge B, as shown. During this virtual displacement, the load P_L produces positive virtual work $P_L \delta y$ while the plastic moment M_P at B produces negative virtual work $M_P \delta y/b$ and those at D, negative work $M_P (\delta y/a + \delta y/b)$. As a condition of equilibrium of the system, the algebraic sum of these virtual works must be zero. Hence

$$P_L \, \delta y - M_P \, \delta y/b - M_P(\delta y/a + \delta y/b) = 0,$$

from which, on canceling δy,

$$P_L = \frac{2M_P}{b} + \frac{M_P}{a}. \tag{b}$$

Substituting $b = l - a$, this gives the same result as expressed by eq. (a) above.

Limit analysis of statically indeterminate beams as illustrated above is often much simpler than elastic analysis, since by such procedure the problem is always reduced to a statically determinate one. It is also gaining favor among structural engineers because it leads to greater economy in design. In dealing with more complicated statically indeterminate beams and frames, the question of selecting, in advance, the proper positions of all plastic hinges is not always a simple one. Sometimes there will be more than one possible mode of collapse for the system and each of these must be investigated separately to find the one that leads to the lowest limit load. These questions belong more properly to special books on plastic analysis and limit design and will not be discussed here.*

EXAMPLE 1. A two-span continuous beam simply supported at A and B and built-in at C carries a uniformly distributed load of intensity w over its entire length (Fig. 9.22a). The beam has a 12WF50 wide-flange section for which the elastic section modulus $Z = 64.7$ in ^{3} Calculate the limit intensity w_L of the load for complete collapse of the system.

SOLUTION. As already worked out on p. 142, the plastic section modulus for this beam is

$$Z_P = 1.10Z = 1.10 \times 64.7 = 71.2 \text{ in.}^3$$

Then assuming a yield stress $\sigma_{y.p.} = 40{,}000$ psi, the corresponding plastic bending moment becomes

$$M_P = 40{,}000 \, \frac{71.2}{12} = 237{,}000 \text{ ft.-lb.}$$

In the completely elastic condition, the bending moment diagram for the beam will be as shown in Fig. 9.22b. From this, we see that plastic hinges are likely to

*See, for example, *Plastic Design in Steel*, Am. Inst. of Steel Construction, 1959.

form first at the supports C and B. After these hinges form, the system, on further increase in load, behaves as two simply supported beams AB and BC, until another plastic hinge forms, either at D or at E. Let us assume that D will become the next plastic hinge. Then the right-hand span becomes a mechanism and the limit load has been reached. At such time, we have for equilibrium

$$M_D = \frac{w_L l^2}{8} - M_P = M_P,$$

from which

$$w_L = \frac{16M_P}{l^2} = \frac{16 \times 237{,}000}{12 \times 12} = 26{,}400 \text{ lb/ft.}$$

With a plastic hinge at B, the reaction at A will be $wl/2 - M_P/l$ and the shear force at any section distance x to the right of A becomes

$$V_x = \frac{wl}{2} - \frac{M_P}{l} - wx.$$

Then to find the section of maximum bending moment in the span AB, we set this expression equal to zero and obtain

$$x = \frac{l}{2} - \frac{M_P}{wl}. \tag{c}$$

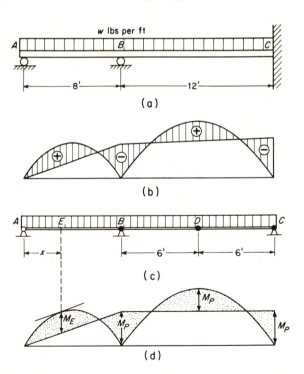

Fig. 9.22

The corresponding bending moment at E is

$$M_E = \left(\frac{wl}{2} - \frac{M_P}{l}\right)x - \frac{wx^2}{2} \tag{d}$$

Substituting the value of x from eq. (c) into eq. (d), we obtain

$$M_E = \frac{wl^2}{8} - \frac{M_P}{2}\left(1 - \frac{M_P}{wl^2}\right).$$

Taking $w = 26{,}400$ lb per ft, $l = 8$ ft, and $M_P = 237{,}000$ ft-lb, this gives $M_E = 109{,}300$ ft-lb. Since this is less than $M_P = 237{,}000$, we conclude that the right-hand span does collapse first and that $w_L = 26{,}400$ lb/ft is the limit load intensity for the continuous beam.

PROBLEM SET 9.5

1. A two-span continuous beam carries a uniformly distributed load as shown in Fig. A. If M_P is the plastic moment for the cross-section, find the intensity w_L of the limit load. *Ans.* $w_L = 11.66M_P/l^2$.

2. The continuous beam in Fig. A has a rectangular cross-section 2 in. wide by 4 in. deep, and $l = 6$ ft. The material is structural steel with a yield stress $\sigma_{\text{y.p.}} = 40{,}000$ psi. What is the numerical value of w_L? *Ans.* $w_L = 8640$ lb/ft.

3. Calculate the limit value P_L for the statically indeterminate beam shown in Fig. B, if the plastic moment M_P for the section is given. *Ans.* $P_L = 6M_P/l$.

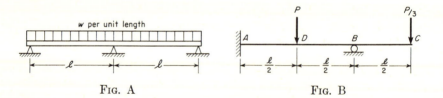

FIG. A   FIG. B

4. Calculate the limiting value P_L for the beam in Fig. B if the load at the end of the overhang is $P/6$ instead of $P/3$. *Ans.* $P_L = 7.2M_P/l$.

5. If the statically indeterminate beam in Fig. B carries a load P at D as shown and a load αP at C, what is the value of α to make the limit value of P a maximum and what is this maximum value? *Ans.* $P_L = 8M_P/l$ when $\alpha = \frac{1}{4}$.

6. A cantilever beam AB built-in at B is supported at A by a tie-rod and loaded with a force P as shown in Fig. C. The beam has a rectangular cross-section 2 in. wide by 1 in. deep, and the tie-rod has a circular cross-section of diameter d. Find the minimum diameter d of the tie-rod required to develop the full collapse load P_L for the beam. Both beam and tie-rod are made of structural steel with a yield stress of 40,000 psi. *Ans.* $d = 0.163$ in.

7. If the load P in Fig. C is applied at A instead of at C, what will be its limit value to produce collapse of the system? The tie-rod has diameter $d = \frac{5}{32}$ in., and all other data are the same as given in the preceding problem. *Ans.* $P_L = 1185$ lb.

8. Calculate the collapse load P_L for a beam with built-in ends and loaded at the middle as shown in Fig. D. The beam has an 8WF17 section and a span $l = 8$ ft. The material is structural steel with a yield stress of 36,000 psi. *Ans.* $P_L = 46{,}700$ lb.

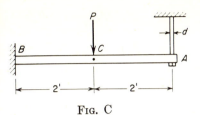

FIG. C

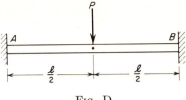

FIG. D

9. If the beam shown in Fig. D carries a uniformly distributed load of intensity w instead of the concentrated load P, find the limit value of w. Use all data as given in the preceding problem. *Ans.* $w_L = 11,700$ lb/ft.

10. A continuous frame ABC built-in at A and supported by a roller at C is subjected to the action of a horizontal force P at B as shown in Fig. E. Calculate the limit value P_L of this force if the uniform section of the frame can develop a plastic moment M_P. *Ans.* $P_L = 2M_P/l$.

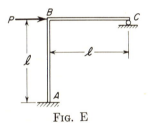

FIG. E

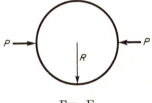

FIG. F

11. Referring to the continuous frame shown in Fig. E, assume that instead of the concentrated force P at B there is a uniformly distributed horizontal load of intensity w acting to the right on the vertical member AB. In such case, find the limit value of w necessary to produce complete collapse of the frame. *Ans.* $w_L = 4M_P/l^2$.

12. A thin steel ring of mean radius R is subjected to the action of two forces P as shown in Fig. F. The cross-section of the ring is such that the plastic moment is M_P. Find the limit load P_L for the ring. *Ans.* $P_L = 4M_P/R$.

13. If the horizontal force P in Fig. E acts at a height h above the ground level A instead of at B as shown, what will be its limit value? *Ans.* $P_L = 2M_P/h$.

14. In Fig. 9.19, p. 256, the beam AB is a steel bar with a 1-in. × 1-in.-square cross-section, the post CD is the same, and the cable ADB is a steel wire having diameter $d = \frac{1}{8}$ in. The angle $\alpha = 20°$ and the span $l = 4$ ft. Calculate the limit value of the load P if both beam and cable have a yield stress of 36,000 psi. *Ans.* $P_L = 1050$ lb.

10

Theory of Columns

10.1 Eccentric Loading of a Short Strut

Consider in Fig. 10.1a a short post or *strut* subjected to an eccentrically applied compressive force P at its upper end. If such a strut is comparatively short and stiff, the deflection due to bending action of the eccentric load will be negligible compared with the eccentricity e, and the principle of supperposition applies. The strut is assumed to have a plane of sym-

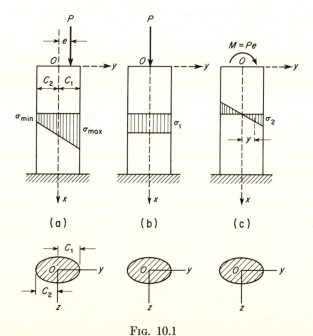

Fig. 10.1

metry (the xy-plane) and the load P lies in this plane at the distance e from the centroidal axis Ox. Such loading may be replaced by its statical equivalent of a centrally applied compressive force P (Fig. 10.1b) and a couple of moment $M = Pe$ (Fig. 10.1c). The centrally applied load produces uniform compressive stress $\sigma_1 = P/A$ over each cross-section as shown by the stress diagram in Fig. 10.1b. Likewise, the end moment M produces linearly varying bending stress $\sigma_2 = My/I$ as shown in Fig. 10.1c. Then, by superposition, the total compressive stress in any fiber due to combined bending and compression becomes

$$\sigma = \frac{P}{A} + \frac{Pey}{I}, \tag{10.1}$$

where A is the cross-sectional area and I is the moment of inertia of the section about the z-axis. Also, since we are most concerned here with compression, compressive stress is treated as positive. The stress diagram for this superposition of stresses is trapezoidal as shown in Fig. 10.1a.

Taking $y = c_1$, the distance to the extreme fiber on the right, and introducing the notation $r = \sqrt{I/A}$ for radius of gyration of the cross section, eq. (10.1) becomes

$$\sigma_{\max} = \frac{P}{A}\left(1 + \frac{ec_1}{r^2}\right). \tag{10.2a}$$

Similarly, taking $y = -c_2$, for the extreme fiber on the left, eq. (10.1) gives

$$\sigma_{\min} = \frac{P}{A}\left(1 - \frac{ec_2}{r^2}\right). \tag{10.2b}$$

Usually we are only interested in the maximum compressive stress as given by eq. (10.2a). However, it is of interest to note from eq. (10.2b) that if the ratio ec_2/r^2 becomes greater than unity, there will be tensile stresses in some fibers on the left side of the strut. For example, if the cross-section is a rectangle of width b in the z direction and height h in the y direction, we have $c_1 = c_2 = h/2$ and $r^2 = I/A = h^2/12$. Then from eq. (10.2b), we conclude that no tensile stress will be produced by eccentric compression so long as the condition

$$-\frac{h}{6} < e < \frac{h}{6} \tag{a}$$

is satisfied. If condition (a) is not fulfilled so that tensile stress does occur, it will always be smaller than the compressive stress obtained from eq. (10.2a), but even so, it may become important in the case of such materials as brick or concrete, which are weak in tension.

Let us consider now the more general case of a strut under the action of a compressive load P which does not lie in either of the two principal planes of bending. In Fig. 10.2, let Oy and Oz be the principal axes through

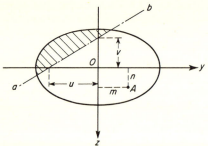

FIG. 10.2

the centroid of the cross-section and let A be the point of application of the load P. Denoting the coordinates of this point by m and n, the moments of P about the axes Oy and Oz will be Pn and Pm, respectively. Then, by superposition, the compressive stress σ for any point in the cross-section defined by coordinates z and y becomes

$$\sigma = \frac{P}{A} + \frac{Pmy}{I_z} + \frac{Pnz}{I_y}, \qquad (10.3)$$

analogous to eq. (10.1). Equating the right-hand side of this equation to zero, we obtain an equation for the locus of points of zero stress in the cross-section. With the notations $r_z = \sqrt{I_z/A}$ and $r_y = \sqrt{I_y/A}$ for radii of gyration, this equation of the line of zero stress becomes

$$\frac{my}{r_z^2} + \frac{nz}{r_y^2} + 1 = 0, \qquad (10.4)$$

which is seen to be the equation of a straight line, shown in Fig. 10.2 as the line ab. All longitudinal fibers situated in the unshaded portion of the cross-section will be in compression, while all those situated in the shaded portion will be in tension. The intercepts u and v of this line on the y- and z-axes are found from eq. (10.4) by setting first z and then y equal to zero. Thus

$$u = -\frac{r_z^2}{m}; \quad v = -\frac{r_y^2}{n}. \qquad (b)$$

For a given cross-section, it is of interest to define a region around the centroid within which the load P will induce compression over the entire section. This region is called the *core* of the section. The determination of this core for a rectangular section is discussed in the following example.

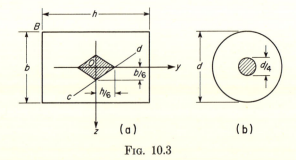

(a)   (b)

FIG. 10.3

EXAMPLE 1. Find the equation of the boundary of the core for the case of an eccentrically loaded strut of rectangular cross-section as shown in Fig. 10.3a.

SOLUTION. Let the point of application of the compressive force P be confined to the lower right-hand quadrant (m and n both positive). Then to make the line of zero stress pass through point B, we set $y = -h/2$ and $z = -b/2$ in eq. (10.4). This gives

$$\frac{mh}{2r_z^2} + \frac{nb}{2r_y^2} = 1,$$

which, with $r_z^2 = h^2/12$ and $r_y^2 = b^2/12$, becomes

$$\frac{6m}{h} + \frac{6n}{b} = 1. \tag{c}$$

This is the equation of a straight line cd having intercepts $h/6$ and $b/6$ on the y- and z-axes, respectively, as shown in Fig. 10.3a. As long as the point of application of the load P is situated between this line and point B, the fiber stress at B will be compression. Similar arguments, when the load P is confined to each of the other three quadrants, can be made and we conclude that the core of the section is the shaded rhombus shown in Fig. 10.3a.

A similar argument can be made for the case of a circular cross-section of diameter d and we find that the core of the section is a circle of diameter $d/4$ as shown in Fig. 10.3b.

PROBLEMSET 10.1

1. A short strut of square cross-section has a notch cut in its side as shown in Fig. A. Calculate the maximum compressive stress σ at the section mn due to a centrally applied load P. Ans. $\sigma_{max} = 8P/a^2$.

2. Solve the preceding problem if the cross-section of the strut is a circle of diameter a. Ans. $\sigma_{max} = 9.1P/a^2$.

3. At the cross-section mn, the C-clamp shown in Fig. B has a rectangular cross-section $1 \times \frac{1}{2}$ in. Determine the maximum tensile stress σ at n if the load $P = 400$ lb and the dimension $b = 3$ in. Ans. $\sigma = 15,200$ psi.

4. A short piece of 10$\lfloor$ 20 channel carries a compressive load the line of action of which passes through the centroid of the web. If the allowable maximum compressive stress is 16,000 psi, calculate the safe load P. Ans. $P = 61,500$ lb.

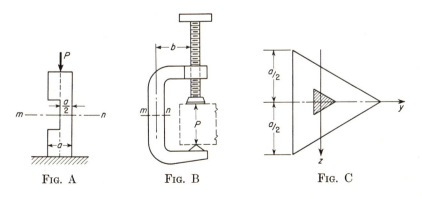

FIG. A   FIG. B   FIG. C

5. A short piece of $6 \times 6 \times 1$-in. angle iron carries a compressive load the line of action of which coincides with the intersection of the middle planes of the legs. If the maximum compressive stress is not to exceed 16,000 psi, what is the safe load P? *Ans.* $P = 37,500$ lb.

6. Solve the preceding problem if the angle iron has an $8 \times 4 \times 1$-in. cross-section. *Ans.* $P = 37,300$ lb.

7. Show that the core of a circular cross-section is circular and find its radius if d is the diameter of the section. *Ans.* $d/8$.

8. A short strut loaded in compression has an equilateral triangular cross-section with sides of length a (Fig. C). Prove that the core of the section is an equilateral triangle with sides of length $a/4$ as shown in Fig. C.

9. A short strut has a hollow circular cross-section of outside diameter d_1 and inside diameter d_2. Prove that the core of the section is a circle having radius $r_c = \frac{1}{8}(d_1 + d_2{}^2/d_1)$.

10.2   Long Columns; Euler's Column Formula

Let us consider now the case of a long slender column AB of length l which is built-in at its lower end A and subjected to a centrally applied compressive load P at its upper end B as shown in Fig. 10.4. This column is assumed to be perfectly straight and of uniform cross-section. It is also assumed that the material is homogeneous and that it behaves elastically.

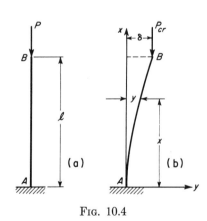

Experience shows that when the vertical load P is small, such a compressed column is laterally stable. That is, if the upper end B is pushed slightly to one side by a lateral force, the column will return to its straight form as soon as this lateral force is removed. However, as P is gradually increased, we observe that at a certain value of this load, the straight form of equilibrium becomes unstable and the column, if pushed to one side, stays there even after the lateral force is removed. This instability phenomenon is called *lateral buckling* and the value of the load at which it occurs is called the *critical load*, denoted by P_{cr}.

Fig. 10.4

To find the load P_{cr} which will cause buckling, we consider the column in the slightly bent configuration shown in Fig. 10.4b and calculate the magnitude of the vertical load necessary to hold it there. Choosing coordinate axes through point A, as shown in Fig. 10.4b, we denote the deflection at any point x on the elastic line by y and the deflection of point B by δ. Then the bending moment at the cross-section defined by x becomes

$$M_x = P(\delta - y). \tag{a}$$

Since we consider only a very small lateral deflection, the relation between bending moment and curvature is (see eq. 8.1, p.198)

$$EI \frac{d^2y}{dx^2} = \pm M.$$

For the chosen axes x and y, d^2y/dx^2 is positive in this case and we use the plus sign in this expression. Hence the differential equation of the elastic line becomes

$$EI \frac{d^2y}{dx^2} = P(\delta - y). \tag{b}$$

Dividing both sides of eq. (b) by EI and introducing the notation

$$k^2 = \frac{P}{EI}, \tag{c}$$

it may be written in the form

$$\frac{d^2y}{dx^2} + k^2y = k^2\delta. \tag{d}$$

This is a linear differential equation with constant coefficients. Its solution $y = f(x)$ consists of two parts :(1) the general solution of the corresponding homogeneous equation (zero on the right-hand side) and (2) a particular solution of the complete equation. It can easily be verified by substitution that a general solution of the homogeneous equation is

$$y_1 = C_1 \sin kx + C_2 \cos kx,$$

where C_1 and C_2 are arbitrary constants, while a particular solution of the complete equation is

$$y_2 = \delta.$$

Thus the complete solution becomes $y = y_1 + y_2$ or

$$y = C_1 \sin kx + C_2 \cos kx + \delta. \tag{e}$$

To find the integration constants C_1 and C_2, we have the boundary conditions at the built-in end A:

$$x = 0, \quad y = 0, \quad \frac{dy}{dx} = 0. \tag{f}$$

One differentiation of eq. (e) with respect to x gives

$$\frac{dy}{dx} = C_1 k \cos kx - C_2 k \sin kx. \tag{g}$$

Now substituting conditions (f) into eqs. (e) and (g), we find

$$C_1 = 0, \quad C_2 = -\delta,$$

and the general solution (e) becomes

$$y = \delta(1 - \cos kx). \tag{h}$$

This shows that the column bends so that its elastic line takes the form of a cosine curve, the amplitude δ of which is as yet undetermined.

To examine this question further, we note that when $x = l$, $y = \delta$. Substituting this condition into eq. (h), we obtain

$$\delta = \delta(1 - \cos kl)$$

from which

$$\delta \cos kl = 0. \tag{i}$$

From eq. (i), we conclude that either $\delta = 0$ or $\cos kl = 0$. If $\delta = 0$, the column stands in the straight vertical configuration of equilibrium and no limitation is imposed on the magnitude of the load P. On the other hand, if $\cos kl = 0$, i.e., if

$$kl = \frac{n\pi}{2}, \tag{j}$$

where n is an odd integer, then the column can also be in equilibrium with any small deflection δ at the top. The smallest value of P for which this can occur will be obtained by taking $n = 1$ in eq. (j). Then, using the notation (c), we obtain

$$\sqrt{\frac{P_{cr}}{EI}}\, l = \frac{\pi}{2},$$

from which

$$P_{cr} = \frac{\pi^2 EI}{4l^2}. \tag{10.5}$$

This is Euler's column formula for the case of a slender column built-in at its base and free at the top. For all values of $P < P_{cr}$, we must have $\delta = 0$ and the straight configuration of equilibrium shown in Fig. 10.4a is said to be *stable*. If point B is pushed to the side and then released, the column will return to the vertical position. If $P = P_{cr}$, equilibrium can exist for any small value of δ and the vertical configuration is said to be indifferent or *neutral*. In this case, if point B is pushed to the side and then released, the column simply remains in the slightly deflected position. If $P > P_{cr}$, and point B is pushed slightly to the side, the deflection increases indefinitely and the vertical equilibrium configuration is said to be *unstable*. Thus the value P_{cr} as defined by eq. (10.5) represents the maximum load that the column can carry; it is sometimes called the *Euler load* for the column.

Examination of eq. (10.5) shows that the critical load for a given column is proportional to the flexural rigidity EI and inversely proportional to the square of the length l of the column. It will also be noted that the critical load is independent of the compressive strength of the material. Thus two geometrically identical slender columns, one of high-strength alloy steel and the other of ordinary structural steel, will fail by buckling at approximately the same value of the load P, since the moduli of elasticity are approximately the same. It must therefore be realized that the "strength" of a long column is in no way dependent upon the strength of its material in compression, but only on its geometry and the stiffness of the material.

Equation (10.5) shows also that the strength of a long column may be increased by increasing the moment of inertia I of its cross-section. Without increasing the cross-sectional area, this may be done by distributing the material as far as possible from the principal axes of the cross-section. Hence tubular sections are more economical for compression members than solid sections. By diminishing the wall thickness of such sections and increasing the transverse dimensions, the stability of the column may be increased. There is a lower limit for the wall thickness, however, below which the wall itself may become unstable and, instead of buckling of the column as a whole, there will be buckling of its longitudinal elements which brings about a corrugation of the wall, called local buckling.

If we do not limit our attention to the case where $n = 1$, eq. (j) is seen to define an infinite number of values for the critical load. Thus, in general,

$$P_{\text{cr}} = \frac{n^2 \pi^2 EI}{4l^2}. \tag{k}$$

Taking $n = 1, 3, 5, \ldots$, we see that the values of P_{cr} are in the ratio $1{:}9{:}25{:}\ldots$. Correspondingly, eq. (h) for the elastic line becomes

$$y = \delta\left(1 - \cos\frac{n\pi x}{2l}\right), \tag{l}$$

and we see that as n is increased, the deflection curve has more and more inflection points. Two of these deflection curves, for $n = 3$ and $n = 5$, are shown in Fig. 10.5. Although they represent theoretically possible

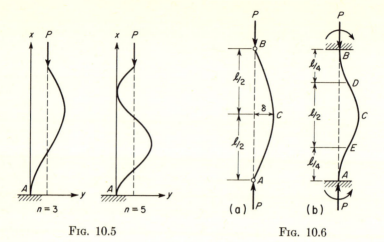

FIG. 10.5   FIG. 10.6

modes of buckling for the columns, they are of no practical interest because the column will always buckle in the first mode, as represented in Fig. 10.4b, as soon as P reaches the first critical value given by eq. (10.5).

Critical loads for long columns having other end conditions than those of the simple cantilever column in Fig. 10.4 can easily be obtained from eq. (10.5). Consider, for example, the long column with pinned ends as shown in Fig. 10.6a. From symmetry, it may be concluded that in the first mode of buckling the elastic line will have a vertical tangent at its mid-point C. Hence, each half of the column is in the same condition as that in Fig. 10.4 and the critical load is obtained from eq. (10.5) simply by replacing l by $l/2$. Thus

$$P_{\mathrm{cr}} = \frac{\pi^2 EI}{l^2}. \tag{10.6}$$

Similarly, for a long column having both ends built-in, Fig. 10.6b, the first buckling mode will be a full cosine wave having inflection points at D and E. Then each end portion behaves as a simple cantilever column of length $l/4$ and we obtain, from eq. (10.5)

$$P_{\mathrm{cr}} = \frac{4\pi^2 EI}{l^2}. \tag{10.7}$$

The case of a column built-in at one end and pinned against lateral deflection at the other end requires some further consideration and is discussed in the following example.

EXAMPLE 1. A slender column of length l is built-in at its lower end A and laterally supported at its upper end B as shown in Fig. 10.7. Find the first critical value of the compressive load P.

SOLUTION. In this case, a horizontal reaction Q will be induced at the pinned end B when buckling takes place. Then the bending moment at section x is

$$M_x = Q(l - x) - Py,$$

and the differential equation of the deflection curve becomes

$$EI\frac{d^2y}{dx^2} = -Py + Q(l - x).$$

The general solution of this equation is

$$y = C_1 \cos kx + C_2 \sin kx + \frac{Q}{P}(l - x),$$

where $$k = \sqrt{\frac{P}{EI}}.$$

For determining the constants C_1 and C_2 and the unknown reaction Q, we have the following conditions at the ends:

$$(y)_{x=0} = 0; \quad (y)_{x=l} = 0; \quad \left(\frac{dy}{dx}\right)_{x=0} = 0.$$

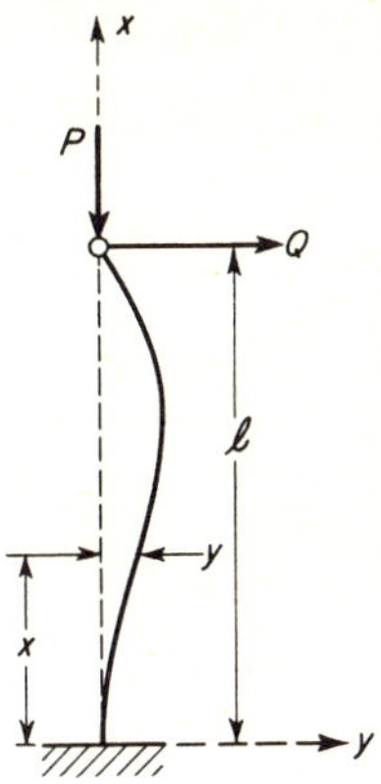

FIG. 10.7

Then, from the above equation for y,

$$C_1 + \frac{Ql}{P} = 0; \quad C_1 \cos kl + C_2 \sin kl = 0; \quad kC_2 - \frac{Q}{P} = 0.$$

Substituting the values of C_1 and C_2 from the first and third of the above expressions into the second one, there results

$$\frac{Q}{P}\left(\frac{1}{k} \sin kl - l \cos kl\right) = 0.$$

As long as the critical load has not been reached, there is no buckling and Q is zero. When the critical load has been reached and buckling is present, Q is no longer zero and hence the expression within the parentheses must be zero. This gives the following transcendental equation for determining the critical load:

$$\tan kl = kl.$$

The smallest value of kl, different from zero, and therefore of P, which satisfies this equation is $kl = 4.49$. Since $k^2 = P/EI$,

$$P_{cr} = k^2 EI = \frac{20.2EI}{l^2} = \frac{\pi^2 EI}{(0.7l)^2}. \tag{m}$$

PROBLEM SET 10.2

1. A 6-ft length of 6I-12.5 standard I-beam is to be used as a pin-ended column. Calculate the critical value of the load. *Ans.* $P_{cr} = 103,000$ lb.

2. A 14WF78 wide-flange steel beam is to be used as a column 45 ft long with the lower end built-in and the upper end pinned. Calculate the maximum compressive load that the column can carry. *Ans.* $P_{max} = 428,500$ lb.

3. The member AB of the simple truss in Fig. A is a 5-ft length of 3-U-4.1 steel channel section with pinned ends and the member BC is a steel rod of circular cross-section having diameter $d = \frac{3}{4}$ in. Find the largest value of the vertical load P that the structure can support if $\sigma_{y.p.} = 40,000$ psi and $E_s = 30 \,(10)^6$ psi. *Ans.* $P_{max} = 13,150$ lb.

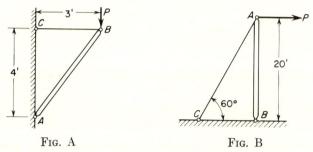

FIG. A   FIG. B

4. The pole AB in Fig. B is a pin-ended Douglas Fir column having a circular cross-section of diameter $d = 8$ in. Assuming that the guy wire AC is strong enough, find the maximum value of the horizontal force P that can be applied at the top of the pole without causing the structure to collapse. *Ans.* $P_{max} = 31,800$ lb.

5. For the simple pin-connected truss shown in Fig. C, the members AC and BC are slender pin-ended steel bars having identical cross-sections. Find the value of the angle θ, defining the direction of the applied force P, required to make the critical value of this force as large as possible. Assume buckling in the plane of the figure. *Ans.* $\theta = 18° \, 26'$.

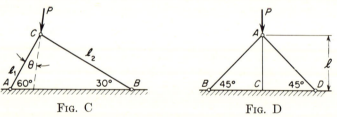

FIG. C   FIG. D

6. In Fig. D, the bars AB, AC, and AD are slender steel rods of circular cross-section which all have the same flexural rigidity EI. They have pinned ends at A, B, and D and a built-in end at C. Calculate the critical value of the vertical load P applied at A. Assume buckling of the bars to take place in the plane of the figure. *Ans.* $P_{cr} = 2.75(\pi^2 EI/l^2)$.

7. A so-called beam-column carries a transverse load $2Q$ at mid-span and a compressive end load P as shown in Fig. E. For $P < P_{cr} = \pi^2 EI/l^2$, set up the dif-

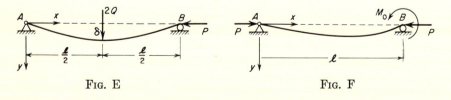

FIG. E   FIG. F

ferential equation of the elastic line and calculate the vertical deflection δ at midspan. The beam has uniform flexural rigidity EI and length l. Assume pinned ends at A and B. *Ans.*

$$\delta = \frac{Ql}{2P}\left(\frac{\tan kl/2}{kl/2} - 1\right), \quad \text{where} \quad k = \sqrt{P/EI}.$$

8. A simply supported beam-column of uniform flexural rigidity EI is subjected to compressive loads P at both ends and to a couple of moment M_0 at the end B as shown in Fig. F. Set up the differential equation of the deflection curve and find its solution $y = f(x)$ for $P < P_{cr}$. *Ans.*

$$y = \frac{M_0}{P}\left(\frac{\sin kx}{\sin kl} - \frac{x}{l}\right), \quad \text{where} \quad k = \sqrt{P/EI}.$$

10.3 Further Discussion of Euler's Column Formula

The case of a column with pinned ends (Fig. 10.6a) is most frequently encountered and is called the *fundamental case.* Dividing both sides of eq. (10.6) for this case by the cross-sectional area A and introducing the notation $r = \sqrt{I/A}$, for radius of gyration, we obtain

$$\sigma_{cr} = \left(\frac{P}{A}\right)_{cr} = \frac{\pi^2 E}{\left(\frac{l}{r}\right)^2}, \tag{10.8}$$

where the ratio l/r is called the *slenderness ratio* of the column and σ_{cr}, the *critical compressive stress.* The curve ABC in Fig. 10.8, plotted from eq. (10.8) is called *Euler's curve.* For a column of any given slenderness ratio, plotted as abscissa, this curve shows the corresponding value of the average compressive stress $(P/A)_{cr}$, plotted as ordinates, for which the column becomes laterally unstable.

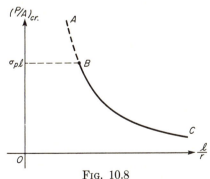

FIG. 10.8

Since it was assumed that the material behaves elastically at the beginning of lateral buckling, we conclude that eq. (10.8) is valid only if it gives a value of σ_{cr} less than the proportional limit $\sigma_{p.l.}$ of the material.

Thus, by setting $\sigma_{cr} = \sigma_{p.l.}$, we obtain from eq. (10.8), a limiting value of l/r below which Euler's formula does not apply. This is represented by point B on the curve in Fig. 10.8, and marks the boundary between so-called short and long columns. Taking structural steel with a proportional limit $\sigma_{p.l.} = 30,000$ psi and a modulus of elasticity $E = 30(10)^6$ psi, for example, we find $l/r = \sqrt{1000\,\pi^2} \approx 100$. Thus for $l/r < 100$, the average compressive stress in a pin-ended steel column will reach the proportional limit before lateral buckling can occur and eq. (10.8) is inapplicable. For this reason, the portion AB of the Euler curve in Fig. 10.8 is shown by a dotted line; only the portion BC is valid.

The Euler curve for the fundamental case of a column with pinned ends can be used for columns with other end conditions by introducing a *modified length* l_1. Comparing eq. (10.5) with eq. (10.6), we see that the modified length of a cantilever column will be $l_1 = 2l$, where l is the actual length of the column. Similarly, from eq. (10.7), we see that the modified length of a column with both ends built-in will be $l_1 = l/2$, while for a column built-in at one end and pinned at the other, $l_1 = 0.7l$ (see p. 273). In general, then, eq. (10.8) may be used in the form

$$\sigma_{cr} = \frac{\pi^2 E}{(l_1/r)^2}. \tag{10.8a}$$

We have seen above that for columns having sufficiently high slenderness ratios, buckling occurs before the average compressive stress can reach the proportional limit of the material. In all such cases, stability of the column governs the selection of a safe working load and we use the portion BC of the Euler curve as shown in Fig. 10.9 as a basis of design. As another extreme, columns having very low slenderness ratios can be expected to

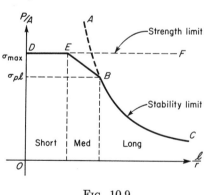

FIG. 10.9

fail due to some kind of weakness of the material before the critical compressive stress can be reached. This failure associated with the strength of the material may result from crushing, as in the case of concrete, or yielding, as in the case of structural steel. In any case, some maximum compressive stress can be set as a limit of strength and the safe working load chosen accordingly. This, we can represent in Fig. 10.9 by a horizontal line DF drawn through the chosen *strength limit*.

In between the ranges of short columns and long columns there will be a range of slenderness ratios too small for true elastic instability to govern and yet too large for strength considerations alone to govern. Columns falling within this range of slenderness ratios require special consideration and are called *medium columns*. Analysis of their behavior under centrally applied compressive loads has been widely treated both experimentally and theoretically, but a discussion of such behavior will not be given here.* A simple and satisfactory procedure, in these cases, is simply to draw a straight line *EB* in Fig. 10.9 and arbitrarily let its ordinates represent the maximum compressive stress for such medium-length columns. In this way we obtain the broken line *DEBC* in Fig. 10.9, which can be used as a basis of design for a column of any length.

As a specific example, let us consider the case of ordinary structural steel for which $E = 30(10)^6$ psi, $\sigma_{p.l.} = 30{,}000$ psi, and $\sigma_{y.p.} = 40{,}000$ psi, Then for all practical purposes, the yield stress $\sigma_{y.p.}$ will represent the strength limit of the material for very short columns. Furthermore, experiments show that $l/r = 60$ is the upper limit of slenderness ratios for which a column can sustain this average compressive stress without buckling. Thus for slenderness ratios from 0 to 60, we take $\sigma_{\max} = \sigma_{y.p.} = 40{,}000$ psi as shown by the line *DE* in Fig. 10.10 For slenderness ratios greater than 100, $\sigma_{\max} = \sigma_{cr}$, and we use the portion *BC* of the Euler curve. For intermediate slenderness ratios ($60 < l/r < 100$), the straight line *EB* is used. Thus the complete line *DEBC* in Fig. 10.10 determines the value of $\sigma_{\max}$ for any given pin-ended mild steel column.

In discussing working stresses for steel columns, we must consider the stress given by the diagram in Fig. 10.10 as an ultimate stress. Such a compressive stress brings the columns to complete failure, either by buckling, yielding, or a combination of both, and the working stress should be

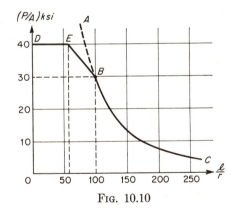

Fig. 10.10

---

*See Timoshenko and Gere, *Theory of Elastic Stability*, McGraw-Hill Book Co., Inc., New York, 1961, p. 175.

taken as $\sigma_{\max}/n$ where n is the desired factor of safety. The value to be chosen for this factor of safety depends upon unforeseen or accidental increases in the load P and also on possible errors in the central application of this load, as well as possible initial crookedness of the column. A common value for n in structural work is 2.5.

Experiments show that both errors in central application of the load and initial crookedness of the column tend to increase with increasing slenderness ratios. This suggests the use of a variable factor of safety which is usually increased linearly with l/r. For example, some structural specifications recommend using a factor of safety $n = 2.0 + 0.015\ l/r$ for $0 < l/r < 100$, and $n = 3.5$ for $l/r > 100$.

In the whole realm of structural design, there is perhaps no situation that presents so troublesome a question in regard to the proper choice of a factor of safety as that of columns. As already indicated, this stems from the fact that inherent inaccuracies in centering the load and in attaining assumed end conditions, as well as initial crookedness of the column, have a pronounced effect upon its behavior under compression. Many thousands of columns have been tested over the years in an attempt to resolve this question. The results of such tests when plotted as shown in Fig. 10.11 form a wide band and leave the question of proper factor of

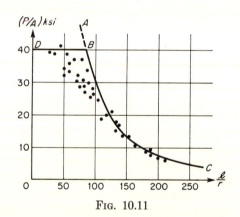

FIG. 10.11

safety very much unsettled. One point, however, is clearly brought out by these test results; namely, the broken line DBC, constructed as already explained, represents an upper bound for the test results. This substantiates the theory and leaves the pronounced but indeterminate influence of imperfections simply to be adequately provided for by the use of a generous factor of safety.

1. A steel bar of rectangular cross-section 1×2 in. is to be used as a column with pinned ends. What is the shortest length l for which Euler's equation applies if $E = 30\ (10)^6$ psi and $\sigma_{p.l.} = 30{,}000$ psi? *Ans.* $l = 28.7$ in.

2. Calculate the critical compressive stress for the column described in the preceding problem if it is 4 ft long. *Ans.* $\sigma_{cr} = 10{,}700$ psi.

3. An extruded nickel steel tube having outside diameter $d = 2$ in. and wall thickness $t = \frac{1}{8}$ in. is used as a pin-ended column 12 ft long. Calculate the critical compressive stress if $E = 30\ (10)^6$ psi. *Ans.* $\sigma_{cr} = 6300$ psi.

4. What is the shortest length l of the column described in the preceding problem for which Euler's formula can be applied if $\sigma_{p.l.} = 160{,}000$ psi? *Ans.* $l = 28.5$ in.

5. Calculate the critical compressive stress for the column described in Problem 3, if the material is extruded magnesium for which $E = 6.5\ (10)^6$ psi. *Ans.* $\sigma_{cr} = 1360$ psi.

6. A steel column with built-in ends is 6 ft long and has a standard 6I-12.5 section. Using the broken line $DEBC$ in Fig. 10.10 and a factor of safety $n = 3$, find the safe load P_w. *Ans.* $P_w = 48{,}200$ lb.

7. Solve the preceding problem if the column is built-in at one end and pinned at the other. *Ans.* $P_w = 45{,}200$ lb.

8. Construct a curve similar to the line $DEBC$ in Fig. 10.10 to be used as a basis of design for columns of extruded aluminum alloy having $E = 10\ (10)^6$ psi, $\sigma_{p.l.} = 20{,}000$ psi, and $\sigma_{max} = 30{,}000$ psi. Assume that $l/r = 30$ is the upper short-column limit and $l/r = 70$ is the lower long-column limit.

10.4 The Secant Formula

Referring to Fig. 10.12, let us consider the case of a perfectly straight cantilever column subjected to a compressive load P applied with ec-

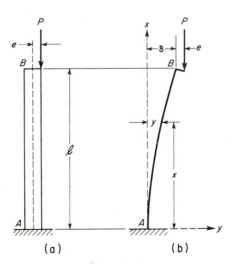

Fig. 10.12

centricity e. This case differs from that discussed in Art. 10.1 only in the fact that we now assume the column to be slender so that the lateral deflections cannot always be considered as small compared with the eccentricity e of the applied load. As the load P is gradually increased, the column will start to bend slightly, and for any value of P the deflection curve will be as shown in Fig. 10.12b.

Choosing coordinate axes x and y as shown and denoting by δ the deflection of the upper end, the bending moment at any cross-section defined by x will be

$$M_x = P(\delta + e - y). \tag{a}$$

Then for small deflections within the elastic limit of the material, the differential equation of the elastic line becomes

$$EI \frac{d^2y}{dx^2} = P(\delta + e - y). \tag{b}$$

Dividing both sides of this equation by EI and using the notation

$$k^2 = \frac{P}{EI}, \tag{c}$$

it may be written in the form

$$\frac{d^2y}{dx^2} + k^2y = k^2(\delta + e). \tag{d}$$

This equation is similar to eq. (d) on p. 269 and its solution is

$$y = C_1 \sin kx + C_2 \cos kx + \delta + e. \tag{e}$$

Noting that when $x = 0$: $y = 0$, $dy/dx = 0$, we find

$$C_1 = 0, \quad C_2 = -(\delta + e)$$

and the solution (e) becomes

$$y = (\delta + e)(1 - \cos kx). \tag{f}$$

This shows that the deflection curve AB has the form of a quarter cosine wave.

At the upper end B we have $y = \delta$ when $x = l$. Substituting these simultaneous values into eq. (f), we obtain

$$\delta = e \left(\frac{1 - \cos kl}{\cos kl} \right). \tag{10.9}$$

For a given value of e, this equation expresses the relation between the lateral deflection δ and the load P, indirectly defined by $k = \sqrt{P/EI}$.

Taking $e = 0$, we see that $\delta = 0$ for any value of $kl < \pi/2$, i.e., any value of $P < P_{cr}$ (see eq. 10.5, p. 270). But when $kl = \pi/2$, i.e., when $P = P_{cr}$, the deflection becomes indeterminate; it can have any small value. These observations agree with those already reached in Art. 10.2 for the case of a centrally loaded column. However, for any small value of $e > 0$, eq. (10.9) gives a definite value for δ so long as $P < P_{cr}$ and shows that δ increases without limit as P approaches P_{cr}. The load-deflection relationship expressed by eq. (10.9) is seen to be nonlinear.

For a chosen value of the eccentricity e and a given column, eq. (10.9) can be represented graphically as shown in Fig. 10.13. For $e = 0$, the load-deflection curve is represented by the two straight lines OAB. No

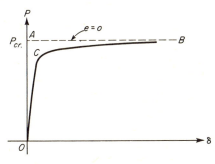

Fig. 10.13

deflection takes place for $P < P_{cr}$ and at $P = P_{cr}$ the deflection is indeterminate. For $e > 0$, but small, the load-deflection relation will be represented by the curve OCB. For small loads, the deflection increases slowly but as the load approaches the critical value, it increases rapidly and tends toward infinity.

For small values of $kl \ll \pi/2$, i.e., for values of $P \ll P_{cr}$, we may take

$$\cos kl \approx 1 - \frac{(kl)^2}{2},$$

in eq. (10.9). Then

$$\delta \approx e\,\frac{(kl)^2/2}{1 - (kl)^2/2} \approx e\,\frac{(kl)^2}{2},$$

or, since $k = \sqrt{P/EI}$,

$$\delta \approx \frac{(Pe)l^2}{2EI}. \tag{g}$$

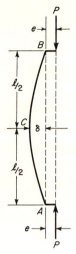

FIG. 10.14

This is seen to agree with the deflection of a cantilever beam loaded by a couple $M = Pe$ at the free end (see Case 5 of Table 8.1, p. 212). This justifies the treatment of eccentrically loaded short struts as discussed in Art. 10.1.

In the present case of an eccentrically loaded long column, it is important to note that there is no proportionality between the load P and the deflection δ that it produces; hence the method of superposition cannot be used. An axially applied load P when acting alone produces no lateral deflection, but when acting in company with a bending couple Pe, it produces also some bending and the resulting deflections are greater than the bending couple alone would produce. This magnification is especially pronounced if the axial force is near the Euler value.

All of the foregoing discussion can be applied also to the fundamental case of a pin-ended column eccentrically loaded as shown in Fig. 10.14. From symmetry conditions, the tangent to the elastic line at C remains vertical and each half of the column is in the same condition as the column in Fig. 10.12. Hence, upon replacing l by $l/2$ in eq. (10.9), we obtain, for the deflection δ of the mid-point C of the elastic line,

$$\delta = e\left(\frac{1 - \cos\dfrac{kl}{2}}{\cos\dfrac{kl}{2}}\right) = e\left[\sec\frac{kl}{2} - 1\right], \tag{h}$$

where l is now the full length of the pin-ended column.

The maximum bending moment in Fig. 10.14 occurs at the mid-section C and has the magnitude

$$M_{\max} = P(\delta + e). \tag{i}$$

Substituting expression (h) for δ, this becomes

$$M_{\max} = Pe \sec\frac{kl}{2}. \tag{j}$$

The corresponding maximum compressive stress on the concave side of the column at C is

$$\sigma_{\max} = \frac{P}{A} + \frac{M_{\max}c}{I}. \tag{k}$$

Substituting expression (j) for $M_{\max}$ and using the notation $r = \sqrt{I/A}$ for radius of gyration of the cross-section, this becomes

$$\sigma_{\max} = \frac{P}{A}\left[1 + \frac{ec}{r^2}\sec\left(\frac{kl}{2}\right)\right]. \tag{10.10a}$$

If k is replaced by $\sqrt{P/EI}$, we have

$$\sigma_{\max} = \frac{P}{A}\left[1 + \frac{ec}{r^2}\sec\left(\frac{l}{2r}\sqrt{\frac{P}{AE}}\right)\right]. \qquad (10.10b)$$

Equation (10.10) represents the so-called *secant formula* for an eccentrically loaded slender column. It gives the maximum compressive stress in terms of the *average compressive stress* P/A, the *eccentricity ratio* ec/r^2, and the so-called *Euler angle* $kl/2$. If the column is short, and the load P is small, we have sec $kl/2 \approx 1$, and eq. (10.10) reduces to the same form as eq. (10.2a) of Art. 10.1.

If we set some limit on $\sigma_{\max}$, say $\sigma_{\max} = \sigma_{\text{y.p.}}$ in the case of structural steel, we may then calculate the corresponding average compressive stress which will first produce yielding in the most severely stressed fibers. For this purpose, eq. (10.10) is written in the form

$$\frac{P_{\text{y.p.}}}{A} = \frac{\sigma_{\text{y.p.}}}{1 + \dfrac{ec}{r^2}\sec\left(\dfrac{l}{2r}\sqrt{\dfrac{P_{\text{y.p.}}}{AE}}\right)}. \qquad (10.11)$$

For any chosen eccentricity ratio, we may solve eq. (10.11) by trial and error and plot a curve which shows, as a function of l/r, the value of P/A at which yielding first begins in the most stressed fibers. A family of such curves for several values of the eccentricity ratio ec/r^2 and for $\sigma_{\text{y.p.}} =$ 40,000 psi is shown in Fig. 10.15.

We usually consider the value of P/A for which yielding in an extreme fiber begins as the limit load for a steel column. Having found this value for a given column from the graph in Fig. 10.15, or from eq. (10.11), the

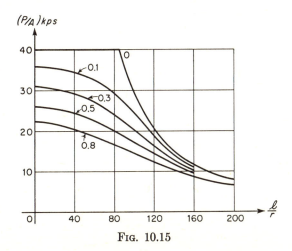

FIG. 10.15

safe average compressive stress will be obtained by dividing the obtained value by a suitable factor of safety n.

If desired, the factor of safety can be incorporated directly into eq. (10.11) as follows. Let P_w denote the safe working load, and n the required factor of safety. Then the compressive load which would bring the extreme fiber stress to the yield point will be nP_w. Substituting this for P in eq. (10.11), we obtain

$$\frac{P_w}{A} = \frac{\sigma_{y.p.}/n}{1 + \frac{ec}{r^2}\sec\left(\frac{l}{2r}\sqrt{\frac{nP_w}{AE}}\right)} \qquad (10.12)$$

Comparison of this equation with eq. (10.10) shows that the introduction of nP_w in place of P in the denominator of eq. (10.10) takes proper account of the fact that σ_{max} does not increase linearly with P.

The transcendental character of the secant formula makes it rather difficult to solve, trial and error procedures usually being necessary. Such procedure is illustrated in the following example.

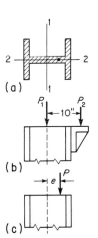

(a)

(b)

(c)

FIG. 10.16

EXAMPLE 1. A wide-flange steel section (Fig. 10.16a) is to be selected for a column 25 ft long and having pinned ends. The column is to carry centrally applied loads $P_1 = 100$ kips and bracketed loads $P_2 = 50$ kips, the latter applied on the principal axis 2-2 of the section at a distance of 10 in. from axis 1-1, Fig. 10.16b. Select a suitable section such that the column will have a factor of safety $n = 2.5$ against failure by yielding ($\sigma_{y.p.} = 40,000$ psi).

SOLUTION. The two loads applied as shown in Fig. 10.16b can be replaced by a single load $P = 150$ kips applied with eccentricity $e = 3.33$ in. as shown in Fig. 10.16c. Then to use the secant formula as a basis of selecting the required section, we write it in the form

$$\frac{\sigma_{y.p.}}{n} = \frac{P}{A} + \frac{Pe}{Z}\sec\left(\frac{l}{2}\sqrt{\frac{nP}{EI}}\right), \qquad (1)$$

where Z is the section modulus and I the moment of inertia of the section. By trial and error, we now look for a WF section which will make the right side of eq. (1) equal to or less than the left side, i.e., equal to or less than $\sigma_{y.p.}/n = 40,000 \div 2.5 = 16,000$ psi. Using Table B.2 of Appendix B, we find

For a 14WF68 section: Right side = 13,500 psi.
For a 14WF61 section: Right side = 15,270 psi.
For a 14WF53 section: Right side = 18,160 psi.

This indicates that the 14WF61 section is suitable so far as bending in the principal plane of loading is concerned. We assume that buckling in the plane of smaller flexural rigidity is prevented.

PROBLEM SET 10.4

1. A 14WF78 section steel column with pinned-end conditions is 25 ft long. It carries at its ends centrally applied compressive loads $P_1 = 200$ kips and bracketed compressive loads $P_2 = 20$ kips on axis 2-2, 12 in. from axis 1-1 (see Fig. 10.16). Using the secant formula, calculate the maximum extreme fiber stress in compression. *Ans.* $\sigma_{max} = 11{,}800$ psi.

2. If the yield stress for steel is 40,000 psi, what factor of safety n does the column in the preceding problem have against failure due to yielding in the most stressed extreme fiber? *Ans.* $n = 3.22$.

3. Select a WF section for the column described in the two preceding problems so that the factor of safety against yielding in the most stressed extreme fiber will be $n = 2.5$. Assume $\sigma_{y.p.} = 40{,}000$ psi. *Ans.* 14WF61.

4. A pin-ended column AB of length l is subjected to axial compression as shown in Fig. A. The load P at A is centrally applied, while that at B is applied with eccentricity e. For small values of P, the maximum bending moment in the column occurs at B and is simply $M_{max} = Pe$. Find the value of the average compressive stress P/A above which the maximum bending moment will occur at some intermediate point between A and B and will, in consequence, be greater than Pe. *Hint:* See answer to Problem 10.2-8 for the equation of the elastic line in this case. *Ans.* $P/A = \pi^2 E/4(l/r)^2$.

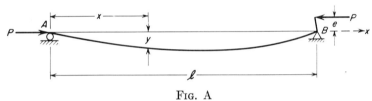

FIG. A

5. An 8WF40 steel section is used for a pin-ended column 20 ft long which is to carry compressive loads $P = 150$ kips. The load P is centrally applied at the lower end of the column and bracketed at the upper end with an eccentricity $e = 1$ in. from axis 2-2 on axis 1-1. Calculate the maximum extreme fiber stress induced in the column. *Ans.* $\sigma_{max} = 31{,}600$ psi.

10.5   Representation of Imperfections by Equivalent Eccentricity

In Art. 10.3 it has been pointed out that the behavior of a column under compressive load is very much affected by *imperfections* such as slight crookedness of the column axis, errors in centering the load, uncertainty as to end conditions, etc. The extent of these imperfections varies sufficiently from one column to another to produce the rather wide scatter of test results when represented as shown in Fig. 10.11. To appreciate the influence of imperfections on the strength of a column, it must be observed that they result in the line of action of the load not coinciding exactly with the axis of the column, thus introducing more or less bending action in addition to direct compression. On this basis it seems logical to conclude that the behavior of a real imperfect column under load will be similar to that of a perfectly straight ideal column loaded with a suitable eccentricity e.

This suggests that the secant formula derived in Art. 10.4 could be used also as a basis of design for supposedly straight centrally loaded columns, simply by choosing an appropriate value of the eccentricity ratio ec/r^2 to account for the effect of imperfections. While such a procedure still leaves unanswered the question of the proper value of ec/r^2, it offers a rational means to represent the effect of imperfections rather than to simply allow for them by an overlarge factor of safety as suggested on p. 278.

To select a suitable value of ec/r^2 to account for imperfections, recourse must be had to the results of tests made on actual columns. As a result of such experiments, a value $ec/r^2 = 0.25$ is commonly recommended for pin-ended columns as used in ordinary structural work. This does not mean necessarily that every column is expected to be this imperfect, but rather that no column is likely to exceed this degree of imperfection. That is, if the results for the columns tested were plotted on the diagram in Fig. 10.15, the points would all lie above the curve labeled $ec/r^2 = 0.3$. The single curve for $ec/r^2 = 0.25$ and for $\sigma_{\text{y.p.}} = 40,000$ psi is shown again in Fig. 10.17a. This curve may be used as a basis for the design of structural steel columns. For such a column of any given slenderness ratio l/r, the corresponding ordinate of the curve shows the value of the average compressive stress P/A for which yielding of the most stressed fibers in compression can be expected to begin. The corresponding safe value would be this P/A divided by the desired factor of safety n. Since the curve already takes account of the effect of imperfections, a safety factor $n = 2$ may be considered as adequate.

Since lack of straightness of a column can be expected to increase with the length, it is sometimes preferred to allow for imperfections by using an equivalent eccentricity e which increases linearly with the length l of the column. On the basis of test results on steel columns, a commonly recommended value is $e/l = 1/400$. In such case the secant formula becomes

$$\frac{P_{\text{y.p.}}}{A} = \frac{\sigma_{\text{y.p.}}}{1 + \dfrac{1}{400} \cdot \dfrac{l}{r} \cdot \dfrac{c}{r} \sec\left(\dfrac{1}{2}\dfrac{l}{r}\sqrt{\dfrac{P_{\text{y.p.}}}{AE}}\right)}, \tag{a}$$

where the ratio c/r depends on the shape of the cross-section. For a solid circular cross-section $c/r = 2$, for a rectangular cross-section $c/r = \sqrt{3}$, for a thin-walled hollow circular cross-section $c/r = \sqrt{2}$, and for an I-section bending in its plane of greatest flexural rigidity, $c/r \approx 1$. Two curves DE, plotted from eq. (a) for $c/r = 1$ and $c/r = 2$, are shown in Fig. 10.17b where the yield point $\sigma_{\text{y.p.}} = 40,000$ psi. These curves may be used as a basis for design of steel columns in the same manner as was explained above in connection with the curve in Fig. 10.17a.

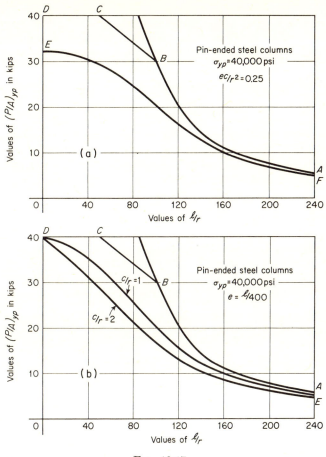

FIG. 10.17

The use of such curves as those in Fig. 10.17 based on the secant formula represents the most rational approach to the design of centrally loaded steel columns. We have seen that the load which such a column can safely carry is influenced by the imperfections, the strength of the material, and the inherent aspect of instability. Each of these factors is rationally represented in the secant formula. The effect of imperfections is accounted for by the chosen equivalent eccentricity of load; the strength of the material itself, by the chosen value of $\sigma_{y.p}$; and the inherent aspect of instability, by the appearance of P/A in the secant term. Thus the secant formula blends the cases of short, medium, and long columns into one continuous range and obviates the necessity to make these classifications as was done in Art. 10.3.

The strongest objection to the secant formula is its transcendental character which necessitates solution by trial and error. However, if such curves as those in Fig. 10.17 are at hand, the selection of a safe load for a given column or of a suitable column to carry a given load becomes a very simple matter.

EXAMPLE 1. The column shown in Fig. 10.7, p. 273, has length $l = 8$ ft and a square box section 4×4 in. outside dimensions and wall thickness $t = \frac{1}{4}$ in. The material is steel with $\sigma_{y.p.} = 40,000$ psi. Using the curves in Fig. 10.17b and a factor of safety $n = 2$, find the safe value of the compressive load P.

SOLUTION. The geometric properties of the cross-section are $A = 3.75$ in.^{2}, $I = 8.84$ in.^{4}, $r = \sqrt{I/A} = 1.54$ in., $c = 2.0$ in. The effective length of the column is $l_1 = 0.7 \times 96 = 67.2$ in. Thus $l/r = 43.6$ and $c/r = 1.3$. Interpolating between the curves DE in Fig. 10.17b for $c/r = 1.3$, we find $(P/A)_{y.p.} = 33,500$ psi, from which $P_{y.p.} = 33,500 \times 3.75 = 125,500$ lb. Then with a factor of safety $n = 2$, we have $P_w = 62,750$ lb.

PROBLEMSET 10.5

1. Using the design curves in Fig. 10.17b and a factor of safety $n = 2$, find the safe load P_w for a pin-ended steel column 4 ft long and having a solid circular cross-section of diameter $d = 2$ in. *Ans.* $P_w = 28,300$ lb.

2. A $3 \times 3 \times \frac{1}{4}$-in. steel angle section is used as a column 10 ft long with built-in ends. Using the design curves in Fig. 10.17b with a factor of safety $n = 1.5$, find the allowable magnitude of the compressive load. *Ans.* $P_w = 15,900$ lb.

3. A 12-ft length of steel tubing has a circular cross-section of 4 in. outside diameter and wall thickness $t = \frac{1}{16}$ in. Assuming pinned ends and using the curve in Fig. 10.17a, find the limiting average compressive stress P/A that the column can carry. *Ans.* $(P/A)_{y.p.} = 20,000$ psi.

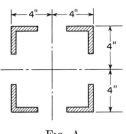

FIG. A

4. A structural steel column 40 ft long is built up of four $2 \times 2 \times \frac{1}{4}$-in. angle irons held together with diagonal lacing to form the square cross-section shown in Fig. A. Assuming that the lacing is adequate to make the column perform as one with a solid section, and that it has pinned ends, find the safe compressive load P_w on the basis of the curves in Fig. 10.17b with a fact or of safety $n = 2.5$. *Ans.* $P_w = 18,800$ lb.

5. Using the curve EF in Fig. 10.17a and a factor of safety $n = 2$, find the safe compressive load P_w for a pin-ended steel column if it has a standard 8I-23 section and is 12 ft long. *Ans.* $P_w = 28,200$ lb.

6. What is the maximum length l that the pin-ended column in the preceding problem can have to carry a compressive load $P = 40,000$ lb with a factor of safety of 2? *Ans.* $l = 10$ ft.

10.6 Empirical Column Formulas

Because of the objections to the transcendental nature of the secant formula, many simpler but wholly empirical formulas have been proposed as substitutes for it. Such formulas usually give the allowable average compressive stress $(P/A)_w$ as a function of the slenderness ratio l/r, without specifically indicating the factor of safety that is used.

Straight Line Formula. One of the most commonly used empirical column formulas is the *straight line formula*

$$\left(\frac{P}{A}\right)_w = \sigma_w + \alpha \frac{l}{r}, \tag{10.13}$$

where σ_w is an allowable working stress for the material in compression and α is a numerical factor. For pin-ended, centrally loaded structural steel columns, a Chicago Building Code recommends $\sigma_w = 16,000$ psi and $\alpha = 70$ so that eq. (10.13) becomes

$$\left(\frac{P}{A}\right)_w = 16,000 - 70 \frac{l}{r}. \tag{10.13a}$$

The use of this formula is limited to $(30 < l/r < 120)$. For $l/r < 30$, $(P/A)_w = 14,000$ psi and for $l/r > 120$, the Euler formula (eq. 10.8) should be used with a factor of safety $n = 2.7$.

The Parabolic Formula. In order to make greater allowance for the effect of imperfections for the more slender columns, a *parabolic formula* is often used, such that

$$\left(\frac{P}{A}\right)_w = \sigma_w - \beta\left(\frac{l}{r}\right)^2, \tag{10.14}$$

where again σ_w is an allowable working stress in compression and β is a numerical factor. For centrally loaded pin-ended structural steel columns, the American Institute of Steel Construction (1947) recommends $\sigma_w = 17,000$ psi and $\beta = 0.485$ so that

$$\left(\frac{P}{A}\right)_w = 17,000 - 0.485\left(\frac{l}{r}\right)^2. \tag{10.14a}$$

The use of this formula is limited to $(0 < l/r < 120)$. For $l/r > 120$, the Euler formula (eq. 10.8) should be used with a suitable factor of safety.

Rankine-Gordon Formula. Still another empirical formula proposed for pin-ended, centrally loaded columns is the *Rankine-Gordon formula*

$$\left(\frac{P}{A}\right)_w = \frac{\sigma_w}{1 + \gamma\left(\frac{l}{r}\right)^2}, \tag{10.15}$$

where σ_w is an allowable working stress in compression and γ a numerical factor. A common form of this formula for structural steel columns is

$$\left(\frac{P}{A}\right)_w = \frac{18{,}000}{1 + \frac{1}{18{,}000}\left(\frac{l}{r}\right)^2}, \qquad (10.15a)$$

to be used only for $0 < l/r < 120$.

For purposes of comparison, each of the foregoing empirical formulas for steel columns is represented graphically in Fig. 10.18, together with the Euler curve ($n = 2$) and the secant formula ($\sigma_{y.p.} = 36{,}000$ psi, $ec/r^2 = 0.2$, and $n = 2$). Taking the secant curve as a basis of comparison, we see that the straight line formula, eq. (10.13a), is somewhat more conservative while both the parabolic curve and the Rankine-Gordon curve are less conservative.

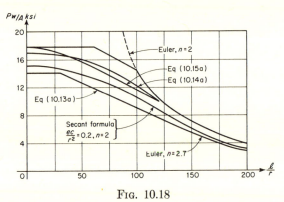

FIG. 10.18

Many empirical formulas for columns of materials other than structural steel are available for design purposes. For example, a formula of the straight line type intended for design of pin-ended aluminum alloy columns in the aircraft industry is

$$\left(\frac{P}{A_w}\right) = \frac{1}{n}\left(43{,}800 - 350\frac{l}{r}\right). \qquad (10.13b)$$

Use of this formula is limited to the range $0 < l/r < 83$. For $l/r > 83$, the Euler formula is to be used.

For pin-ended cast-iron columns, the New York Building Code (1917) specifies an allowable average compressive stress

$$\left(\frac{P}{A}\right)_w = 9000 - 40\frac{l}{r}, \qquad (10.13c)$$

for $0 < l/r < 70$. Owing to the weakness of cast-iron in tension due to bending, it should not be used as a material for columns having $l/r > 70$.

For wood columns of rectangular cross-section, the Forest Products Laboratory has proposed a special formula based on the results of many tests. For *short columns*, having an unsupported length not greater than 10 times the smaller lateral dimension, the allowable average compressive stress shall be

$$\left(\frac{P}{A}\right)_w = S, \tag{10.16a}$$

where values of S for different varieties of timber are given in Table A.2 of Appendix A, p. 343. For *medium columns*,

$$\left(\frac{P}{A}\right)_w = S\left[1 - \frac{1}{3}\left(\frac{l}{Kb}\right)^4\right], \tag{10.16b}$$

where $K = \frac{1}{2}\pi\sqrt{E/6S}$, E being the modulus of elasticity in bending as given in the table. Use of this formula is limited to the range $(10 < l/b < K)$. When $l/b = K$, we see that $(P/A)_w = 2S/3$. For long columns $(l/b > K)$, Euler's formula is to be used with a factor of safety $n = 3$. Thus

$$\left(\frac{P}{A}\right)_w = \frac{\pi^2 E}{36(l/b)^2}. \tag{10.16c}$$

Values of S, E, and K for common structural grades of timber, kept dry, are to be found in the table mentioned above.

The empirical formulas discussed above are all intended for centrally loaded columns, the reduction in allowable average compressive stress with increasing l/r being intended only to allow for imperfections. In the case of columns carrying bracketed loads or, in general, intentional bending as well as compression, the American Institute of Steel Construction has proposed an empirical procedure which avoids the difficulties connected with the trial and error solution of the secant formula. To explain this, we refer to the pin-ended column in Fig. 10.19, where P is the axial load and M is an end moment. Neglecting the effect of the lateral deflection δ on the bending moment at C, the maximum compressive stress would be

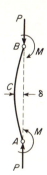

Fig. 10.19

$$\sigma_{max} = \frac{P}{A} + \frac{M}{Z},$$

where Z is the section modulus with respect to the principal axis about which bending takes place. Dividing both sides of this equation by σ_{max}, we obtain

$$\underset{\sigma_{max}}{\frac{P/A}{}} + \underset{\sigma_{max}}{\frac{M/Z}{}} = 1. \tag{a}$$

Now if only the centrally applied loads P were acting, i.e., if $M = 0$, we could logically take $\sigma_{max} = \sigma_c$. the allowable average compressive stress given by one of the empirical column formulas, say the parabolic formula, eq. (10.14a). On the other hand, if only the end moments M were acting, i.e., if $P = 0$, we would take $\sigma_{max} = \sigma_b$, the allowable working stress for the material in pure bending. For the combined loading, we arbitrarily replace σ_{max} in the first term on the left side of eq. (a) by σ_c, and σ_{max} in the second term by σ_b, and then require that the sum of the two terms shall not exceed unity. Thus

$$\underset{\sigma_c}{\underline{P/A}} + \underset{\sigma_b}{\underline{M/Z}} \lesssim 1. \qquad (10.17)$$

The use of the lower working stress σ_c in the first term is presumed to correct for the additional bending moment $P\delta$. In calculating the value of σ_c from the parabolic or other empirical formula, the least radius of gyration of the cross-section should be used regardless of the plane of the end moments M.

EXAMPLE 1. Repeat the solution of Example 1 of Art. 10.4, using condition (10.17) instead of the secant formula. The working stress in bending is $\sigma_b = 20,000$ psi.

SOLUTION. The total compressive load $P = 150$ kips and the end moments $M = 500$ in.-kips. For a suitable section, we try first the 14WF68 section already considered on p. 284. For this section, $A = 20.0$ in.^{2}, $Z = 103$ in.^{3}, and $r_{min} = 2.46$ in. Then $l/r = 300 \div 2.46 = 122$ and the parabolic formula, eq. (10.14a), gives

$$\sigma_c = 17,000 - 0.485\,(122)^2 = 9800 \text{ psi.}$$

Accordingly the left side of expression (10.17)becomes

$$\frac{7500}{9800} + \frac{4850}{20,000} = 1.01.$$

This being so nearly equal to unity, we many consider the 14WF68 section satisfactory.

EXAMPLE 2. Using the Forest Products Laboratory formulas, select a Douglas Fir timber of rectangular cross-section to be used as a pin-ended column of length $l = 14$ ft if the axial load $P = 60,000$ lb.

SOLUTION. From Table A.2 of Appendix A, p. 343, we find for Douglas fir: $S = 880$ psi and $K = 27.3$. If the column were very short, the required cross-sectional area would be $A = P/S = 60,000/880 = 68.2$ sq in. This then would require an 8 × 8-in. cross-section, which suggests that for a long column we try first a 10 × 10-in. section. Then from eq. (10.16b)

$$\left(\frac{P}{A}\right)_w = 880\left[1 - \frac{1}{3}\left(\frac{14 \times 12}{27.3 \times 10}\right)^4\right] = 838 \text{ psi.}$$

With this working stress, the safe load $P = 838 \times 100 = 83,800$ lb, which is too large. For an 8×10-in. section, eq. (10.16b) gives

$$\left(\frac{P}{A}\right)_w = 880\left[1 - \frac{1}{3}\left(\frac{14 \times 12}{27.3 \times 8}\right)^4\right] = 777 \text{ psi,}$$

and the safe load is $777 \times 80 = 62,200$ lb. This indicates that the 8×10-in. section is adequate.

PROBLEM SET 10.6

1. Using the A.I.S.C. formula, eq. (10.14a), select a WF-beam section to be used as a fixed-end column 25 ft long if the axial load $P = 200,000$ lb. *Ans.* 12WF50.

2. Using the A.I.S.C. column formula, eq. (10.14a), find the safe axial load P for a steel pipe 16 ft long and having pinned ends if the outside diameter is 12.75 in. and the inside diameter is 12.00 in. *Ans.* $P = 234,000$ lb.

3. Solve the preceding problem, using the straight line formula, eq. (10.13a). *Ans.* $P = 188,500$ lb.

4. A 12WF45 beam section is to be used for a cantilever column built-in at the lower end, free at the top, and 8 ft long. Using the Rankine-Gordon formula, find the safe load P axially applied. *Ans.* $P = 154,000$ lb.

5. What is the safe load P for a 10×12-in. shortleaf pine column 25 ft long if it has pinned ends? *Ans.* $P = 58,500$ lb.

6. What is the safe axial load P for an 8×10-in. spruce column if it is 12 ft long? *Ans.* $P = 48,200$ lb.

7. A 12WF72 beam section serves as a pin-ended column 12 ft long. It carries an axial load $P_1 = 100$ kips and an eccentric load P_2 acting on axis 1-1, 4 in. from axis 2-2. Using condition (10.17), find the safe magnitude of the eccentric load P_2. Calculate σ_c from eq. (10.14a) and use $\sigma_b = 20,000$ psi. *Ans.* $P_2 = 76,900$ lb.

8. Select a 12-in. WF steel section to serve as a pin-ended column 16 ft long if it is to carry a load $P = 80,000$ lb with eccentricity $e = 6$ in., bending in the plane of the web. Use condition (10.17) with σ_c determined from eq. (10.14a) and $\sigma_b = 20,000$ psi. *Ans.* 12WF36.

11
Mechanical Properties of Materials

11.1 The Tensile Test

The preceding chapters dealt with the methods of analyzing the stress distribution produced by various kinds of forces on structures. Knowing the stresses, the designer must then select the material and the dimensions of the structure in such a way that it will safely withstand various loading conditions in service. For this purpose it is necessary to have information regarding the elastic properties and strength characteristics of structural materials under various stress conditions. The designer must know the limits under which the material can be considered as perfectly elastic for various stress conditions, and also the behavior of the material beyond those limits. Information of this type can be obtained only by experimental investigations. Materials-testing laboratories are equipped with testing machines* which produce certain typical deformations of test specimens, such as tension, compression, torsion, and bending.

Experiments show that test results are sometimes affected by the size and shape of the test specimen. Thus to make the results of tests comparable, certain proportions for test specimens have been established and are recognized as standard. The most widely used of all mechanical tests of structural materials is undoubtedly the tension test. The standard tensile test specimen in the United States is circular, with $\frac{1}{2}$-in. diameter and 2-in. gage length, so that

$$\frac{l}{d} = 4 \quad \text{or} \quad l = 4.51\sqrt{A},$$

where $A = \pi d^2/4$ is the cross-sectional area of the specimen.

*For a description of materials-testing machines and a bibliography on the subject see the article by J. Marin in M. Hetenyi ed., *Handbook of Experimental Stress Analysis*, New York, 1950.

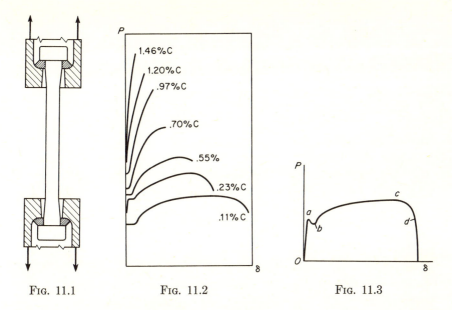

FIG. 11.1   FIG. 11.2   FIG. 11.3

The length of the cylindrical portion of the specimen is always somewhat greater than the gage length l and is usually at least $l + d$. The ends of the specimen are generally made with a larger cross-section in order to prevent the specimen from breaking in the grips of the testing machine, where stress conditions are more severe because of local irregularities in stress distribution. A cylindrical specimen with $l = 10d$ is shown in Fig. 11.1, which also shows the spherical seats in the grips of the machine, used to insure central application of the load.

Tensile test machines are usually provided with a device which automatically draws a *tensile test diagram* representing the relation between the load P and the extension δ of the specimen. Such a diagram exhibits important characteristics of the material. Fig. 11.2, for example, shows a series of tensile test diagrams for carbon steel with various contents of carbon. It can be seen that as the carbon content increases, the ultimate strength of the steel also increases, but at the same time the elongation before fracture decreases and the material has less ductility. High-carbon steel is relatively brittle. It follows Hooke's law up to a high value of stress and then fractures at a very small elongation. On the other hand, a mild steel with a small carbon content is ductile and stretches considerably before fracture.

Fig. 11.3 represents the tensile test diagram for mild structural steel. From this diagram the important characteristics such as *yield point*, *ultimate strength*, and amount of *plastic elongation* can be obtained.

In determining the *proportional limit*, sensitive extensometers are necessary in order to detect the slightest deviation from a straight line in the tensile test diagram. Obviously the position found for this limit depends considerably on the sensitivity of the instruments. In order to obtain greater uniformity in results, a specified amount of permanent set or a certain deviation from proportionality is often taken as the basis for determining the proportional limit. The International Congress for Testing Materials at Brussels (1906) defined the proportional limit as the tensile stress at which the permanent set is 0.001 per cent.

The *yield point* is a very important characteristic for structural steel. At the yield point stress, the specimen elongates a considerable amount without any increase in load. In the case of mild steel this elongation may be more than 2 per cent. Sometimes yielding is accompanied by an abrupt decrease in load, and the tensile test diagram has the shape shown in Fig. 11.3. In such a case the upper and lower limits of the load at a and b, divided by the initial cross-sectional area, are called the *upper* and *lower* yield points, respectively. The position of the upper yield point is affected by the speed of testing, by the form of the specimen and by the shape of the cross section. The lower yield point as already discussed in Art. 2.2 is usually considered a true characteristic of the material and therefore is used as a basis for determining working stresses.

Owing to the relatively large stretching of the material at the yield point it is not necessary to use sensitive extensometers to determine this point. It can be determined with the simplest instruments or can be taken directly from the tensile test diagram. For structural carbon steel the stress at the yield point is about 55–60 per cent of the ultimate strength. Structural steel with about 1 per cent silicon has a yield point stress about 70–80 per cent of the ultimate strength. The ultimate strength for the silicon steel is about the same as for the carbon steel. Such a high value for the yield point justifies the usual practice of taking higher working stresses for silicon steel.

A sharply defined yield point is a characteristic not only of structural steel but also of materials such as bronze and brass. There are other materials, however, which do not have a pronounced yield point. For these materials the stress at which the permanent set reaches the value 0.2 per cent is sometimes arbitrarily called the yield point. It must be kept in mind that the yield point defined in this manner does not represent a definite physical characteristic of the material but depends upon the arbitrarily chosen permanent set.

The *ultimate strength* is usually defined as the stress obtained by dividing the maximum load on the specimen (point c in Fig. 11.3) by the initial cross-sectional area. This quantity also is often taken as a basis for determining the working stresses.

The area under the tensile test diagram *Oacde* (Fig. 11.3) represents the
work required to produce fracture. This quantity is also used as a characteristic property of the material and depends not only on the strength but
also on the ductility of the material.

The *ductility* of a metal is usually considered to be characterized by the
elongation of the gage length of the specimen during a tensile test and by the
reduction in area of the cross-section where fracture occurs. In the first
stage of plastic elongation, from *a* to *c* in Fig. 11.3, the specimen elongates
uniformly along its length. This uniform elongation is accompanied by a
uniform lateral contraction, so that the volume of the specimen remains
practically constant.* At point *c* the tensile force reaches a maximum
value, and further extension of the specimen is accompanied by a decrease
in the load. At this stage of plastic elongation the deformation becomes
localized and necking begins, the specimen taking the shape shown in
Fig. 11.4. It is difficult to determine accurately the moment when
necking begins and thereby establish separately the magnitude of
the uniform stretching and the magnitude of the elongation due
to necking. It is therefore customary to measure the total increase
in the gage length after the specimen has fractured. The *elonga-*
tion is then defined as the ratio of this total elongation of the gage
length to its initial length. In practice the elongation at fracture
is usually given in percentage. If *l* is the original gage length and
δ the total elongation, the elongation at failure in percentage is

$$\epsilon = \frac{\delta}{l} \cdot 100. \qquad\qquad (a)$$

Fig. 11.4

This elongation is usually taken as a measure of the ductility of the material.
Elongation obtained in this manner depends on the proportions of the
specimen. The increase in the gage length due to necking is a large part of
the total increase and is practically the same for a short gage length as for a
long gage length. Hence the elongation defined by eq. (a) becomes larger as
the gage length decreases. For steel, the elongation obtained for specimens
with $l = 5d$ is about 1.22 times the elongation for a specimen of the same
material with $l = 10d$. Experiments also show that the shape of the cross-
section affects the local deformation at the neck and hence affects the
elongation of the specimen. This shows that comparable results with
respect to elongation can be obtained only by using geometrically similar
specimens.

*The small elastic deformation in which the volume does change can be neglected in
comparison with the comparatively large plastic deformation.

The *reduction in area* at the cross-section of fracture is expressed as follows:

$$q = \frac{A_0 - A_1}{A_0},$$ (b)

in which A_0 is the initial cross-sectional area and A_1 the final cross-sectional area at the section where fracture occurs.

11.2 Yield Point

The early portion of the tensile test diagram in Fig. 11.3 is shown to a larger scale in Fig. 11.5, as a *stress-strain diagram*. The shape of this diagram at the yield point depends noticeably on the mechanical arrangement of the testing machine. If extension of the specimen is produced by an increase of distance between the grips of the machine moving at a uniform speed, the sudden plastic stretching will somewhat decrease the tensile force in the specimen, and a sharp peak A in the diagram will be obtained. If an elastic spring is inserted in series with the specimen, the slope of the curve AB of the diagram can be reduced as shown by the broken line AB_1. On the other hand if the tensile load is applied directly to the specimen, the tensile force at yielding will be affected by the inertia of the load in sudden motion, and small vibrations may appear on the diagram.

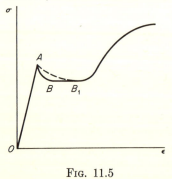

FIG. 11.5

In order to study in more detail the deformations which occur at the yield point, specimens with polished surfaces have been used. Such experiments show that at the time the tensile stress drops from point A to point B (Fig. 11.5) fine, dull lines begin to appear on the surface of the specimen. These lines are inclined about 45° to the direction of tension and are called *Lueders' lines** (see Fig. 2.2, p. 27). With further stretching, the lines increase in width and in number, and during stretching from B to B_1 they cover the entire surface of the specimen. Instead of polishing, sometimes special paints (called *stress coats*) are used to indicate Lueders' lines. The paints are brittle and cannot sustain large deformations; hence they crack during loading and indicate the pattern of Lueders' lines.

Studies with a microscope show that Lueders' lines represent the inter-

*These lines were first described by W. Lueders, *Dinglers Polytech. J.*, 1854

sections with the lateral surface of the specimen of thin layers of material in which plastic deformation has occurred while the adjacent portions of the material remain perfectly elastic. By cutting the specimen and using a special etching, the thin plastic layers in the interior of the specimen can be made visible. Under a microscope it is seen that these layers consist of crystals which have been distorted by sliding.

Experiments show that the values of the yield point stress and the yield point strain depend upon the rate of strain. The curves in Fig. 11.6 show stress-strain diagrams for mild steel for a wide range of rates of strain ($u = de/dt = 9.5 \times 10^{-7}$ per second to $u = 300$ per second). It is seen that not only the yield point but also the ultimate strength and the total elongation depend greatly upon the rate of strain. In general, these quantities increase as the rate of strain increases.

To explain the sudden stretching of steel at its yield point, it has been suggested* that the boundaries of the grains consist of a brittle material and form a rigid skeleton which prevents plastic deformation of the grains at low stress. Without such a skeleton the tensile test diagram would be like that indicated in Fig. 11.7 by the broken line. Owing to the presence

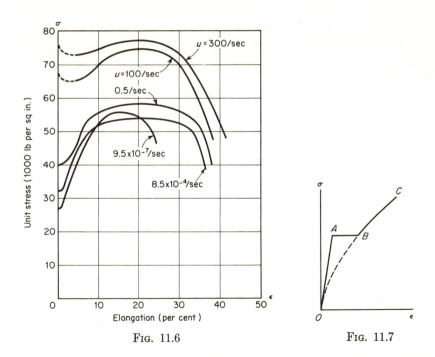

Fig. 11.6

Fig. 11.7

*See P. Ludwik and Scheu, "Werkstoffanschuss," *Ver. deut. Ing. Ber.*, No. 70, 1925.

of the rigid skeleton, the material remains perfectly elastic and follows
Hooke's law up to point A, where the skeleton breaks down. Then the
plastic grain material suddently obtains the permanent strain AB, after
which the material follows the usual curve BC for a plastic material. This
theory explains the condition of instability of the material at the upper
yield point. It also accounts for the fact that materials with small grain
size usually show higher values for the yield stress. As a result, such
materials undergo more stretching at the yield point, as defined by the
length of the horizontal line AB in Fig. 11.7. In addition, the theory
explains the fact that in high-speed tests the increase in yield point stress
is accompanied by an increase in the amount of stretching at yielding,
as shown by the curves in Fig. 11.6.

11.3 Stretching of Steel Beyond the Yield Point

During stretching of a steel specimen beyond the yield point, the material
hardens and the stress required for stretching the bar increases as shown
by the portion BC of the stress-strain diagram. in Fig. 11.8. Elongation
of the specimen is combined with uniform reduction of the cross-sectional

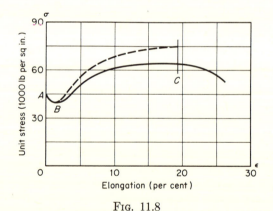

Fig. 11.8

area so that the volume of the specimen remains practically constant.
The work done during stretching is transformed largely into heat, and
the specimen becomes hot. Calorimetric measurements show that not all of
the mechanical energy is transformed into heat, however; part of it remains
in the specimen in the form of strain energy. Owing to differences in
orientation of the crystals, the stresses are not uniformly distributed

over the cross sections, and after unloading, some residual stress and a certain amount of strain energy remain in the specimen.

If after unloading we load the specimen a second time, we will find that its yield point stress is raised. This characteristic is shown in Fig. 11.9, which represents a tensile test diagram for mild steel. After stretching the bar to the point C it was unloaded. During unloading, the material followed approximately a straight-line law, as shown by the line CD on the diagram. When the load was applied to the bar a second time, the material again followed approximately Hooke's law and the line DF was obtained. At point F, which corresponds to the previous loading at C, the curve abruptly changed character and traced the portion FG, which can be considered a prolongation of the curve BC. This represents a raising of the yield point due to previous stretching of the material. If several days are allowed to elapse after the first unloading, then upon reloading a still higher yield point may be obtained, as indicated by the dotted line at F'. Fig. 11.10 shows the results of a tensile test of die-cast aluminum.

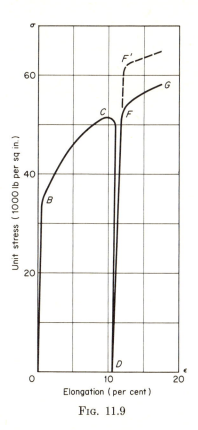

Fig. 11.9

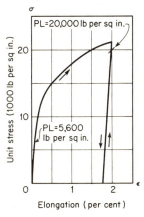

Fig. 11.10

The initial proportional limit of the material was 5600 psi. After stretching the specimen 2 per cent, the proportional limit upon reloading was found to be 20,000 psi and the yield point about 21,000 psi.

More complete investigations show that the time which elapses between unloading and reloading has a great influence on the stress-strain curve during reloading. If reloading begins immediately after unloading, accurate measurements show that there are deviations from the straight-line law at very low stress, and the proportional limit is greatly lowered. But if a considerable interval of time elapses between unloading and reloading, the material recovers its elastic properties completely. Fig. 11.11 shows curves obtained for mild steel which indicate that if reloading follows in ten minutes after overstrain, the material does not follow Hooke's law, but after five days it has partially recovered its elasticity and after twenty-one days it has almost completely recovered it.

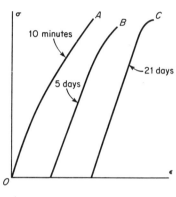

FIG. 11.11

Experiments also show that if the material is subjected to mild heat treatment after unloading, say in a bath of 100°C, the recovery of elastic properties occurs in a much shorter interval of time. Fig. 11.12 shows the results of tests made on a steel bar. The initial tensile test is represented by the curve A. Curve B represents reloading of the same bar ten minutes after unloading, and considerable deviation from Hooke's law is noticeable.

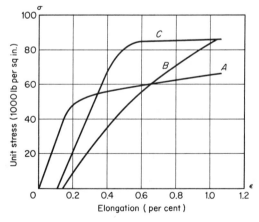

FIG. 11.12

Curve C is the diagram obtained with the same bar after a second unloading and after heat treating at 100°C for four minutes. In this case the material has completely recovered its elastic properties.

The phenomenon of strain hardening due to plastic deformation is encountered in many technological processes such as rolling bars and drawing tubes and wires at low temperature, cutting sheet metal by shears and drawing, and punching holes. In all these cases the part of the material which undergoes plastic deformation becomes harder, and its ductility is greatly reduced. To eliminate this undesirable effect of strain hardening it is customary to anneal the material, which restores the initial ductility.

Sometimes the strain hardening of ductile materials is of practical use in manufacturing. It is common practice to subject the chains and cables of hoisting machines to a certain amount of overstrain in order to eliminate undesirable stretching of these parts in service. The cylinders of hydraulic presses are sometimes subjected to an initial internal pressure sufficient to produce permanent deformation of the walls. The strain hardening and the residual stresses produced by this pressure prevent any permanent set in service. Overstraining of the metal is sometimes used in the manufacture of guns. By stretching the metal in the wall of a gun beyond the initial yield point and afterwards subjecting it to a mild heat treatment, the elastic properties of the material are improved and at the same time initial stresses are produced which combine with the stresses produced by the explosion to give a more favorable stress distribution. Turbine discs and rotors are sometimes given an analogous treatment. By running these parts at overspeed, a permanent set is obtained around the central hole, which raises the yield point of the material and produces initial stresses which are in a favorable direction. Die-cast aluminum fans are sometimes subjected to overstrain at the bore to prevent any possibility of their loosening on the shaft in service. Considerable plastic flow of the metal is sometimes produced in pressing the hubs of locomotive wheels onto their axles, and this has proved to have a favorable effect. Copper bars in the commutators of electric machinery are subjected to considerable cold work by drawing in order to give them the required strength.

In using overstrain in this manner to raise the yield point and improve the elastic properties of a structure, it is necessary to keep in mind: (1) that the hardening disappears if the structure is subjected to annealing temperatures and (2) that stretching the metal in a certain direction, while making it stronger with respect to tension in that direction, does not proportionately improve the mechanical properties with respect to compression in the same direction.

The fact that stretching a metal in a certain direction does not improve the mechanical properties in compression in the same proportion as it does in tension must not be overlooked in cases in which the material is subjected to reversal of stresses. It should also be mentioned that there are indications that material which has yielded in a particular region is more sensitive in that region to chemical action, and there is a tendency for corrosion to enter the metal along the surfaces of sliding. This phenomenon is of particular importance in the case of boilers and other containers subjected simultaneously to stress and to chemical action.

In constructing a tensile test diagram such as curve ABC in Fig. 11.8 the tensile load is usually divided by the initial cross-sectional area A_0 of the specimen in order to obtain the conventional unit stress. But for large stretching there will be a considerable reduction in cross-sectional area; and to obtain the true stress the actual area A, instead of A_0, should be used. From the constancy of the volume of the specimen we have

$$l_0 A_0 = lA, \quad A = \frac{A_0 l_0}{l} = \frac{A_0}{1 + \epsilon}, \tag{a}$$

and the true stress is

$$\sigma = \frac{P}{A} = \frac{P}{A_0}(1 + \epsilon). \tag{b}$$

To obtain the true stress diagram the ordinates of the conventional diagram must be multiplied by $1 + \epsilon$. In Fig. 11.8 such a diagram is shown by the broken line. It extends as far as a vertical through point C, where the load reaches its maximum value. On further stretching of the specimen, local reduction of the cross-section (necking) begins and ϵ is no longer constant along the specimen. Then eq. (b) is no longer applicable, since the stresses over the minimum cross-section are not uniformly distributed. In such a case, eq. (b) gives only an average value of σ. The average unit elongation ϵ at the minimum section may be found from eq. (a), which gives

$$\epsilon = \frac{A_0}{A} - 1. \tag{c}$$

Using the symbol q for the unit reduction of the cross-sectional area, eq. (b), p. 298, we obtain

$$A = A_0 (1 - q);$$

and eq. (c) gives

$$\epsilon = \frac{q}{1 - q}. \tag{d}$$

From eq. (d) the unit elongation at the minimum section can be readily calculated if the reduction in area of that section is measured. This quantity is called the *effective elongation* and is much larger than the elongation $\epsilon = \delta/l$ determined from the total elongation δ of the gage length.

11.4 Types of Fractures in Tension

In discussing fractures, we distinguish between (1) *brittle fracture*, as in the case of cast iron or glass, and (2) *shear fracture*, as in the case of mild steel, aluminum and other metals. In the first case, fracture occurs practically without plastic deformation over a cross-section perpendicular to the axis of the specimen. In the second case fracture occurs after considerable plastic stretching and has the familiar *cup-cone* form, shown in Fig. 11.13.

FIG. 11.13

In discussing these two kinds of fracture, the theory has again been forwarded that the strength of the material can be described by two characteristics, the resistance to separation and the resistance to sliding. If the resistance to sliding is greater than the resistance to separation, we have a brittle material, and fracture will occur as a result of overcoming the cohesive forces without any appreciable deformation. If the resistance to

Table 11.1 ULTIMATE STRENGTH OF CYLINDRICAL AND GROOVED SPECIMENS

δ (inches)	Ultimate Strength (lb per sq in.)			
	Carbon Steel		Nickel Chrome Steel	
	Computed from Original Area	Computed from Reduced Area	Computed from Original Area	Computed from Reduced Area
1/32	163,000	176,000	193,000	237,000
1/16	164,000	177,000	184,000	232,000
1/8	143,000	158,000	154,000	199,000
Normal specimen	102,000	227,000	108,000	348,000

separation is larger than the resistance to sliding, we have a ductile material. Then sliding along inclined planes begins first, and the cup-cone fracture occurs only after considerable uniform stretching and subsequent local reduction of the cross-sectional area (necking) of the specimen.

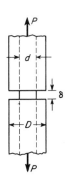

FIG. 11.14

The preceding discussion refers only to tensile tests of standard circular specimens of cylindrical shape. The results obtained with other shapes of specimens are quite different, as illustrated by the grooved specimen shown in Fig. 11.14. During a tensile test, reduction of the cross-sectional area at the grooved section is partially prevented by the presence of the portions of larger diameter D. It is natural that this action should increase as the width δ of the groove decreases. Table 11.1 gives the results of tests obtained with two different materials: (1) carbon steel with proportional limit 56,000 psi, yield point 64,500 psi, ultimate strength 102,000 psi, elongation $26\frac{1}{2}$ per cent, reduction in area 55 per cent; and (2) nickel chrome steel with proportional limit, 80,000 psi, yield point 85,000 psi, ultimate strength 108,000 psi, elongation 27 per cent, reduction in area 69 per cent. These figures were obtained from ordinary tensile tests on normal cylindrical specimens with 1/2-in-diameter and 2-in.-gage length. The original cross-sectional area was used in calculating the stresses. The grooved specimens of the type shown in Fig. 11.14 had $d = \frac{1}{2}$ in. and $D = 1\frac{1}{2}$ in.

The table shows that in all cases the breaking load for the grooved specimens was larger than for the corresponding cylindrical specimens. With the grooved specimens only a small reduction in area took place, and the appearance of the fracture was like that of brittle materials. The true ultimate strength of the cylindrical specimens was larger than for the grooved specimens because fracture of the cylindrical specimens occurred after considerable plastic flow. This resulted in strain hardening and increased not only the resistance to sliding but also the resistance to separation.

Similar conditions are sometimes encountered in engineering practice. An effect analogous to that of the narrow groove in Fig. 11.14 may be produced by internal cavities in large forgings, such as turborotors. Thermal stresses and residual stresses may combine with the effect of the stress concentration at the cavity to produce a crack. The resulting fracture will have the characteristics of a brittle failure without appreciable plastic flow, although the material may prove ductile in the usual tensile tests.

Because most of the grooved specimen remains elastic during a tensile test to failure, it will have a very small elongation, and hence only a small amount of work is required to produce fracture. A small impact force can easily supply the work required for failure. The specimen is brittle because of its shape, not because of any mechanical property of the material. In machine parts subjected to impact all sharp changes in cross-section are dangerous and should be avoided.

11.5 Compression Tests

The compression test is commonly used for testing brittle materials such as stone, concrete, and cast-iron. The specimens used in the tests are usually made in either cubic or cylindric shape. In compressing the specimens between the plane surfaces of the testing machine it is normally assumed that the compressive force is uniformly distributed over the cross-section. The actual stress distribution is much more complicated, even if the surfaces are in perfect contact and the load is centrally applied. Owing to friction on the surfaces of contact between the specimen and the heads of the machine, the lateral expansion which accompanies compression is prevented at these surfaces and the material in this region is in a more favorable stress condition. As a result, the type of fracture obtained in a compression test of a cubic specimen of concrete is as shown in Fig. 11.15. The material in contact with the machine remains unaffected while the material at the sides is crushed out.

Fig. 11.15

In order to obtain the true resistance to compression of a material such as concrete, the influence of friction at the surfaces of contact must be eliminated or minimized. For this purpose A. Föppl covered the surfaces of contact with paraffin and found that the ultimate strength was then greatly reduced. The type of failure was completely different, and cubic specimens failed by subdividing into plates parallel to one of the lateral sides. Another method of eliminating the effect of friction forces is to use specimens in the form of prisms having a length in the direction of compression several times larger than the lateral dimensions. The middle portion of the prism then approaches the condition of uniform compression. A very interesting method of producing uniform compression on cylindrical specimens as developed in the Kaiser-Wilhelm Institut is shown in Fig. 11.16. The head pieces of the testing machine and the ends of the cylindri-

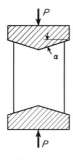

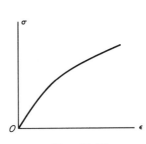

Fig. 11.16   Fig. 11.17

cal specimen are machined to conical surfaces with the angle α equal to the angle of friction. Thus the effect of friction is compensated for by the wedging action, and uniform compression results.

Compression tests of materials such as concrete, stone, and cast-iron show that these materials have a very low proportional limit. Beyond the proportional limit the deformation increases at a faster rate relative to the load, and the compression test diagram has the shape shown in Fig. 11.17. Sometimes it is desirable to have an analytical expression for such a diagram. For these cases Bach proposed an exponential law given by the equation

$$\epsilon = \frac{\sigma^n}{E},$$ (a)

in which n is a number depending on the properties of the material. Bach found the values $n = 1.09$ for pure cement and $n = 1.13$ for granite.

Compression tests of ductile materials show that the shape of the diagram depends on the proportions of the specimen. As the dimension in the direction of compression decreases, the effect of friction at the ends becomes more pronounced and the compression test diagram becomes steeper. For example, Fig. 11.18 shows the results of compression tests on copper cylinders with various ratios d/h of the diameter to the height of the specimen. In compression tests of such ductile materials as copper, fracture is seldom obtained. Compression is accompanied by lateral expansion and a compressed cylinder ultimately assumes the shape of a flat disc.

11.6 Tests of Materials Under Combined Stresses

Leaving the discussion of simple tension and compression tests, let us now consider cases in which the materials are tested under combined stresses. We begin with a discussion of materials tested under uniform hydrostatic pressure.* Such tests show that under uniform pressure, homogeneous materials can sustain enormous compressive stresses and remain elastic. Tests also show that large pressures produce only small changes in volume.

Several attempts have been made to produce uniform hydrostatic tension of materials, but up to now there has not been a satisfactory solution to this interesting problem.

Tensile tests of various steels combined with lateral pressure have shown that the pressure has a great effect on the shape of the neck and on the reduction in area at the minimum cross-section. Fig. 11.19 shows the yoke arrangement by which tension was applied to specimens within the

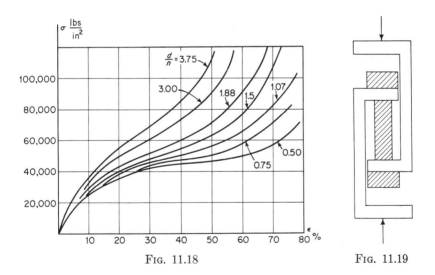

FIG. 11.18 FIG. 11.19

---

*The most comprehensive tests of this kind were made by P. W. Bridgman, who developed a technique for obtaining enormous pressures; see his books, *The Physics of High Pressure*, New York, 1931, and *Studies in Large Plastic Flow and Fracture*, New York, 1952. A new triaxial stress-testing machine was described by H. A. B. Wiseman and Joseph Marin, *Proc. Am. Soc. Test. Mat.*, Vol. 54, 1954.

pressure vessel. Figs. 11.20a and b illustrate fractures of medium carbon steel (0.45 per cent carbon) at atmospheric pressure and at a lateral pressure of 145,000 psi. In the first case the average true stress was 114,000 psi. In the second case the corresponding value was 474,000 psi. It was also found that with an increase in lateral pressure the relative extent of the flat part at the bottom of the cup-cone fracture diminishes; at a certain pressure it entirely disappears, and the fracture becomes entirely shear fracture.

(a)

(b)

FIG. 11.20

The combination of axial compression and lateral pressure was used by Th. v. Kármán in compression tests of marble. These tests showed that with increasing lateral pressure marble becomes more and more plastic, and initially cylindrical specimens may obtain barreled forms, as shown in Fig. 11.21.

In studying two-dimensional stress conditions, thin-walled cylindrical tubes have been tested. By subjecting a tube to axial tension combined with internal pressure, the yield point stress for various ratios of the two principal stresses was established for several materials, including iron, copper, and nickel. The results obtained in this way were in satisfactory agreement with the maximum distortion energy theory to be discussed in Art. 11.7.

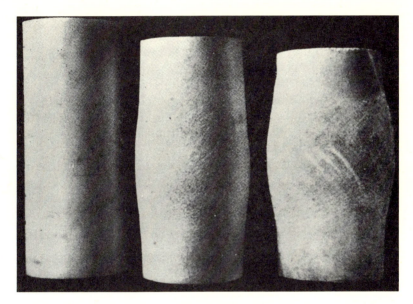

Fig. 11.21

In practical applications not only the yield point stress but also the ductility and strain hardening are of great importance in cases of combined stresses. Unusual cases of failure, such as explosions of large spherical storage tanks and sudden cracks in the hulls of welded cargo ships, have recently called attention to these subjects. In both these types of failure, low-carbon steel plates were used which showed satisfactory strength and ductility in ordinary tensile tests. But the fractured surfaces of the plates in the exploded pressure vessels and in the damaged ships did not show plastic deformation and had a brittle character. Most of these failures occurred at low atmospheric temperatures and under two-dimensional stress conditions.

In order to determine the influence of temperature and two-dimensional stress on the strength and ductility of low-carbon steel, a considerable amount of experimental work has been done in recent times in various laboratories. Thin-walled tubes were used to produce two-dimensional stress conditions. These tubes were subjected simultaneously to axial tension and internal hydrostatic pressure, so that tensile stresses σ_t in the circumferential direction and σ_a in the axial direction could be produced in any desired ratio $n = \sigma_t/\sigma_a$. Using tubes of medium-carbon steel (0.23 per cent carbon) with 1.450-in. outside diameter and 0.100-in. wall thickness, E. A. Davis made tests* with five different values of the ratio n. Fig. 11.22 shows the types of fractures obtained. For the small values of the ratio n

FIG. 11.22

*See E. A. Davis (Westinghouse Reasearch Laboratories), *J. Appl. Mech.*, Vol. 12, p. 13, 1945, and Vol. 15, p. 216, 1948.

the cracks were circumferential, and for the larger values they were longitudinal. By making an additional series of tests it was established that the transition from one type of failure to the other occurred at the value $n = 0.76$. It was found that in the case of circumferential cracks the fracture occurred along the planes of maximum shearing stress and at true stresses of about the same magnitude as in the case of flat specimens prepared from the same material as the tubes. In the case of the longitudinal cracks, rupture appeared to be more brittle. Failure usually started along the planes of maximum shearing stress, but owing to high stress concentration at the crack ends, it continued as brittle fracture in the axial plane without substantial plastic deformation. The maximum shearing stress at which the longitudinal cracks began was always much smaller than in the case of circumferential cracks. It seems that the differences in the two fractures were due largely to the shape of the specimens. In the case of circumferential cracks the material was much more free to neck down than in the case of longitudinal cracks and therefore the latter occurred with smaller local deformation and smaller decrease in load beyond the ultimate strength.

In experiments at the University of California* tests were made at two different temperatures using thin-walled tubes of low-carbon steel. The diameter of the tubes was $5\frac{1}{4}$ in. and the temperatures were $70°$ F and $-138°$ F. The tests at room temperature always gave a shear type of fracture with considerable plastic deformation. The tests at low temperature (with $n = 1$) showed brittle fracture with very small plastic deformation. This brittleness was attributed to the local stresses at the welded junctions of the tubes with the end connections.

After these tests with small tubes, large-size tubular specimens of 20-in. outside diameter and 10-ft length, made of $\frac{3}{4}$-in. ship plate, were tested at $70°$ F and at $-40°$ F. The tests at low temperature, especially with the ratio $n = 1$, showed brittle fracture at stresses much smaller than those obtained from tensile tests of ordinary cylindrical specimens made of the same material.

11.7 Strength Theories

The mechanical properties of structural materials are normally determined by tests which subject the specimen to comparatively simple stress conditions. The strength of materials under more complicated stress conditions has only been investigated in a few exceptional cases, such as those discussed in the preceding article.

*See H. E. Davis and E. R. Parker, *J. Appl. Mech.*, Vol. 15, p. 201, 1948.

In order to determine suitable allowable stresses for the complicated stress conditions which occur in practical design, various *strength theories* have been developed. The purpose of these theories is to predict failure conditions under combined stresses, assuming that the behavior in a simple tension or compression test is known. By failure of the material is meant either yielding or actual rupture, whichever occurs first.

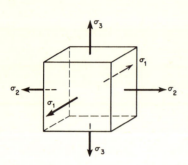

FIG. 11.23

The most general state of stress which can exist in a body is always completely determined by specifying the principal stresses σ_1, σ_2, and σ_3 (Fig. 11.23). In the following discussion, tension is considered positive and compression negative, and the axes in Fig. 11.23 are chosen so that the relations between the algebraic values of the principal stresses are

$$\sigma_1 > \sigma_2 > \sigma_3. \tag{a}$$

The *maximum stress theory* considers the maximum or minimum principal stress as the criterion for strength. For ductile materials this means that yielding begins in an element of a stressed body when either the maximum stress reaches the yield point in simple tension or the minimum stress reaches the yield point in simple compression. Thus the conditions for yielding are

$$(\sigma_1)_{\text{y.p.}} = \sigma_{\text{y.p.}}, \quad \text{or} \quad |(\sigma_3)_{\text{y.p.}}| = \sigma'_{\text{y.p.}} \tag{11.1}$$

in which $\sigma_{\text{y.p.}}$ and $\sigma'_{\text{y.p.}}$ are the yield point stresses in simple tension and compression, respectively. There are many examples which contradict the maximum stress theory. It has already been pointed out that in simple tension sliding occurs along planes inclined at 45° to the axis of the specimen For these planes neither the tensile nor the compressive stresses are maximum, and failure is caused by shear stresses instead. It has also been pointed out that a homogeneous and isotropic material, even though weak in simple compression, may sustain very large hydrostatic pressures without yielding. This indicates that the magnitude of the maximum stress is not sufficient to determine the conditions for yielding of the material or its fracture.

A second strength theory is the *maximum strain theory*. In this theory it is assumed that a ductile material begins to yield either when the maximum strain (elongation) equals the yield point strain in simple tension or when the minimum strain (shortening) equals the yield point strain in simple compression.

Observing that stress in one direction produces lateral deformation in the other two perpendicular directions, and using superposition, we find for the three principal strain components of the element in Fig. 11.23, the following expressions:

$$\epsilon_1 = \frac{1}{E}[\sigma_1 - \mu(\sigma_2 + \sigma_3)],$$

$$\epsilon_2 = \frac{1}{E}[\sigma_2 - \mu(\sigma_1 + \sigma_3)], \qquad (11.2)$$

$$\epsilon_3 = \frac{1}{E}[\sigma_3 - \mu(\sigma_1 + \sigma_2)].$$

With condition (a), the first of these equations represents the maximum strain and the third, the minimum strain. Substituting the yield point strains $\sigma_{y.p.}/E$ in tension and $\sigma'_{y.p.}/E$ in compression for ϵ_1 and ϵ_3, the criterion of failure according to the maximum strain theory becomes

or

$$\sigma_1 - \mu(\sigma_2 + \sigma_3) = \sigma_{y.p.}$$

$$|\sigma_3 - \mu(\sigma_1 + \sigma_2)| = \sigma'_{y.p.} \qquad (11.3)$$

There are many cases in which the maximum strain theory may also be shown to be invalid. For example, if a plate is subjected to equal tensions in two perpendicular directions, the maximum strain theory indicates that the tensile stress at yielding will be higher than the yield point in simple tension. This result is obtained because the elongation in each direction is decreased by the tension in the perpendicular direction. However, this conclusion is not supported by experiments. Tests of materials under uniform hydrostatic pressure also contradict this theory. For this case, the second of eqs. (11.3) gives

$$|\sigma_3|_{y.p.} = \frac{\sigma'_{y.p.}}{1 - 2\mu},$$

in which σ_3 represents the hydrostatic pressure. Experiments show that homogeneous materials under uniform compression can withstand much higher stresses and remain elastic.

The *maximum shear theory* gives better agreement with experiments, at least for ductile materials which have $\sigma_{y.p.} = \sigma'_{y.p.}$. This theory assumes that yielding begins when the maximum shear stress in the material becomes equal to the maximum shear stress at the yield point in a simple tension test. Since the maximum shear stress in the material is equal to half the difference between the maximum and minimum principal stresses, and since the maximum shear stress, in a tension test, is equal to half the normal stress, the condition for yielding becomes

$$\sigma_1 - \sigma_3 = \sigma_{y.p.}. \tag{11.4}$$

This theory is in good agreement with experiments and is widely used in machine design for ductile materials.

To compare the preceding strength theories let us consider the case of pure shear. For this special case of two-dimensional stress the maximum tensile, compressive, and shearing stresses are all numerically equal (see p. 63) and we have

$$\sigma_1 = -\sigma_3 = \tau, \quad \sigma_2 = 0.$$

Assuming that the material has the same yield point in tension and compression, the conditions for yielding according to the maximum stress theory, maximum strain theory, and maximum shear theory, respectively, are

$$\left. \begin{array}{l} \tau_{y.p.} = \sigma_{y.p.}, \\[2mm] \tau_{y.p.} = \dfrac{\sigma_{y.p.}}{1 + \mu}, \\[2mm] \tau_{y.p.} = \dfrac{\sigma_{y.p.}}{2}. \end{array} \right\} \tag{b}$$

Taking $\mu = 0.3$, we find the following results for pure shear:

Maximum stress theory..................$\tau_{y.p.} = \sigma_{y.p.}$

Maximum strain theory..................$\tau_{y.p.} = 0.77\sigma_{y.p.}$

Maximum shear theory..................$\tau_{y.p.} = 0.50\sigma_{y.p.}$

It is seen that the difference between the various theories is considerable in this particular case. In the design of a circular shaft in torsion, for example, it is first necessary to assume an allowable value of working stress in shear $\tau_w = \tau_{max} = \tau_{y.p.}/n$. Then the diameter of the shaft may be found from eq. (4.5), p. 73. Using the three theories discussed above, the following ratios of the diameters are obtained:

$$1 : 1.09 : 1.26.$$

In more recent times, consideration of the strain energy of deformation per unit volume of the material has been used as a basis of selecting working stresses in machine design.* Considering the element in Fig. 11.23 and applying the same reasoning as for simple tension, we find that the strain energy per unit volume is

$$u = \tfrac{1}{2}(\epsilon_1 \sigma_1 + \epsilon_2 \sigma_2 + \epsilon_3 \sigma_3).$$

Substituting the values of the strain components from eqs. (11.2), this becomes

$$u = \frac{1}{2E}[\sigma_1{}^2 + \sigma_2{}^2 + \sigma_3{}^2 - 2\mu(\sigma_1\sigma_2 + \sigma_2\sigma_3 + \sigma_1\sigma_3)]. \qquad (11.5)$$

As already pointed out, we know that materials can withstand very large hydrostatic pressure without failure. Hence it is reasonable to resolve the total strain energy into two parts: (1) the strain energy of uniform tension or compression and (2) the strain energy of distortion and use, for determining the limiting stress condition, only this later part of the strain energy, i.e., the *distortion energy*. To accomplish this separation, we use eqs. (11.2). Adding these equations we obtain

$$\epsilon_1 + \epsilon_2 + \epsilon_3 = \frac{1 - 2\mu}{E}(\sigma_1 + \sigma_2 + \sigma_3), \qquad (c)$$

which states that the unit volume change is proportional to the summation of the three principal stresses. If this summation is zero, the volume change vanishes and the material is subjected only to the deformation of distortion. If $\sigma_1 = \sigma_2 = \sigma_3 = p$, as in the case of hydrostatic pressure, we have

$$\epsilon_1 = \epsilon_2 = \epsilon_3 = \epsilon = \frac{1 - 2\mu}{E}p. \qquad (d)$$

There will be no distortion in this case and uniform tension or compression exists alone.

For the general case, we introduce the notation

$$\frac{\sigma_1 + \sigma_2 + \sigma_3}{3} = p \qquad (e)$$

and then divide each of the three principal stresses into two parts as follows:

$$\sigma_1 = p + \sigma'_1, \quad \sigma_2 = p + \sigma'_2, \quad \sigma_3 = p + \sigma'_3. \qquad (f)$$

Summing up these three quantities and using eq. (e) we obtain

$$\sigma'_1 + \sigma'_2 + \sigma'_3 = 0.$$

*See R. E. Peterson, *Stress Concentration Design Factors*, New York, 1953.

Since the summation of σ'_1, σ'_2, and σ'_3 vanishes, these stresses produce only distortion, and eqs. (f) provide a means for dividing the given system of stresses σ_1, σ_2, and σ_3 into two systems;(1) uniform tension or compression p, producing only change of volume, and (2) the system of stresses σ'_1, σ'_2, and σ'_3, producing only distortion.

As an example of the application of eqs. (f) let us consider the case of simple tension, Fig. 11.24a. Substituting $\sigma_2 = \sigma_3 = 0$ into eqs. (e) and (f) we obtain

$$p = \frac{\sigma_1}{3}, \quad \sigma'_1 = \frac{2\sigma_1}{3}, \quad \sigma'_2 = \sigma'_3 = -\frac{\sigma_1}{3}.$$

Simple tension in the x direction can thus be resolved into uniform tension (Fig. 11.24b) and a combination of pure shear in the xy- and xz-planes (Fig. 11.24c). It can be seen that the work of the stresses producing only distortion (Fig. 11.24c) on the displacements produced by uniform tension (Fig. 11.24b) vanishes. The strain energies of cases (b) and (c) are thus independent of each other, and the total strain energy in simple tension is obtained by adding together the strain energy of uniform tension and the strain energy of distortion.

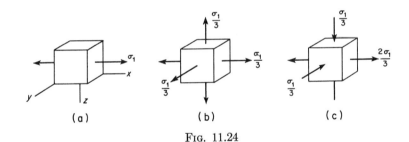

FIG. 11.24

This conclusion also holds in the general case when all three principal stresses σ_1, σ_2, and σ_3 are acting. From this it follows that the strain energy of distortion is obtained by subtracting the strain energy of uniform tension from the total strain energy. Substituting

$$\sigma_1 = \sigma_2 = \sigma_3 = \frac{\sigma_1 + \sigma_2 + \sigma_3}{3}$$

into eq. (11.5) we obtain, for the strain energy of uniform tension alone, the expression

$$\frac{1 - 2\mu}{6E}(\sigma_1 + \sigma_2 + \sigma_3)^2.$$

Thus the strain energy of distortion in the general case is

$$u_1 = \frac{1}{2E}[\sigma_1{}^2 + \sigma_2{}^2 + \sigma_3{}^2 - 2\mu(\sigma_1\sigma_2 + \sigma_2\sigma_3 + \sigma_1\sigma_3)]$$

$$- \frac{1 - 2\mu}{6E}(\sigma_1 + \sigma_2 + \sigma_3)^2$$

$$= \frac{1 + \mu}{6E}[(\sigma_1 - \sigma_2)^2 + (\sigma_2 - \sigma_3)^2 + (\sigma_1 - \sigma_3)^2]. \tag{11.6}$$

This equation may now be taken as the basis for predicting failure of ductile materials having a pronounced yield point stress $\sigma_{\text{y.p.}}$ in simple tension. According to this theory, for the general case of stresses σ_1, σ_2, and σ_3, yielding begins when the distortion energy (eq. 11.6) reaches the value of the distortion energy at the yield point in a simple tension test. This latter quantity is obtained from eq. (11.6) by substituting

$$\sigma_1 = \sigma_{\text{y.p.}}, \quad \sigma_2 = \sigma_3 = 0,$$

which gives

$$u_1 = \frac{1 + \mu}{3E}\sigma_{\text{y.p.}}{}^2. \tag{11.7}$$

Then the condition for yielding based on the distortion energy theory is

$$(\sigma_1 - \sigma_2)^2 + (\sigma_2 - \sigma_3)^2 + (\sigma_1 - \sigma_3)^2 = 2\sigma_{\text{y.p.}}{}^2. \tag{11.8}$$

In the particular case of two-dimensional stress we put $\sigma_3 = 0$ in eq. (11.8) and the condition for yielding becomes

$$\sigma_1{}^2 - \sigma_1\sigma_2 + \sigma_2{}^2 = \sigma_{\text{y.p.}}{}^2. \tag{11.9}$$

Considering, for example, combined axial tension and torsion of thin tubes and denoting by σ and τ the corresponding stresses, the principal stresses will be (see eq. 7.7, p. 183)

$$\sigma_1 = \frac{\sigma}{2} + \sqrt{\frac{\sigma^2}{4} + \tau^2}, \quad \sigma_2 = \frac{\sigma}{2} - \sqrt{\frac{\sigma^2}{4} + \tau^2},$$

and the condition of yielding (eq. 11.9) becomes

$$\sigma^2 + 3\tau^2 = \sigma_{\text{y.p.}}{}^2. \tag{11.10}$$

In the case of torsion alone we have $\sigma = 0$ and eq. (11.10) gives

$$\tau_{\text{y.p.}} = \frac{\sigma_{\text{y.p.}}}{\sqrt{3}} = 0.577\sigma_{\text{y.p.}}, \tag{11.11}$$

which is in good agreement with experimental results.

The condition of yielding given by eq. (11.8) is currently accepted as valid for ductile materials, and it is assumed that the material begins to yield when the strain energy of distortion reaches a definite value (see eq. 11.7).

11.8 Impact Tests

Impact tests are used in studying the *toughness* of materials, i.e., the ability of the material to absorb energy during plastic deformation. In static tensile tests this energy is represented by the area under the tensile test diagram, and it can be concluded that in order to have high toughness the material must have high strength and at the same time large ductility. Brittle materials have low toughness since they have only small plastic deformation before fracture. The use of such materials in structures is dangerous since fracture may occur suddenly without any noticeable deformation.

In discussing various kinds of fractures (see Art. 11.4) it was indicated that the same material may behave as a brittle or a plastic material, depending on the external conditions. Considering, for example, such an important case as mild structural steel, we find that under ordinary tensile tests, it may have large plastic deformation, while if tested at some lower temperature it may fracture entirely as a brittle material. Disastrous examples of such fractures occurred during World War II in the numerous failures of welded cargo ships.* Subsequent research work showed that the brittleness temperature of the steel plates used in the hulls of the ships was in the same range as the service temperature.

To explain the transition from brittle to plastic fracture, A. F. Joffe distinguished between two kinds of tensile stresses, (1) tensile stress σ_n producing brittle fracture by separation and (2) tensile stress σ_s corresponding to the beginning of sliding. In Fig. 11.25 these two quantities are represented as functions of the specimen temperature t. In Joffe's experiments the resistance to separation remained practically independent of temperature, and in Fig. 11.25 the diagram for σ_n is given by the horizontal line. At the same time the resistance to sliding was influenced considerably by the temperature of the specimen, and the ordinates of the curve for σ_s decrease as the temperature increases. The point of intersection C of the two curves defines the critical value t_{cr} of the temperature. If the temperature of testing is higher than t_{cr}, the resistance to sliding is smaller than the resistance to separation and the specimen will yield plastically. For temperatures lower than t_{cr} we have $\sigma_n < \sigma_s$, and the specimen will fail by a separation fracture without plastic deformation.

*See the paper by Finn Jonassen in W. M. Murray (ed.), *Fatigue and Fracture of Metals*, 1952.

There are other important conclusions which can be obtained on the basis of the diagram in Fig. 11.25. Let us consider the effect of speed of loading on the test results. It is known that with an increase of speed the resistance of the material to sliding increases while its resistance to separation remains practically constant. As a result of this the ordinates of the σ_s curve will increase, and the curve will move to the new position A_1B_1 (Fig. 11.25) while the line σ_n remains stationary. Thus the intersection point of the two curves is

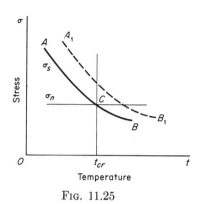

FIG. 11.25

displaced to the right, indicating that with an increase in the speed of loading the critical temperature increases. This conclusion is verified by impact tests, which give brittle fractures at higher temperatures than in static tests.

Now assume that the specimen is subjected to torsion. Yielding of the specimen in shear will begin at about the same value of shearing stress as in the tension tests, but the corresponding value of the maximum normal stress σ_n, equal in this case to the maximum shearing stress, will be about one-half of the value of σ_n in a tension test. Hence in constructing, for torsion tests, a diagram similar to Fig. 11.25, we must take values of the ordinates of the σ_s curve about one-half the values for tension tests. As a result the intersection point C of the curves will be displaced to the left, and we conclude that in torsion tests the critical temperature must be lower than in tensile tests. This conclusion is also in agreement with experiments.

Considering further the influence of the state of stress on the value of the critical temperature, let us assume that a uniform tension in all three directions is superposed on simple tension, so that we obtain a three-dimensional stress condition. It is known that such a superposition does not affect the value of the maximum shearing stress at which yielding begins. The value of σ_s increases, however, and the ordinates of the σ_s curve in Fig. 11.25 increase and the intersection point C moves to the right. Thus the critical temperature for the assumed three-dimensional stress condition will be higher than for simple tension. Similar three-dimensional stress conditions are produced at the notch in a grooved specimen. Such specimens have higher values of t_{cr} than in the case of smooth specimens.*

*For more details on stresses at grooves see E. Orowan's article in W. M. Murray (ed.), *Fatigue and Fracture of Metals*, 1952.

The fundamental ideas regarding the critical temperature at which the transition occurs from brittle to plastic fracture were extended by N. N. Davidenkov and applied to various kinds of steel. Using a diagram similar to Fig. 11.25, he was able to predict the influence of various factors on the value of the critical temperature and showed by his experimental work that the predictions were in satisfactory agreement with the experimental facts.* For determining the critical temperature, impact tests were used. Since in the case of brittle fracture the amount of work required to produce failure is many times smaller than for plastic fracture, the tests showed at the critical temperature a sharp change in the amount of energy absorbed. Fig. 11.26 represents the results of impact tensile tests of smooth cylindrical steel specimens. It can be seen that a sharp change occurs in the energy absorbed in the interval $-130°C$ to $-110°C$.

By changing the process of heat treatment the grain size of steel can be varied considerably, and it is of practical interest to investigate the influence of grain size on the magnitude of the critical temperature. It is known that with an increase in grain size the resistance of steel to separation diminishes. Hence for coarse-grained steels the horizontal line for σ_n (Fig. 11.25) will be lowered and the critical temperature will be higher than for fine-grained steels. To verify this conclusion, specimens of coarse- and fine-grained medium-carbon steel (0.23 per cent carbon) were tested in impact and the results are shown in Fig. 11.27. It is seen that the critical temperature for the coarse-grained steel was about $-95°C$, while in the case of the fine-grained steel it was $-160°C$.

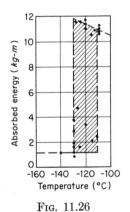

FIG. 11.26

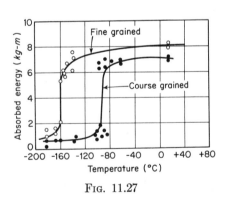

FIG. 11.27

*See Davidenkov's books, *Dynamical Testing of Metals*, 1936, and *Problems of Impact in Metal Study*, Ed. Acad. Science, Moscow, 1938 (in Russian). The results given in the following discussion, if not specifically noted, are taken from the latter book.

The effect of the size of the specimen on the value of the critical temperature has also been investigated. But here the simple diagram of Fig. 11.25 does not give a clear interpretation of the experimental results. With an increase in dimensions we may expect a decrease in the resistance to brittle fracture, since the probability of having critical imperfections increases with volume. Hence for larger volumes the horizontal line for σ_n will be lowered in Fig. 11.25, producing a displacement of point C to the right. But it also appears that an increase in volume reduces the value of σ_s and the corresponding lowering of the σ_s curve in Fig. 11.25 results in a displacement of C in the opposite direction. Thus the final result depends on the relative importance of the two factors. Experiments with smooth cylindrical specimens indicate that the lowering of the σ_n line is more important and point C moves to the right, showing that the critical temperature increases with an increase in volume of the specimen. This factor must be considered when applying the results of tests on small specimens to the design of large-sized structures.

In the preceding discussion we have considered only tensile tests of cylindrical specimens in which the stress distribution was uniform. In practice, however, notched specimens are used in impact tests and stress concentrations are present. To investigate the effect of non-uniform stress distribution on the magnitude of the critical temperature, let us begin with the case of bending of a smooth cylindrical specimen. Experiments in bending with static loads indicate that yielding of the steel begins at a much higher stress than in the case of uniform tension. The yield point stress is first reached in the thin layer of fibers at the farthest distance from the neutral axis, and the formation of planes of yielding in those fibers is prevented by the presence of the adjacent material at lower stress. The resulting increase in the value of the yield stress must be considered in applying the diagram of Fig. 11.25 to bending tests, The ordinates of the σ_s curve must be increased, which results in a displacement of the intersection point C to the right. The critical temperature, as obtained from bending tests, will then be higher than the value obtained from tensile tests. This conclusion agrees with experimental results.

Similar reasoning can be applied to cases of stress concentration produced by grooves and notches (see p. 46) and we may expect an increase in t_{cr} for notched bars.

After this general discussion let us consider the type of impact test which should be used in practice to determine t_{cr}. The correct determination of t_{cr} is important in order to avoid the dangerous situation in which the critical temperature of the material is the same as the service temperature of the structure. It is apparent that impact tests at room temperature are not sufficient and in important situations a series of tests over a range of temperatures should be made. A transition curve, similar to Fig. 11.27, should be constructed and from it t_{cr} determined. When the critical temperature has been determined, and knowing the service temperature t_0 of the structure, Davidenkov recommends that the measure of safety be taken as the ratio

$$\frac{T_0 - T_{cr}}{T_0} \tag{a}$$

in which T_0 and T_{cr} are the absolute temperatures corresponding to t_0 and t_{cr}. This ratio diminishes and approaches zero as T_0 approaches T_{cr}. The result is a very dangerous situation in which small external impulses may produce brittle fracture of the structure. On the other hand, the ratio approaches unity as T_{cr} approaches absolute zero. In this case brittle fractures will not occur and it is only necessary to select the dimensions of the structure so that it will be strong enough to carry the loads without plastic deformation.

In selecting a reasonable value of the ratio (a) for use in design, the conditions which actually exist in the structure must be considered. Such stress raisers as sharp reentrant corners and imperfections in welding contribute to an increase in t_{cr}. An increase in size of the structure has the same effect. To have sufficient safety and to keep the ratio (a) as large as possible, materials with low values of t_{cr} should be used. The critical temperature can be lowered not only by changing the chemical content of the material but also by proper heat treatment. A fine-grained steel has a lower value of t_{cr} than a coarse-grained steel. Considerable interest in the brittle character of metals at low temperatures has developed recently in this country, and we can expect an improvement in our knowledge of this important subject.

11.9   Fatigue of Metals

Machine parts are frequently subjected to varying stresses and it is important to know the strength of materials under such conditions. It is well known that materials fail under repeated loading and unloading, or under reversal of stress, at stresses smaller than the ultimate strength of the material under static loads. The magnitude of the stress required to produce failure decreases as the number of *cycles* of stress increases. This phenomenon of the decreased resistance of a material to repeated stresses is called *fatigue*, and the testing of a material by the application of such stresses is called an *endurance test*.

If σ_{max} and σ_{min} are the maximum and minimum values of the repeated stress, then the algebraic difference

$$R = \sigma_{max} - \sigma_{min} \tag{a}$$

is called the *range of stress*. The cycle is completely defined if the range and the maximum stress are given. The average or mean stress is

$$\sigma_m = \tfrac{1}{2}(\sigma_{max} + \sigma_{min}). \tag{b}$$

In the particular case of *reversed stress* $\sigma_{min} = -\sigma_{max}$, $R = 2\sigma_{max}$ and $\sigma_m = 0$. Any cycle of varying stress can be obtained by superposing a cycle of reversed stress on a steady average stress. The maximum and minimum values of the varying stress are then given by the following formulas:

$$\sigma_{max} = \sigma_m + \frac{R}{2}; \quad \sigma_{min} = \sigma_m - \frac{R}{2}. \tag{c}$$

There are various methods of applying the load in an endurance test. The specimen can be subjected to direct tension and compression, to bending, to torsion, or to some combination of these. The simplest way is by reversed bending. A common cantilever form of fatigue test bar is shown in Fig. 11.28. The cross-section of the specimen is varied along the

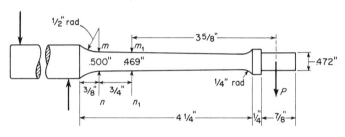

Fig. 11.28

length in such a manner that the maximum stress occurs between cross-sections mn and m_1n_1 and is practically constant within that region. The effect of stress concentrations is eliminated by using a large fillet radius and by increasing the diameter of the bar near the fillet. The load P is always downward and the specimen rotates at constant speed. The stress therefore changes sign every half-revolution, and the number of cycles of stress is equal to the number of revolutions of the machine. The stress is a completely reversed stress, the average stress being zero and the range of stress twice σ_{max}.

By taking several specimens and testing them at various loads P, a curve such as is shown in Fig. 11.29a can be obtained. Here σ_{max} is represented as a function of the number of cycles n required to produce fracture. The curve shown was obtained with mild steel. At the beginning σ_{max} decreases rapidly as n increases, but after about 4 million cycles there is no longer any appreciable change in σ_{max}, and the curve approaches asymptotically the horizontal line $\sigma_{max} = 27,000$ psi. The stress corresponding to such an asymptote is called the *endurance limit* of the material. It is now the usual practice in endurance tests to plot σ_{max} against log n. In this manner the magnitude of the endurance limit is disclosed by a definite discontinuity in the curve. An example of such a curve is shown in Fig. 11.29b.

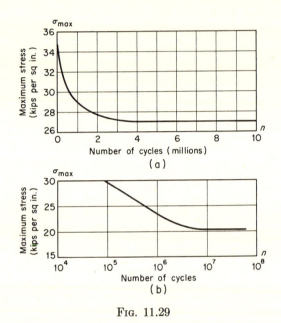

Fig. 11.29

There is a great difference between the fractures of mild steel specimens tested statically and those tested by alternating stresses. In the first case considerable plastic flow precedes fracture, and the surfaces at the ruptured section show a silky, fibrous structure due to the great stretching of the crystals. A fatigue crack, however, appears entirely different. A crack begins at some point in the material owing to a local defect or to a stress concentration produced by an abrupt change in the cross-section. Once formed, the crack spreads owing to the stress concentrations at its ends. This spreading progresses under the action of the alternating stress until the cross-section becomes so reduced in area that the remaining portion fractures suddenly under the load.

Two zones can usually be distinguished in a fatigue fracture, one due to the gradual development of the crack and the other due to sudden fracture. The latter zone resembles the fracture of a tensile test specimen with a deep, narrow groove (see p. 306) in which the shape of the specimen prevents sliding, and therefore fracture occurs as a result of overcoming the cohesive forces. This fracture is of the brittle type even though the material is ductile. In the case of cantilever test specimens (Fig. 11.28) the maximum stresses are at the outer fibers. Hence the fatigue crack usually starts at the circumference and spreads towards the center. Where there are stress concentrations due to fillets, grooves, or holes, the crack usually starts at the most highly stressed portion and spreads outward from this point. In such cases the fracture surface shows concentric rings with respect to this starting point. This is a very common type of fracture in machine parts which have been subjected to alternating stresses. It is thus evident that the brittle type of fatigue fracture is due to the peculiar mechanism of fracture, not to crystallization of the material as was once thought.

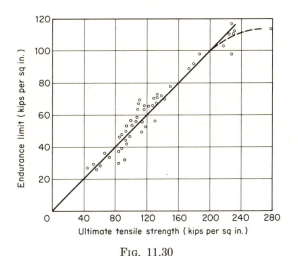

FIG. 11.30

It is evident from the above discussion that the determination of the endurance limit for a particular material requires a large number of tests and considerable time. Hence it is of practical interest to establish relations between the endurance limit and other mechanical properties which can be determined by static tests. The large amount of experimental data accumulated has not yet made it possible to establish such a correlation. As a rough estimate, the endurance limit for ferrous metals under reversal of stresses can be taken equal to 0.40 to 0.55 times the ultimate strength obtained in the usual way from a tensile test. When working with materials whose mechanical characteristics are very well known, such as carbon steels, estimates of this type can be considered reliable. Otherwise such estimates are likely to be misleading, and direct endurance tests should be used instead. Some results of endurance tests of steels are given in Fig. 11.30 and also in Table A.4 of Appendix A.

In the majority of cases, endurance tests are carried out for completely reversed stresses ($\sigma_{max} = -\sigma_{min}$), while in many cases in machine design the stresses vary but are not completely reversed. It is necessary to know the endurance limits under these varying stresses. Wöhler was the first experimenter who studied the phenomenon of fatigue systematically.[*] He showed that the range of stress R necessary to produce fracture decreases as the mean stress σ_m increases. On the basis of these tests and of Bauschinger's work, Gerber proposed a parabolic law relating the range of stress R and the mean stress σ_m. This is illustrated by the parabolic curves in Fig. 11.31, in which the mean stress and the range of stress are expressed as fractions of the ultimate strength. The range is a maximum when the stress is completely reversed ($\sigma_m = 0$) and it approaches zero when the mean stress approaches the ultimate strength. If the endurance limit for reversed stress and the ultimate strength are known, the endurance limit for any varying stress can be obtained from such curves. Other investigations show that there is no general law connecting the mean stress and the range of stress. For instance, there are materials for which the relation between R and σ_m is represented more accurately by the broken lines (Goodman law) in Fig. 11.31 than by parabolas.

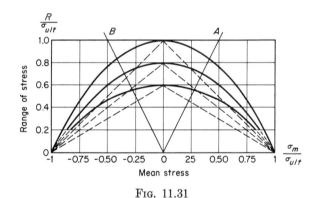

FIG. 11.31

The straight lines OA and OB in Fig. 11.31 have a slope of 2 and determine the region AOB in which the stress changes sign during a cycle. Outside this region the stress always remains tension or compression. Experimentally determined values within the region AOB usually lie

*A. Wöhler, Z. Bauwesen, Vols. 8, 10, 13, 16, and 20, 1858–70. An account of this work in English is given in Engineering, Vol. 11, 1871; see also Unwin, The Testing of Materials of Construction, 3d ed., 1910.

between the parabolas and the corresponding straight lines. When the stress is always tension or always compression, the values of the range R, as found by test, are sometimes not only below Gerber's parabolas but also below the corresponding straight lines.

11.10   Fatigue Under Combined Stresses

Most of our experimental information on the fatigue strength of materials has been obtained under conditions of uniaxial stress, as in rotating bending-test specimens. But in practical problems we frequently encounter cases of combined stress, and it is important to know the fatigue strength for such conditions. To obtain the fatigue strength of various ductile materials in pure shear, torsion tests were made in which the angle of twist was reversed. The results of some of these tests are shown* in Fig. 11.32. For purposes of comparison, the endurance limit in bending is taken as the abscissa and the endurance limit in shear is plotted as the ordinate. It is seen that the ratio of these limits for all the materials tested is very nearly

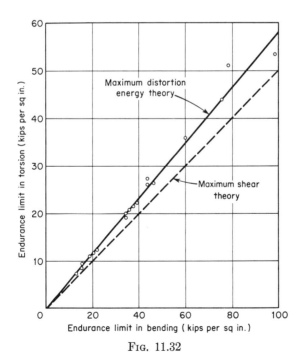

Fig. 11.32

*See, R. E. Peterson, *Stress Concentration Design Factors*, New York, 1953.

equal to $\sqrt{3}$. This is the value given by the maximum distortion energy theory for the ratio of the yield point stresses in bending and shear (see eq. 11.11).

Fatigue tests under combined stresses produced by the simultaneous action of alternating bending and torsion have also been made and the results are shown in Fig. 11.33. Here again the test results are in good agreement with the maximum distortion energy theory, as might be expected, since slip generally precedes the development of a fatigue crack.

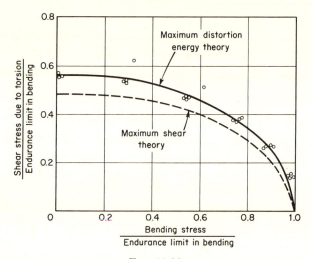

FIG. 11.33

To obtain an equation for calculating the endurance limit for combined bending and torsion, we have only to substitute into the corresponding equation for yielding (eq. 11.10) the value of the endurance limit σ_E for reversed bending, in place of $\sigma_{y.p.}$, which gives

$$\sigma^2 + 3\tau^2 = \sigma_E^2. \tag{a}$$

The corresponding ellipse is shown in Fig. 11.33, and it is apparent that the test results are in good agreement with the equation.

Other fatigue tests with biaxial tension or tension and compression, and with the ratio σ_1/σ_2 remaining constant during a cycle, are also in satisfactory agreement with the maximum distortion energy theory. Thus we can use for determining the fatigue limit in the case of complete reversal of stresses the following equation (see eq. (11.9) p. 319):

$$\sigma_1^2 - \sigma_1\sigma_2 + \sigma_2^2 = \sigma_E^2 \tag{b}$$

in which σ_E is the endurance limit for uniaxial stress conditions. Assuming that $\sigma_1 > \sigma_2$ and using the notation $\sigma_2 = \alpha\sigma_1$, we obtain from eq. (b)

$$\sigma_1\sqrt{1 - \alpha + \alpha^2} = \sigma_E. \tag{c}$$

In the case of pulsating stresses in which the stress varies from zero to some maximum value, the corresponding uniaxial pulsating stress σ_{max} should be substituted for σ_E in eq. (c).

In conclusion, it should be pointed out that the endurance limit for various materials can be affected by many extraneous factors. For example, moderate cold-stretching of steel has been found to produce some increase in its endurance limit. However, when this cold-working is overdone, the endurance limit may be lowered.

Most fatigue tests are made at room temperature. Some experiments with fatigue of steel specimens made at lower temperatures ($-20°C$) have shown a slight increase in the endurance limit at this lower temperature. However, tests at higher than room temperature (up to 300°C) showed no appreciable effect of temperature on the endurance limit.

Fatigue tests on specimens in the presence of various corrosive agents such as salt water have shown that the endurance limit may be greatly reduced by the combined action of fatigue with corrosion. There are many known cases of failures in service which can be attributed to such *corrosion fatigue* such as marine propeller shafts, turbine blades, oil-well pump rods, etc. For this reason, special corrosion-resistant materials are frequently used in such cases. Protective coatings and surface cold-working have also been used successfully in guarding against such failures.

11.11 Fatigue and Stress Concentrations

In discussing the stress concentrations produced by sharp variations in the cross-sections of bars and shafts (see Art. 2.5) it was indicated that such stress concentrations are especially damaging in the case of varying stresses. In machine parts, stress concentrations are always present due to fillets, grooves, holes, keyways, etc., and experience shows that most fatigue cracks in service begin at points of stress concentration.

Figure 11.34 shows the torsional fatigue failure of a shaft of a large motor-generator set which unfortunately operated near resonance.* The crack started at the keyway, where a high stress concentration took place, and gradually developed along the helical path. Figure 11.35 represents a torsion failure of the shaft of a Diesel-driven generator. A high stress con-

*These figures are taken from a paper by R. E. Peterson presented at the Conference on Strength of Material Problems in Industry, at the Massachusetts Institute of Technology, July 1937.

FIG. 11.34

centration at the small fillet resulted in several helical cracks, which when joined together produced the saw-toothed appearance. Finally, Fig. 11.36 represents a characteristic fatigue failure of a heavy helical spring. The crack started from the inside, as theory predicts (see p. 78), and again followed the direction of one of the principal stresses. All these pictures clearly demonstrate the damaging action produced by stress concentration, and it is clear that this factor must be seriously considered in the design of machine parts subjected to alternating stress.

Early fatigue tests made with specimens having sharp changes of cross section showed that there was a reduction in strength due to the stress concentration, but this reduction was usually smaller than expected from the magnitude of the calculated stress concentration factor. For instance, in the case of flat steel specimens with small circular holes subjected to direct stress, the theoretical factor of stress concentration is 3 (see p. 48). If the magnitude of the peak stress is the controlling factor in endurance tests, it would be expected that the tension-compression load required to produce fatigue failure of a specimen with a hole would be about one-third

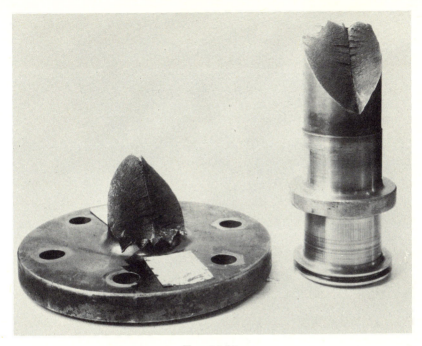

FIG. 11.35

of the load for a specimen without a hole. However, experiments showed that in this case the reduction in strength due to the stress concentration is small as compared with the calculated effect.

To explain this discrepancy and to give the necessary information for designers, a very extensive series of tests were made by R. E. Peterson at the Westinghouse Research Laboratories.* Geometrically similar cantilever test specimens varying in diameter from 0.1 in. to 3 in., with a fillet or with a transverse circular hole and of different materials were tested in special fatigue-testing machines. The results of these tests for specimens with fillets are given in Fig. 11.37. The smaller diameters of the specimens are taken as abscissas while the ordinates represent the ratios k_f of the endurance test loads for plain specimens to the endurance test loads for the corresponding specimens with stress concentrations. Similar results were obtained for specimens with transverse holes.

The horizontal lines in Fig. 11.37 give the values of the stress concentration factors obtained for each fillet size by a direct measurement of strain at

*R. E. Peterson, *J. Appl. Mech.*, Vol. 1, pp.79 and 157, 1933; and R. E. Peterson and A. M. Wahl, *ibid.*, Vol. 3, p. 15, 1936.

FIG. 11.36

the points of maximum stress concentration. These values are designated by k_t and are called *theoretical values* of stress concentration in the following discussion. If the fatigue strength of the specimen depends only on the peak stress, then k_t must evidently be equal to k_f.

On a basis of his tests, Peterson came to the following conclusions:

(1) In some cases fatigue results are quite close to theoretical stress concentration values. This conclusion is of great practical importance, since a general idea seems to exist, based on some early experiments, that fatigue data for stress concentration cases are always well below theoretical values, i.e., on the safe side for design purposes.

(2) Fatigue results for alloy steels and quenched carbon steels are usually closer to the theoretical values than are the corresponding fatigue results for carbon steels not quenched. It was expected in these tests that the theoretical values of k_t would be reached for all steels provided the specimens were made large enough, but Fig. 11.37 shows that the curves

for normalized 0.45 per cent carbon steel are apparently asymptotic to values considerably below the theoretical.

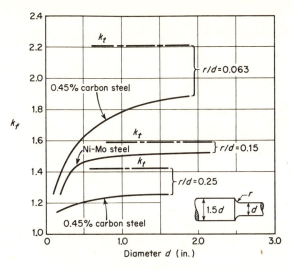

FIG. 11.37

(3) With a decrease in the size of the specimen, the reduction in fatigue strength due to a fillet or hole becomes somewhat less; and for very small fillets or holes the reduction in fatigue strength is comparatively small. This can be clearly seen from the curves in Fig. 11.37.

It can be appreciated that the problem of reducing the damaging effect of stress concentrations is of primary importance to designers. Some lowering of stress concentrations can be obtained by a suitable change in design. For example, a design can be improved considerably by eliminating sharp reentrant corners and introducing fillets of generous radius, by designing fillets of proper shape, by introducing relieving grooves, etc. In Fig. 11.38 are shown methods for reducing the stress concentration at a shoulder of a

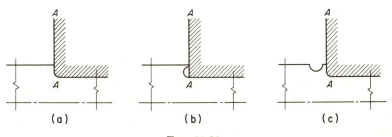

FIG. 11.38

shaft, while maintaining the positioning line AA. The stress can be reduced by cutting into the shoulder and introducing a fillet of larger radius without developing interference with the fitted member, as shown in Fig. 11.38b. If the shoulder height is too small, a relief groove may be used as shown in Fig. 11.38c.

In Fig. 11.39 two different bolt-and-nut designs are shown. In Fig. 11.39a the nut is in compression while the bolt is in tension. High stress concentration takes place at the bottom of the thread in the face of the nut, and under the action of variable forces, fatigue fracture occurs in that plane. In the lip design, Fig. 11.39b, the peak stress is somewhat relieved because the lip is stressed in the same direction as the bolt. Fatigue tests show the lip design to be about 30 per cent stronger.

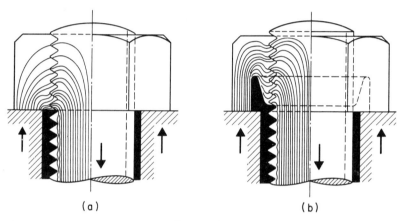

(a)　　　　　　　　　　　(b)

Fig. 11.39

Sometimes these relieving measures are not sufficient to eliminate fatigue failures. As an important example let us consider the typical failures which occur at the wheel seats of locomotive and railroad-car axles, at the wheel or bearing seats of automobile axles, at the pressed or fitted bits of long drill rods in oil-well operations, etc. All these cases of fitted members subjected to the action of variable stresses have been a constant source of fatigue failures. Considering, for example, the case of a wheel hub pressed on an axle, Fig. 11.40a, we can see that a high stress concentration combined with friction is produced at the reentrant corners m and n. During rolling of the axle a reversal of stress at points m and n takes place, and finally a fatigue failure over the cross-section mn, may occur. Stress concentrations can be somewhat reduced by introducing raised seats and fillets as shown in Fig. 11.40b. A further improvement is obtained by introducing the relief groove a. Although such changes are an improvement, they are not

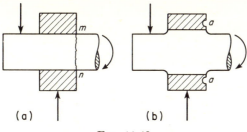

FIG. 11.40

sufficient in this case. Experience shows that the mere press fit of a hub on an axle, reduces the fatigue strength of the axle to less than half of its initial strength, while the changes shown in Fig. 11.40b raise the fatigue strength of the axle perhaps no more than 20 per cent. To improve this condition and eliminate fatigue failures, surface cold-rolling of the axle in the region of stress concentration has been successfully applied.

11.12 Physical Properties of Metals at High Temperatures*

There are many cases in which parts of engineering structures are subjected simultaneously to the action of stresses and of high temperatures. Such conditions are found, for instance, in power plants, chemical industries, and in the missile industry. Owing to the modern tendency to increase temperature and pressure in steam power plants and in the oil-refining industries, the question of the strength of materials at high temperature has become of practical importance and a considerable amount of research work has been done in this field. Experiments show that the yield point and ultimate strength of metals in tension depend very much on the temperature. Fig. 11.41 shows how these as well as other common mechanical properties of a medium carbon steel vary with the temperature.

For loads acting over a long period of time and at high temperatures as, for instance, the weight of a structure or steam pressure in power plants, we need additional information regarding the time effect. Experience shows that under such conditions a continuous deformation, called *creep*, may take place which is the most important factor to be considered in design. Although a considerable amount of research work in this direction has been done and much more is now in progress, the question of the behavior of metals under high temperature and prolonged loading cannot be considered completely cleared.

*For further information, see papers and bibliography presented in *Symposium on the Effect of Temperature on Properties of Metals*, issued jointly by the A.S.T.M. and the A.S.M.E., 1931.

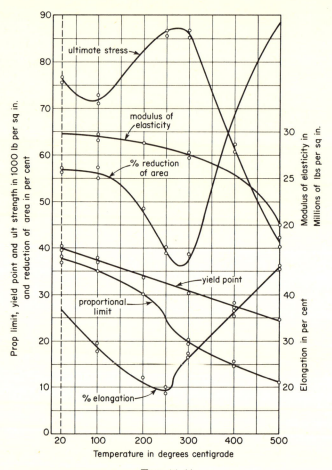

Fig. 11.41

In most experiments of this kind, the gradual elongation of materials under prolonged tension is studied. Tensile test specimens at high temperature are subjected to a certain constant load and temperature, and the progressive creep under this load is investigated. The results of such an experiment when plotted as a time-extension diagram give a curve of the shape shown in Fig. 11.42. When the load is first applied, there is an immediate elastic extension OA. The specimen then begins to stretch at a decreasing rate as shown by the portion AB of the curve. At B the rate of extension reaches a value which remains substantially constant for some time, that is, along the portion BC of the diagram. At C, the rate of extension begins to increase and fracture finally takes place at point D. For the stresses encountered in practice, the portion OB represents a comparatively short time, while the entire lifetime of the specimen lies within the range BC. The slope of the portion BC, representing the rate of extension at a certain stress and temperature, is therefore of the utmost practical interest because the life of the structure depends on the rate of this extension. If the tensile stress is decreased, the slope of BC decreases, but there is no conclusive evidence that it will ever become horizontal, that is, that there is a limiting stress at which the specimen can indefinitely resist the stress and high temperature. Hence in such cases the design must be based on the assumption of a certain duration of service of the structure and of a certain amount of distortion which can be considered permissible. The working stresses are chosen so that the distortion of the structure during its lifetime will not exceed a definite limit, depending on the type of structure. For example, in the design of moving parts such as steam turbines, the creep should never exceed 1 per cent in 100,000 hr (about 11 years) and generally is limited to a fraction of 1 per cent. Creep rates as high as 1 per cent in 10,000 hr may be used for steam piping and boiler tubes.

The comparative working stresses or creep strengths of certain alloy steels over a wide range of temperature are shown in Fig. 11.43. These values, however, may be much affected by variations in grain size, by heat treatment, and by previous strain hardening and should be used with caution.

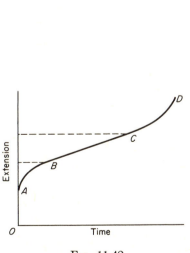

Fig. 11.42

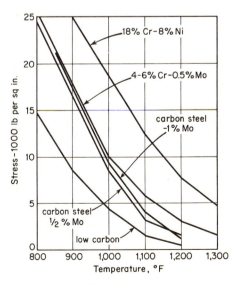

Fig. 11.43

APPENDIX A
Physical Properties for Common Structural Materials

TABLE A.1
Average Physical Properties of Common Metals *

Metal	Density, lb per cu in.	Temp. Coef. of Linear Expansion per Degree Fahr.	Proportional Limit, psi †		Ultimate Strength, psi ‡			Mod. Elasticity, psi †		Per Cent Elongation in 2 In.
			Tension	Shear	Tension	Comp'n	Shear	Tension E	Shear G	
Steel, hot rolled (0.2% carbon)	0.283	6.5×10^{-6}	35,000	21,000	60,000	90,000	45,000	30×10^6¶	12×10^6	30
Steel, cold rolled (0.2% carbon)	0.283	....	60,000	36,000	80,000	....	60,000	30×10^6	12×10^6	18
Steel, hot rolled (0.8% carbon)	0.283	7.3×10^{-6}	70,000	42,000	120,000	....	105,000	30×10^6	12×10^6	10
Steel, oil quenched (0.8% carbon)	0.283	....	120,000	72,000	180,000	....	150,000	30×10^6	12×10^6	2
Nickel steel (oil quenched) (3½% Ni: 0.4% C)	....	....	160,000	96,000	285,000	....	....	30×10^6	12×10^6	5
Wrought iron	0.278	6.7×10^{-6}	30,000	18,000	50,000	60,000	40,000	27×10^6	10×10^6	30
Gray cast iron	0.260	6.0×10^{-6}	6,000	....	20,000	80,000	§	15×10^6	6×10^6	Slight
Malleable cast iron	0.264	6.6×10^{-6}	36,000	23,000	54,000	....	48,000	25×10^6	12.5×10^6	18
Copper, cast	0.322	9.3×10^{-6}	8,000	....	30,000	45,000	27,000	13×10^6	6×10^6	4
Copper, hard-drawn	0.322	9.3×10^{-6}	38,000	23,000	55,000	....	....	17×10^6	6×10^6	20
Aluminum, cast (99% Al)	0.095	12.8×10^{-6}	9,000	....	13,000	....	10,500	10×10^6	4×10^6	4
Aluminum, hard-drawn	0.097	12.8×10^{-6}	20,000	....	30,000	....	....	10×10^6	4×10^6	7
Magnesium (extruded)	0.064	14.5×10^{-6}	17,000	....	32,000	....	17,000	6.5×10^6	2.4×10^6	
Brass, cast (60% Cu: 40% Zn)	0.300	10.4×10^{-6}	20,000	....	45,000	....	....	13×10^6	5×10^6	20
Common brass, rolled	0.310	10.4×10^{-6}	25,000	15,000	60,000	....	50,000	14×10^6	5×10^6	30
Bronze, cast (90% Cu: 10% Sn)	0.295	10.0×10^{-6}	20,000	....	33,000	56,000	....	12×10^6	....	10

* These physical properties will be greatly modified by variations in composition, heat-treatment, previous cold working or drawing, etc.
† The compressive proportional limit and modulus of elasticity may be assumed same as for tension except for cast iron where prop. limit in compression is 25,000 psi.
‡ The ultimate compressive strength of a ductile material is an uncertain quantity. When no value is stated in the above table, for practical purposes this strength may be assumed the same as the proportional limit in tension.
§ In pure shear, cast iron fails by diagonal tension.
¶ For steel, many engineers use $E = 29 \times 10^6$ psi.

TABLE A.2
AVERAGE PHYSICAL PROPERTIES AND WORKING STRESSES OF STRUCTURAL TIMBER
Common Structural Grade—Kept Continuously Dry

Species	Density (dry), lb per cu ft	Bending, psi				Shearing, psi			Compression, psi						
						Parallel to Grain		Long. Shear in Beams	Parallel to Grain					Perpendicular to Grain	
		Prop. Limit	Modulus of Rupture	Working Str.	Modulus of Elasticity	Ult. Str.	Working Str.	Working Str.	Prop. Limit	Ult. Str.	S Working Str.	E Modulus of Elasticity	K (See Art. 10.6)	Prop. Limit	Working Str.
Red Cedar	23	5,500	6,500	720	1.0×10^6	400	100	64	4,000	5,000	560	1.0×10^6	27.1	700	200
Bald Cypress	30	6,500	7,500	1,040	1.2×10^6	500	125	80	4,500	6,000	880	......	....	800	350
Douglas Fir	30	6,500	10,500	1,200	1.6×10^6	500	125	72	3,500	5,000	880	1.6×10^6	27.3	800	325
Western Hemlock	28	6,300	10,000	1,040	1.4×10^6	400	75	60	4,500	5,000	720	1.4×10^6	28.3	600	300
White Oak	50	9,500	13,000	1,100	1.7×10^6	900	225	110	5,000	8,500	880	1.65×10^6	27.8	2,200	500
Longleaf Pine	34	6,500	11,000	1,400	1.6×10^6	800	130	103	4,000	5,800	1,025	1.6×10^6	25.3	900	380
Shortleaf Pine	32	5,000	8,000	1,200	1.6×10^6	700	150	88	.....	.....	880	1.6×10^6	27.3	.....	325
Soft Pine	24	6,300	8,000	720	1.0×10^6	400	80	68	3,500	5,000	700	1.4×10^6	....	700	250
Redwood	22	4,500	7,500	960	1.2×10^6	350	90	56	3,800	5,000	800	1.2×10^6	24.8	550	250
Spruce	24	8,000	10,000	880	1.2×10^6	800	200	68	5,500	7,000	640	1.2×10^6	27.8	700	250
Tamarack	31	7,500	13,000	960	1.3×10^6	400	100	76	3,500	4,500	.....	1.4×10^6	....	700	300

The physical properties of a given species of timber vary widely, depending upon the quality, seasoning, etc., of the specimen. The above values may be increased some 20 or 25 per cent if selected stock be used. If used where occasionally or continuously wet, most of the above stress values should be reduced from one-fifth to one-third. For more detailed information, see American Society for Testing Materials Specifications, D245–33.

Parts of the above table are reprinted from the 1933 A.S.T.M. Standards by permission of the American Society for Testing Materials.

TABLE A.3

AVERAGE PHYSICAL PROPERTIES OF BUILDING STONE, BRICK AND CONCRETE

Material	Density, lb per cu ft	Temp. Coef. of Linear Expansion per Degree Fahr.	Ultimate Compressive Strength, psi	Ultimate Shearing Strength Across Grain, psi	Modulus of Rupture in Bending, psi	Modulus of Elasticity in Compression psi
Granite..........	165	3.6×10^{-6}	20,000	2,300	1,600	7.5×10^{6}
Limestone........	160	3.0×10^{-6}	10,000	1,400	1,200	8.4×10^{6}
Marble..........	170	4.0×10^{-6}	12,000	1,300	1,500	8.2×10^{6}
Sandstone........	135	5.2×10^{-6}	10,000	1,700	1,500	3.3×10^{6}
Slate...........	175	....	15,000	.....	8,000	14.0×10^{6}
Common brick........	125	4.0×10^{-6}	4,000	.....	800	2.0×10^{6}
Stone or gravel concrete *						
7½ gals water per sack of cement....	150	6.0×10^{-6}	3,700	1,000 †	550	3.1×10^{6}
6¾ gals water per sack of cement....	150	6.0×10^{-6}	4,300	1,250 †	600	3.3×10^{6}
6 gals water per sack of cement......	150	6.0×10^{-6}	5,200	1,500 †	700	3.5×10^{6}

* Strength values are for concrete 28 days old. Working strengths may be taken as one-quarter of these 28-day strengths.
† This direct shearing strength must not be used in a beam involving diagonal tension where the concrete may break with a shearing stress equal to from 5 to 10 per cent of its compressive strength

TABLE A.4 Mechanical Properties of Steels

Source	Material, %	State	Prop. Limit lb/in.²	Yield Point lb/in.²	Ult. Strength lb/in.²	Elong. (2-in. gage) %	Reduction in Area %	Endurance Limit lb/in.²	Remarks
1	.37 C, .55 Mn	Annealed at 850° C Normalized at 850° C Heat-treated, water at 850°C, temp. at 550° C	36,500 38,000 65,000	37,800 41,500 69,000	70,000 79,200 105,000	32 29 22	49 46 56	±29,000 ±29,300 ±51,000	Bar 2¼ in. diameter
2	.49 C, .46 Mn	Normalized at 910° C Heat-treated, oil at 790° C, temp. at 430° C	44,700 75,800	47,100 78,800	91,500 121,800	27 11.3	40 51	±33,000 ±64,000	Bar 15/16 in. sq
1	.35 C, .45 Mn, 3.4 Ni	Rolled Annealed at 840° C Normalized at 840° C, temp. at 730° C Heat-treated, water at 800° C, temp. at 600° C	47,000 52,000 52,000 72,000	60,000 60,000 56,000 77,000	105,000 104,000 94,000 107,000	21 22 25 23	42 49 48 56	±41,000 ±44,000 ±47,500 ±52,000	Plate 3½ ft × 2 ft × 2 in.
2	.24 C, .37 Mn, 3.3 Ni, .87 Cr	Annealed at 780° C Heat-treated, oil 830° C	56,700 115,000	60,000 128,000	87,000 138,000	33 18	67 62	±49,000 ±68,000	Bar 2 in. × 1 in.
3	.30 C, .56 Mn, 4.3 Ni, 1.4 Cr	Air-hardened from 800° C Air-hardened from 800° C, temp. 600° C	45,000 92,000	177,000 142,000	244,000 157,000	10.8 17.5	37 55	±102,000 ±80,000	Bar 1⅛ in. diameter
1	.32 C, .74 Mn, .32 Si	Cast Annealed at 925° C Normalized at 925° C	20,000 37,000 40,500	33,500 41,000 46,000	76,000 80,000 85,000	26 27 28	34 40 46	±30,500 ±35,000 ±35,000	Bar 2¼ in. × 1¼ in.

Source: 1. Research Laboratory, Westinghouse Elec. Corp. See S. Timoshenko and J. M. Lessells, *Applied Elasticity,* p. 522, 1924
2. H. F. Moore and T. Jasper, *Univ. of Illinois Eng. Exp. Sta. Bull.,* No. 136, p. 33.
3. L. Aitchison, *Engineering Steels,* p. 209, 1921.

APPENDIX B

Moments of Inertia of Plane Areas

B-1.   Moment of Inertia of a Plane Area with Respect to an Axis in Its Plane

In discussing the bending of beams, one encounters integrals of the type

$$I_x = \int_A y^2 dA \tag{1}$$

in which each element of area dA is multiplied by the square of its distance from the x-axis and integration is extended over the cross-sectional area A of the beam (Fig. B1). Such an integral is called the *moment of inertia* of the area A with respect to the x-axis.

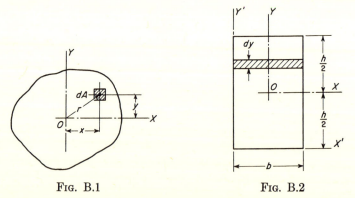

FIG. B.1   FIG. B.2

In simple cases, moments of inertia can readily be calculated analytically. Take, for instance, a rectangle (Fig. B2). In calculating the moment of inertia of this rectangle with respect to the horizontal axis of symmetry x, we can divide the rectangle into infinitesimal elements such as shown in the figure by the shaded area. Then

$$I_x = 2\int_0^{h/2} y^2 b \, dy = \frac{bh^3}{12}. \tag{2}$$

346

In the same manner, the moment of inertia of the rectangle with respect to the y-axis is

$$I_y = 2\int_0^{b/2} x^2 h \, dx = \frac{hb^3}{12}.$$

Eq. (2) can also be used for calculating I_x for the parallelogram shown in Fig. B3, because this parallelogram can be obtained from the rectangle shown by dotted lines by a displacement parallel to the axis x of elements such as the one shown. The areas of the elements and their distances from the x-axis remain unchanged during such displacement so that I_x is the same as for the rectangle.

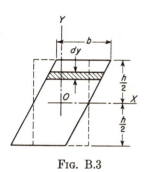

FIG. B.3

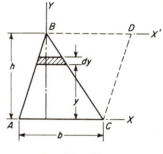

FIG. B.4

In calculating the moment of inertia of a triangle with respect to the base (Fig. B4), the area of an element such as shown in the figure is

$$dA = b\frac{h - y}{h}dy$$

and eq. (1) gives

$$I_x = \int_0^h b\frac{h - y}{h} y^2 \, dy = \frac{bh^3}{12}.$$

The method of calculation illustrated by the above examples can be applied to any area. The moment of inertia is obtained by dividing the figure into infinitesimal strips parallel to the axis and then integrating as in eq. (1).

The calculation can often be simplified if the figure can be divided into portions whose moments of inertia about the axis are known. In such case, the total moment of inertia is the sum of the moments of inertia of all the parts.

From its definition, eq. (1), it follows that the moment of inertia of an area with respect to an axis has the dimensions of a length raised to the fourth power; hence, by dividing the moment of inertia with respect to a

certain axis by the cross-sectional area of the figure, the square of a certain
length is obtained. This length is called the *radius of gyration* with respect
to that axis. For the x- and y-axes, the radii of gyration are

$$r_x = \sqrt{\frac{I_x}{A}}; \quad r_y = \sqrt{\frac{I_y}{A}}. \tag{3}$$

EXAMPLE. Calculate the moments of inertia I_x and I_y for the I-section having
the dimensions shown in Fig. B.5.

SOLUTION. In calculating I_x, we consider the net section (shaded) to consist of
the circumscribed rectangle minus the two rectangular cut-outs on either side of the
web. Then, using eq. (2), we have

$$I_x = \frac{bh^3}{12} - \frac{(b-t)(h-2d)^3}{12}. \tag{a}$$

In calculating the moment of inertia about the y-axis, we consider the net section
to consist of the two flanges plus the web. Thus

$$I_y = \frac{2db^3}{12} + \frac{(h-2d)t^3}{12}. \tag{b}$$

Equations (a) and (b) can be used for an I-section of any proportions. For example, taking $h = 12$ in., $b = 6$ in., $d = 1$ in., and $t = \frac{1}{2}$ in., we have

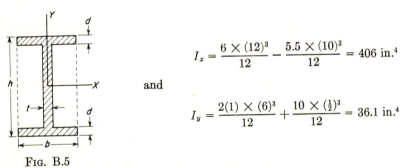

$$I_x = \frac{6 \times (12)^3}{12} - \frac{5.5 \times (10)^3}{12} = 406 \text{ in.}^4$$

and

$$I_y = \frac{2(1) \times (6)^3}{12} + \frac{10 \times (\frac{1}{2})^3}{12} = 36.1 \text{ in.}^4$$

FIG. B.5

B-2.   Polar Moment of Inertia of a Plane Area

The moment of inertia of a plane area with respect to an axis perpendicular to the plane of the figure is called the *polar moment of inertia* with
respect to the point, where the axis intersects the plane (point O in Fig. B1).
It is defined as the integral

$$J = \int_A r^2 \, dA, \tag{4}$$

in which each element of area dA is multiplied by the square of its distance
from the axis and integration is extended over the entire area of the figure.

Referring to Fig. B1, $r^2 = x^2 + y^2$, and from eq. (4)

$$J = \int_A (x^2 + y^2)\, dA = I_x + I_y. \tag{5}$$

That is, the polar moment of inertia of a plane area with respect to any point O is equal to the sum of the moments of inertia with respect to two perpendicular axes x and y through the same point.

Let us consider a *circular cross-section*. We encounter the polar moment of inertia of a circle with respect to its center in discussing the twist of a circular shaft (see Art. 4.1). If we divide the area of the circle into thin elemental rings, as shown in Fig. B6, we have $dA = 2\pi r\, dr$, and from eq. (4)

$$J = 2\pi \int_0^{d/2} r^3\, dr = \frac{\pi d^4}{32}. \tag{6}$$

We know from symmetry that in this case $I_x = I_y$; hence, from eqs. (5) and (6),

$$I_x = I_y = \frac{J}{2} = \frac{\pi d^4}{64}. \tag{7}$$

The moment of inertia of an ellipse with respect to a principal axis x (Fig. B7) can be obtained by comparing the ellipse with the circle shown in the figure by the dotted line.

The height y of any element of the ellipse, such as the element shown shaded, can be obtained by reducing the height y_1 of the corresponding element of the circle in the ratio b/a. From eq. (2), the moments of inertia of these two elements with respect to the x-axis are in the ratio b^3/a^3. The moments of inertia of the ellipse and of the circle are evidently in the same ratio; hence, the moment of inertia of the ellipse is

$$I_x = \frac{\pi(2a)^4}{64} \cdot \frac{b^3}{a^3} = \frac{\pi a b^3}{4}. \tag{8}$$

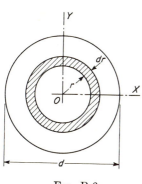

FIG. B.6

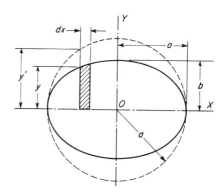

FIG. B.7

In the same manner, for the vertical axis,

$$I_y = \frac{\pi b a^3}{4};$$

the polar moment of inertia of an ellipse is then, from eq. (5),

$$J = I_x + I_y = \frac{\pi a b^3}{4} + \frac{\pi b a^3}{4}. \tag{9}$$

EXAMPLE. Find the polar moment of inertia of the rectangle shown in Fig. B.2 with respect to the centroid O.

SOLUTION. From eq. (5),

$$J_O = I_x + I_y = \frac{bh^3}{12} + \frac{hb^3}{12}.$$

B-3. Parallel-Axis Theorem

If the moment of inertia of an area with respect to an x-axis through the centroid (Fig. B8) is known, the moment of inertia with respect to any parallel x'-axis can be calculated from the equation:

$$I_{x'} = I_x + Ad^2, \tag{10}$$

in which A is the area of the figure and d is the distance between the axes. This can be proved as follows: from eq. (1)

$$I_{x'} = \int_A (y + d)^2 \, dA = \int_A y^2 \, dA + 2 \int_A yd \, dA + \int_A d^2 \, dA.$$

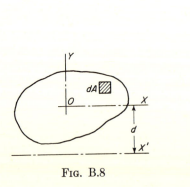

FIG. B.8

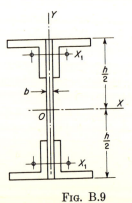

FIG. B.9

The first integral on the right side is equal to I_x, the third integral is equal to Ad^2, and the second integral vanishes because the x-axis passes through the centroid; hence, this equation reduces to (10). Eq. (10) is especially useful in calculating moments of inertia of cross sections of built-up beams (Fig. B9). The positions of the centroids of standard angles and the moments of inertia of their cross sections with respect to an axis through their centroid are given in handbooks. By use of the parallel-axis theorem, the moment of inertia of such a built-up section with respect to the x-axis can readily be calculated.

EXAMPLE. Calculate the moment of inertia I_x for the built-up I-section shown in Fig. B.9 if each angle is a 4 in. × 4 in. × $\frac{1}{2}$ in. section and the web is a 20 in. × $\frac{1}{2}$ in. plate.

SOLUTION. From Table B-5, we find, for one angle, the following data:

$$A = 3.75 \text{ in.}^2, \quad I_1 = 5.6 \text{ in.}^4, \quad \bar{x} = 1.18 \text{ in.}$$

Then, using the parallel-axis theorem, we have for the complete section

$$I_x = \frac{(\frac{1}{2}) \times (20)^3}{12} + 4[5.6 + 3.75(10 - 1.18)^2] = 1522 \text{ in.}^4$$

B-4. Product of Inertia, Principal Axes

The integral

$$I_{xy} = \int_A xy \, dA, \tag{11}$$

in which each element of the area dA is multiplied by the product of its coordinates, and integration is extended over the entire area A of a plane figure, is called the *product of inertia* of the figure. If a figure has an axis of symmetry which is taken for the x- or y-axis (Fig. B10), the product of inertia is equal to zero. This follows from the fact that in this case for any element such as dA with a positive x, there exists an equal and symmetrically situated element dA' with a negative x. The corresponding elementary products $xydA$ cancel each other; hence integral (11) vanishes.

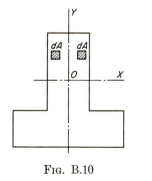

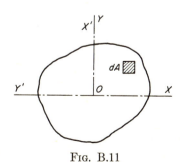

FIG. B.10   FIG. B.11

In the general case, for any point of any plane figure, we can always find two perpendicular axes such that the product of inertia for these axes vanishes. Take, for instance, the axes x and y, Fig. B11. If the axes are rotated 90° about O in the clockwise direction, the new positions of the axes are x' and y' as shown in the figure. There is then the following relation between the old coordinates of an element dA and its new coordinates:

$$x' = y; \quad y' = -x.$$

Hence the product of inertia for the new coordinates is

$$I_{x'y'} = \int_A x'y' \, dA = - \int_A xy \, dA = -I_{xy};$$

thus, during this rotation, the product of inertia changes its sign. As the product of inertia changes continuously with the angle of rotation, there must be certain directions for which this quantity becomes zero. The axes in these directions are called the *principal axes*. Usually the centroid is taken as the origin of coordinates and the corresponding principal axes are then called the *centroidal principal axes*. The centroidal principal axes are of importance since the moments of inertia are maxima or minima with respect to these axes (see the next article). If a figure has an axis of symmetry, this axis and an axis perpendicular to it are principal axes of the figure, because the product of inertia with respect to these axes is equal to zero, as explained above.

If the product of inertia of a figure is known for x- and y-axes (Fig. B12) through the centroid, the product of inertia for parallel x'- and y'-axes can be found from the equation:

$$I_{x'y'} = I_{xy} + Aab. \tag{12}$$

The coordinates of an element dA for the new axes are

$$x' = x + a; \quad y' = y + b.$$

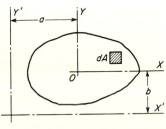

FIG. B.12

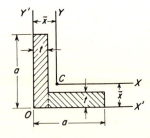

FIG. B.13

Hence,

$$I_{x'y'} = \int_A x'y' \, dA = \int_A (x + a)(y + b)dA$$

$$= \int_A xy \, dA + \int_A ab \, dA + \int_A bx \, dA + \int_A ya \, dA.$$

The last two integrals vanish because C is the centroid so that the equation reduces to (12).

EXAMPLE. Calculate the product of inertia I_{xy} for the angle section shown in Fig. B.13 if the dimensions are $a = 6$ in. and $t = 1$ in.

SOLUTION. Dividing the figure into two rectangles as shown and noting that, with respect to centroidal axes, the products of inertia of the rectangles are both zero, we have

$$I_{x'y'} = at \times \frac{a}{2} \times \frac{t}{2} + (a - t)t \times \frac{a + t}{2} \times \frac{t}{2},$$

which reduces to

$$I_{x'y'} = \frac{a^2 t^2}{2} - \frac{t^4}{4}. \tag{a}$$

With the given values of a and t, this gives $I_{x'y'} = 17.75$ in.[4] Next, from Table B-5, we find

$$A = 11.0 \text{ in.}^2 \quad \text{and} \quad \bar{x} = 1.86 \text{ in.}$$

Hence, from eq. (12),

$$I_{xy} = I_{x'y'} - A\bar{x}^2 = 17.75 - 11(1.86)^2 = -20.3 \text{ in.}^4$$

B-5. Change of Direction of Axes. Determination of Principal Axes

Suppose that the moments of inertia

$$I_x = \int_A y^2 \, dA; \quad I_y = \int_A x^2 \, dA \tag{a}$$

and the product of inertia

$$I_{xy} = \int_A xy \, dA \tag{b}$$

are known, and it is required to find the same quantities for the new axes x_1 and y_1 (Fig. B14). Considering an elementary area dA, the new coordinates from the figure are

$$x_1 = x \cos \phi + y \sin \phi; \quad y_1 = y \cos \phi - x \sin \phi, \tag{c}$$

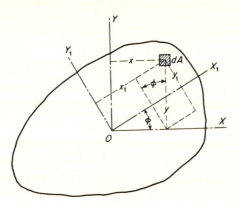

FIG. B.14

in which ϕ is the angle between x and x_1. Then

$$I_{z_1} = \int_A y_1{}^2 \, dA = \int_A (y \cos \phi - x \sin \phi)^2 \, dA = \int_A y^2 \cos^2 \phi \, dA$$

$$+ \int_A x^2 \sin^2 \phi \, dA - \int_A 2xy \sin \phi \cos \phi \, dA,$$

or, by using *(a)* and *(b)*,

$$I_{z_1} = I_x \cos^2 \phi + I_y \sin^2 \phi - I_{xy} \sin 2\phi. \tag{13}$$

In the same manner

$$I_{y_1} = I_x \sin^2 \phi + I_y \cos^2 \phi + I_{xy} \sin 2\phi. \tag{13'}$$

By substituting the trigonometric identities, $\cos^2 \phi = \frac{1}{2}(1 + \cos 2\phi)$ and $\sin^2 \phi = \frac{1}{2}(1 - \cos 2\phi)$, eqs. (13) and (13') become

$$I_{z_1} = \frac{I_x + I_y}{2} + \frac{I_x - I_y}{2} \cos 2\phi - I_{xy} \sin 2\phi \tag{14}$$

and

$$I_{y_1} = \frac{I_x + I_y}{2} - \frac{I_x - I_y}{2} \cos 2\phi + I_{xy} \sin 2\phi. \tag{14'}$$

These equations are very useful for calculating I_{z_1} and I_{y_1}.

The value of ϕ which makes I_{z_1} a maximum or a minimum may be found by differentiating I_{z_1} in eq. (14) with respect to ϕ and setting the derivative equal to zero. This gives

$$\tan 2\phi = \frac{2I_{xy}}{I_y - I_x}. \tag{15}$$

Substituting this value of 2ϕ in eq. (14), one obtains

$$(I_{z_1})_{\substack{\max \\ \min}} = \frac{I_z + I_y}{2} \pm \sqrt{\left(\frac{I_z - I_y}{2}\right)^2 + I_{zy}^2}. \qquad (16)$$

For calculating $I_{x_1y_1}$, we find

$$I_{x_1y_1} = \int_A x_1y_1\,dA = \int_A (x\cos\phi + y\sin\phi)(y\cos\phi - x\sin\phi)dA$$

$$= \int_A y^2 \sin\phi\cos\phi\,dA - \int_A x^2 \sin\phi\cos\phi\,dA + \int_A xy\,(\cos^2\phi - \sin^2\phi)dA$$

By using *(a)*and *(b)*

$$I_{x_1y_1} = \frac{I_z - I_y}{2}\sin 2\phi + I_{xy}\cos 2\phi. \qquad (17)$$

The angle ϕ, locating the two perpendicular axes with respect to which the product of inertia is zero, may be found by setting $I_{x_1y_1} = 0$ in eq. (17). The resulting value of $\tan 2\phi$ is that already stated in eq. (15) and shows that the moment of inertia is a maximum or minimum about the principal axes (the axes about which the product of inertia vanishes).

The radii of gyration corresponding to the principal axes are called *principal radii of gyration.*

EXAMPLE. For the z-section shown in Fig. B.15, find the value of ϕ which locates the centroidal principal axes and determine the corresponding principal moments of inertia. The following numerical data are given: $h = 10$ in., $b = 5$ in.. $t = 1$ in.
SOLUTION. Using the parallel-axis theorem,

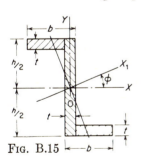

FIG. B.15

$$I_z = \frac{1(10)^3}{12} + 2\left[\frac{4(1)^3}{12} + 4(4.5)^2\right] = 246 \text{ in.}^4$$

$$I_y = \frac{10(1)^3}{12} + 2\left[\frac{1(4)^3}{12} + 4(2.5)^2\right] = 61.5 \text{ in.}^4$$

$$I_{xy} = 0 - 2[4(2.5)(4.5)] = -90.0 \text{ in.}^4$$

Now substitute these values into eq. (15):

$$\tan 2\phi = -\frac{2(90)}{61.5 - 246} = +\frac{180}{184.5} = 0.976.$$

Hence $2\phi = 44° 18'$ and $\phi = 22°09'$.
Finally, using eq. (16), we obtain

$$I_{\substack{\max \\ \min}} = \frac{246 + 61.5}{2} \pm \sqrt{\left(\frac{184.5}{2}\right)^2 + (90)^2}$$

from which $I_{\max} = 282.75$ in.^{4} and $I_{\min} = 24.75$ in.^{4}

1. Find the moment of inertia of the rectangle in Fig. B.2 with respect to its base, i.e., find $I_{x'}$ by direct integration of eq. (1). *Ans.* $I_{x'} = bh^3/3$.

2. Repeat the solution to the preceding problem by using the centroidal moment of inertia, eq. (2), together with the parallel-axis theorem.

3. Find the moment of inertia of the triangle in Fig. B.4 with respect to the axis X' through the apex B and parallel to the base. *Ans.* $I_{x'} = bh^3/4$.

4. Find the moment of inertia of a square having sides of length a with respect to a diagonal. *Ans.* $I_d = a^4/12$.

5. Find the polar moment of inertia of a square having sides of length a with respect to its centroid. *Ans.* $J_c = a^4/6$.

6. Find the polar moment of inertia of an equilateral triangle having sides of length a with respect to its centroid. *Ans.* $J_c = \sqrt{3}\,a^4/48$.

7. Calculate the centroidal moment of inertia I_x of the 4 in. $\times$ 4 in. $\times$ 1 in. T-section shown in Fig. A. *Ans.* $I_x = 9.44$ in.^{4}

8. Using the same dimensions as given in the preceding problem, calculate the moment of inertia I_y for the T-section shown in Fig. A. *Ans.* $I_y = 5.58$ in.^{4}

9. Calculate the centroidal moment of inertia I_x of the 6 in. $\times$ 4 in. $\times$ 1 in. angle section shown in Fig. B. *Ans.* $I_x = 10.75$ in.^{4}

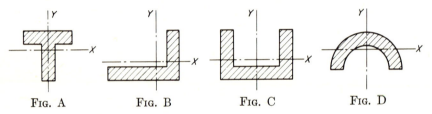

FIG. A   FIG. B   FIG. C   FIG. D

10. Calculate the centroidal moment of inertia I_x of the 6 in. $\times$ 4 in. $\times$ 1 in. channel section shown in Fig. C. *Ans.* $I_x = 17.0$ in.^{4}

11. Calculate the centroidal moment of inertia I_x of the semicircular section shown in Fig. D if the outside diameter is 6 in. and the inside diameter is 4 in. *Ans.* $I_x = 5.09$ in.^{4}

12. Calculate the moment of inertia of a trapezoid having bases a and b ($a > b$) and altitude h with respect to a centroidal axis X parallel to the bases. *Ans.*

$$I_x = \frac{h^3}{36}\left(\frac{a^2 + 4ab + b^2}{a + b}\right).$$

13. Find the product of inertia of the rectangle shown in Fig. B.2 with respect to the axes $x'y'$ coinciding with its edges. *Ans.* $I_{x'y'} = b^2h^2/4$.

14. Given a right triangle with base b and altitude h, find its product of inertia for axes $x'y'$ coinciding with the two orthogonal edges. *Ans.* $I_{x'y'} = b^2h^2/24$.

15. Find the product of inertia of the preceding right triangle with respect to centroidal axes x and y parallel to the orthogonal edges. *Ans.* $I_{xy} = -b^2h^2/72$.

16. For the right triangle described in the two preceding problems, find the value of the angle ϕ which centroidal principal axes make, respectively, with the axes x and y. *Ans.* $\tan 2\phi = bh/(h^2 - b^2)$.

17. Calculate the product of inertia of a quadrant of a circle of radius r with respect to axes x' and y' coinciding with the two radial edges. *Ans.* $I_{x'y'} = r^4/8$.

APPENDIX B-1

Tables

357

TABLE B-1

ELEMENTS OF COMMON SECTIONS

Rectangle	$A = bh$ $c = \dfrac{h}{2}$ $r_x = \dfrac{h}{\sqrt{12}}$	$I_x = \dfrac{bh^3}{12}$ $I_y = \dfrac{hb^3}{12}$ $I_{xy} = 0$
Isosceles Triangle	$A = \dfrac{bh}{2}$ $c = \dfrac{h}{3}$ $r_x = \dfrac{h}{\sqrt{18}}$	$I_x = \dfrac{bh^3}{36}$ $I_y = \dfrac{hb^3}{48}$ $I_{xy} = 0$
Trapezoid	$A = \dfrac{h}{2}(a+b)$ $c = \dfrac{h}{3}\left(\dfrac{a+2b}{a+b}\right)$ $r_x = \dfrac{h\sqrt{a^2 + 4ab + b^2}}{\sqrt{18}\,(a+b)}$	$I_x = \dfrac{h^3}{36}\left(\dfrac{a^2 + 4ab + b^2}{a+b}\right)$ $I_y = \dfrac{h(a+b)(a^2 + b^2)}{48}$ $I_{xy} = 0$
Circle	$A = \pi r^2$ $c = r$ $r_x = \dfrac{r}{2}$	$I_x = \dfrac{\pi r^4}{4}$ $I_y = \dfrac{\pi r^4}{4}$ $I_{xy} = 0$
Ellipse	$A = \pi ab$ $c = b$ $r_x = \dfrac{b}{2}$	$I_x = \dfrac{\pi ab^3}{4}$ $I_y = \dfrac{\pi ba^3}{4}$ $I_{xy} = 0$

TABLE **B**-1 (*Continued*)

Square	$A = a^2$ $c = \dfrac{a}{\sqrt{2}}$ $r_x = \dfrac{a}{\sqrt{12}}$	$I_x = \dfrac{a^4}{12}$ $I_y = \dfrac{a^4}{12}$ $I_{xy} = 0$
Thin-walled Tube	$A = 2\pi rt$ $c = r + \dfrac{t}{2}$ $r_x = \dfrac{r}{\sqrt{2}}$	$I_x = \pi t r^3$ $I_y = \pi t r^3$ $I_{xy} = 0$
Semicircle	$A = \dfrac{\pi r^2}{2}$ $c = \dfrac{4r}{3\pi}$ $r_x = 0.264r$	$I_x = 0.110r^4$ $I_y = \dfrac{\pi r^4}{8}$ $I_{xy} = 0$
Right Triangle	$A = \dfrac{bh}{2}$ $c = \dfrac{h}{3}$ $r_x = \dfrac{h}{\sqrt{18}}$	$I_x = \dfrac{bh^3}{36}$ $I_y = \dfrac{hb^3}{36}$ $I_{xy} = -\dfrac{b^2h^2}{72}$
Circular Quadrant | $A = \dfrac{\pi r^2}{4}$ $c = \dfrac{4r}{3\pi}$ $r_x = 0.264r$ | $I_x = 0.0549r^4$ $I_y = 0.0549r^4$ $I_{xy} = -0.0164r^4$ |

TABLE B-2

ELEMENTS OF WIDE FLANGE SECTIONS

(Abridged List)

Section Index	Nominal Size	Depth of Section In.	Weight per Foot Lb	Area of Section Sq In.	Flange Width In.	Flange Thickness In.	Web Thickness In.	Axis 1-1 I In.4	Axis 1-1 Z In.3	Axis 1-1 r In.	Axis 2-2 I In.4	Axis 2-2 Z In.3	Axis 2-2 r In.
36 WF 300	36 × 16½	36.72	300.0	88.17	16.655	1.680	0.945	20,290.2	1,105.1	15.17	1,225.2	147.1	3.73
36 WF 230	"	35.88	230.0	67.73	16.475	1.260	0.765	14,988.4	835.5	14.88	870.9	105.7	3.59
36 WF 194	36 × 12	36.48	194.0	57.11	12.117	1.260	0.770	12,103.4	663.6	14.56	355.4	58.7	2.49
36 WF 150	"	35.84	150.0	44.16	11.972	0.940	0.625	9,012.1	502.9	14.29	250.4	41.8	2.38
33 WF 240	33 × 15¾	33.50	240.0	70.52	15.865	1.400	0.830	13,585.1	811.1	13.88	874.3	110.2	3.52
33 WF 200	"	33.00	200.0	58.79	15.750	1.150	0.715	11,048.2	669.6	13.71	691.7	87.8	3.43
33 WF 152	33 × 11½	33.50	152.0	44.71	11.565	1.055	0.635	8,147.6	486.4	13.50	256.1	44.3	2.39
33 WF 130	"	33.10	130.0	38.26	11.510	0.855	0.580	6,699.0	404.8	13.23	201.4	35.0	2.29
30 WF 21C	30 × 15	30.38	210.0	61.78	15.105	1.315	0.775	9,872.4	649.9	12.64	707.9	93.7	3.38
30 WF 172	"	29.88	172.0	50.65	14.985	1.065	0.655	7,891.5	528.2	12.48	550.1	73.4	3.30
30 WF 132	30 × 10½	30.30	132.0	38.83	10.551	1.000	0.615	5,753.1	379.7	12.17	185.0	35.1	2.18
30 WF 116	"	30.00	116.0	34.13	10.500	0.850	0.564	4,919.1	327.9	12.00	153.2	29.2	2.12
27 WF 177	27 × 14	27.31	177.0	52.10	14.090	1.190	0.725	6,728.6	492.8	11.36	518.9	73.7	3.16
27 WF 145	"	26.88	145.0	42.68	13.965	0.975	0.600	5,414.3	402.9	11.26	406.9	58.3	3.09
27 WF 114	27 × 10	27.28	114.0	33.53	10.070	0.932	0.570	4,080.5	299.2	11.03	149.6	29.7	2.11
27 WF 94	"	26.91	94.0	27.65	9.990	0.747	0.490	3,266.7	242.8	10.87	115.1	23.0	2.04
24 WF 160	24 × 14	24.72	160.0	47.04	14.091	1.135	0.656	5,110.3	413.5	10.42	492.6	69.9	3.23
24 WF 130	"	24.25	130.0	38.21	14.000	0.900	0.565	4,009.5	330.7	10.24	375.2	53.6	3.13
24 WF 120	24 × 12	24.31	120.0	35.29	12.088	0.930	0.556	3,635.3	299.1	10.15	254.0	42.0	2.68

Section	Size												
24 WF 100		24.00	100.0	29.43	12.000	0.775	0.468	2,987.3	248.9	10.08	203.5	33.9	2.63
24 WF 94	24 × 9	24.29	94.0	27.63	9.061	0.872	0.516	2,683.0	220.9	9.85	102.2	22.6	1.92
24 WF 84		24.09	84.0	24.71	9.015	0.772	0.470	2,364.3	196.3	9.78	88.3	19.6	1.89
24 WF 76		23.91	76.0	22.37	8.985	0.682	0.440	2,096.4	175.4	9.68	76.5	17.0	1.85
21 WF 142	21 × 13	21.46	142.0	41.76	13.132	1.095	0.659	3,403.1	317.2	9.03	385.9	58.8	3.04
21 WF 112		21.00	112.0	32.93	13.000	0.865	0.527	2,620.6	249.6	8.92	289.7	44.6	2.96
21 WF 96	21 × 9	21.14	96.0	28.21	9.038	0.935	0.575	2,088.9	197.6	8.60	109.3	24.2	1.97
21 WF 82		20.86	82.0	24.10	8.962	0.795	0.499	1,752.4	168.0	8.53	89.6	20.0	1.93
21 WF 73	21 × 8¼	21.24	73.0	21.46	8.295	0.740	0.455	1,600.3	150.7	8.64	66.2	16.0	1.76
21 WF 62		20.99	62.0	18.23	8.240	0.615	0.400	1,326.8	126.4	8.53	53.1	12.9	1.71
18 WF 114	18 × 11¾	18.48	114.0	33.51	11.833	0.991	0.595	2,033.8	220.1	7.79	255.6	43.2	2.76
18 WF 105		18.32	105.0	30.86	11.792	0.911	0.554	1,852.5	202.2	7.75	231.0	39.2	2.73
18 WF 96		18.16	96.0	28.22	11.750	0.831	0.512	1,674.7	184.4	7.70	206.8	35.2	2.71
18 WF 85	18 × 8¾	18.32	85.0	24.97	8.838	0.911	0.526	1,429.9	156.1	7.57	99.4	22.5	2.00
18 WF 70		18.00	70.0	20.56	8.750	0.751	0.438	1,153.9	128.2	7.49	78.5	17.9	1.95
18 WF 64		17.87	64.0	18.80	8.715	0.686	0.403	1,045.8	117.0	7.46	70.3	16.1	1.93
18 WF 55	18 × 7½	18.12	55.0	16.19	7.532	0.630	0.390	889.9	98.2	7.41	42.0	11.1	1.61
18 WF 50		18.00	50.0	14.71	7.500	0.570	0.358	800.6	89.0	7.38	37.2	9.9	1.59
16 WF 88	16 × 11½	16.16	88.0	25.87	11.502	0.795	0.504	1,222.6	151.3	6.87	185.2	32.2	2.67
16 WF 78	16 × 8½	16.32	78.0	22.92	8.586	0.875	0.529	1,042.6	127.8	6.74	87.5	20.4	1.95
16 WF 71		16.16	71.0	20.86	8.543	0.795	0.486	936.9	115.9	6.70	77.9	18.2	1.93
16 WF 64		16.00	64.0	18.80	8.500	0.715	0.443	833.8	104.2	6.66	68.4	16.1	1.91
16 WF 58		15.86	58.0	17.04	8.464	0.645	0.407	746.4	94.1	6.62	60.5	14.3	1.88
16 WF 50	16 × 7	16.25	50.0	14.70	7.073	0.628	0.380	655.4	80.7	6.68	34.8	9.8	1.54
16 WF 45		16.12	45.0	13.24	7.039	0.563	0.346	583.3	72.4	6.64	30.5	8.7	1.52
16 WF 36		15.85	36.0	10.59	6.992	0.428	0.299	446.3	56.3	6.49	22.1	6.3	1.45
14 WF 426	14 × 16	18.69	426.0	125.25	16.695	3.033	1.875	6,610.3	707.4	7.26	2,359.5	282.7	4.34
14 WF 342		17.56	342.0	100.59	16.365	2.468	1.545	4,911.5	559.4	6.99	1,806.9	220.8	4.24
14 WF 264		16.50	264.0	77.63	16.025	1.938	1.205	3,526.0	427.4	6.74	1,331.2	166.1	4.14
14 WF 228		16.00	228.0	67.06	15.865	1.688	1.045	2,942.4	367.8	6.62	1,124.8	141.8	4.10
14 WF 202		15.63	202.0	59.39	15.750	1.503	0.930	2,538.8	324.9	6.54	979.7	124.4	4.06
14 WF 167	14 × 16	15.12	167.0	49.09	15.600	1.248	0.780	2,020.8	267.3	6.42	790.2	101.3	4.01
14 WF 142	14 × 14½	14.75	142.0	41.85	15.500	1.063	0.680	1,672.2	226.7	6.32	660.1	85.2	3.97
14 WF 136		14.75	136.0	39.98	14.740	1.063	0.660	1,593.0	216.0	6.31	567.7	77.0	3.77
14 WF 119		14.50	119.0	34.99	14.650	0.938	0.570	1,373.1	189.4	6.26	491.8	67.1	3.75
14 WF 111		14.37	111.0	32.65	14.620	0.873	0.540	1,266.5	176.3	6.23	454.9	62.2	3.73
14 WF 95		14.12	95.0	27.94	14.545	0.748	0.465	1,063.5	150.6	6.17	383.7	52.8	3.71
14 WF 87		14.00	87.0	25.56	14.500	0.688	0.420	966.9	138.1	6.15	349.7	48.2	3.70

TABLE B-2 (*Continued*)
ELEMENTS OF WIDE FLANGE SECTIONS
(Abridged List)

Section Index	Nominal Size	Depth of Section In.	Weight per Foot Lb	Area of Section Sq In.	Flange Width In.	Flange Thickness In.	Web Thickness In.	Axis 1-1 I In.⁴	Axis 1-1 Z In.³	Axis 1-1 r In.	Axis 2-2 I In.⁴	Axis 2-2 Z In.³	Axis 2-2 r In.
14 WF 78	14 × 12	14.06	78.0	22.94	12.000	0.718	0.428	851.2	121.1	6.09	206.9	34.5	3.00
14 WF 74	14 × 10	14.19	74.0	21.76	10.072	0.783	0.450	796.8	112.3	6.05	133.5	26.5	2.48
14 WF 68	"	14.06	68.0	20.00	10.040	0.718	0.418	724.1	103.0	6.02	121.2	24.1	2.46
14 WF 61	"	13.91	61.0	17.94	10.000	0.643	0.378	641.5	92.2	5.98	107.3	21.5	2.45
14 WF 53	14 × 8	13.94	53.0	15.59	8.062	0.658	0.370	542.1	77.8	5.90	57.5	14.3	1.92
14 WF 43	"	13.68	43.0	12.65	8.000	0.528	0.308	429.0	62.7	5.82	45.1	11.3	1.89
14 WF 38	14 × 6¾	14.12	38.0	11.17	6.776	0.513	0.313	385.3	54.6	5.87	24.6	7.3	1.49
14 WF 34	"	14.00	34.0	10.00	6.750	0.453	0.287	339.2	48.5	5.83	21.3	6.3	1.46
14 WF 30	"	13.86	30.0	8.81	6.733	0.383	0.270	289.6	41.8	5.73	17.5	5.2	1.41
12 WF 190	12 × 12	14.38	190.0	55.86	12.670	1.736	1.060	1,892.5	263.2	5.82	589.7	93.1	3.25
12 WF 120	"	13.12	120.0	35.31	12.320	1.106	0.710	1,071.7	163.4	5.51	345.1	56.0	3.13
12 WF 106	"	12.88	106.0	31.19	12.230	0.986	0.620	930.7	144.5	5.46	300.9	49.2	3.11
12 WF 92	"	12.62	92.0	27.06	12.155	0.856	0.545	788.9	125.0	5.40	256.4	42.2	3.08
12 WF 85	"	12.50	85.0	24.98	12.105	0.796	0.495	723.3	115.7	5.38	235.5	38.9	3.07
12 WF 72	"	12.25	72.0	21.16	12.040	0.671	0.430	597.4	97.5	5.31	195.3	32.4	3.04
12 WF 65	"	12.12	65.0	19.11	12.000	0.606	0.390	533.4	88.0	5.28	174.6	29.1	3.02
12 WF 58	12 × 10	12.19	58.0	17.06	10.014	0.641	0.359	476.1	78.1	5.28	107.4	21.4	2.51
12 WF 53	"	12.06	53.0	15.59	10.000	0.576	0.345	426.2	70.7	5.23	96.1	19.2	2.48
12 WF 50	12 × 8	12.19	50.0	14.71	8.077	0.641	0.371	394.5	64.7	5.18	56.4	14.0	1.96
12 WF 45	"	12.06	45.0	13.24	8.042	0.576	0.336	350.8	58.2	5.15	50.0	12.4	1.94
12 WF 40	"	11.94	40.0	11.77	8.000	0.516	0.294	310.1	51.9	5.13	44.1	11.0	1.94
12 WF 36	12 × 6½	12.24	36.0	10.59	6.565	0.540	0.305	280.8	45.9	5.15	23.7	7.2	1.50
12 WF 27	"	11.95	27.0	7.97	6.500	0.400	0.240	204.1	34.1	5.06	16.6	5.1	1.44

Section	Size												
10 WF 112	10 × 10	11.38	112.0	32.92	10.415	1.248	0.755	718.7	126.3	4.67	235.4	45.2	2.67
10 WF 100	"	11.12	100.0	29.43	10.345	1.118	0.685	625.0	112.4	4.61	206.6	39.9	2.65
10 WF 89	"	10.88	89.0	26.19	10.275	0.998	0.615	542.4	99.7	4.55	180.6	35.2	2.63
10 WF 77	"	10.62	77.0	22.67	10.195	0.868	0.535	457.2	86.1	4.49	153.4	30.1	2.60
10 WF 72	"	10.50	72.0	21.18	10.170	0.808	0.510	420.7	80.1	4.46	141.8	27.9	2.59
10 WF 66	"	10.38	66.0	19.41	10.117	0.748	0.457	382.5	73.7	4.44	129.2	25.5	2.58
10 WF 54	"	10.12	54.0	15.88	10.028	0.618	0.368	305.7	60.4	4.39	103.9	20.7	2.56
10 WF 49	"	10.00	49.0	14.40	10.000	0.558	0.340	272.9	54.6	4.35	93.0	18.6	2.54
10 WF 45	10 × 8	10.12	45.0	13.24	8.022	0.618	0.350	248.6	49.1	4.33	53.2	13.3	2.00
10 WF 39	"	9.94	39.0	11.48	7.990	0.528	0.318	209.7	42.2	4.27	44.9	11.2	1.98
10 WF 33	"	9.75	33.0	9.71	7.964	0.433	0.292	170.9	35.0	4.20	36.5	9.2	1.94
10 WF 29	10 × 5¾	10.22	29.0	8.53	5.799	0.500	0.289	157.3	30.8	4.29	15.2	5.2	1.34
10 WF 25	"	10.08	25.0	7.35	5.762	0.430	0.252	133.2	26.4	4.26	12.7	4.4	1.31
10 WF 21	"	9.90	21.0	6.19	5.750	0.340	0.240	106.3	21.5	4.14	9.7	3.4	1.25
8 WF 67	8 × 8	9.00	67.0	19.70	8.287	0.933	0.575	271.8	60.4	3.71	88.6	21.4	2.12
8 WF 58	"	8.75	58.0	17.06	8.222	0.808	0.510	227.3	52.0	3.65	74.9	18.2	2.10
8 WF 48	"	8.50	48.0	14.11	8.117	0.683	0.405	183.7	43.2	3.61	60.9	15.0	2.08
8 WF 40	"	8.25	40.0	11.76	8.077	0.558	0.365	146.3	35.5	3.53	49.0	12.1	2.04
8 WF 35	"	8.12	35.0	10.30	8.027	0.493	0.315	126.5	31.1	3.50	42.5	10.6	2.03
8 WF 31	"	8.00	31.0	9.12	8.000	0.433	0.288	109.7	27.4	3.47	37.0	9.2	2.01
8 WF 28	8 × 6½	8.06	28.0	8.23	6.540	0.463	0.285	97.8	24.3	3.45	21.6	6.6	1.62
8 WF 24	"	7.93	24.0	7.06	6.500	0.398	0.245	82.5	20.8	3.42	18.2	5.6	1.61
8 WF 20	8 × 5¼	8.14	20.0	5.88	5.268	0.378	0.248	69.2	17.0	3.43	8.5	3.2	1.20
8 WF 17	"	8.00	17.0	5.00	5.250	0.308	0.230	56.4	14.1	3.36	6.72	2.6	1.16

MISCELLANEOUS BEAMS AND COLUMNS

Section	Size												
6 WF 20	6 × 6	6.20	20.5	5.90	6.018	0.367	0.258	41.7	13.4	2.66	13.3	4.4	1.50
6 WF 15½	"	6.00	15.5	4.62	6.000	0.269	0.240	30.3	10.1	2.56	9.69	3.2	1.45
5 WF 18½	5 × 5	5.12	18.5	5.45	5.025	0.420	0.265	25.4	9.94	2.16	8.89	3.54	1.28
12 B 22	12 × 4	12.31	22.0	6.47	4.030	0.424	0.260	155.7	25.3	4.91	4.55	2.26	0.84
12 B 19	"	12.16	19.0	5.62	4.010	0.349	0.240	130.1	21.4	4.81	3.67	1.83	0.81
8 B 13	8 × 4	8.00	13.0	3.83	4.000	0.254	0.230	39.5	9.88	3.21	2.62	1.31	0.83
6 B 16	6 × 4	6.25	16.0	4.72	4.030	0.404	0.260	31.7	10.1	2.59	4.32	2.14	0.96
6 B 12	"	6.00	12.0	3.53	4.000	0.279	0.230	21.7	7.24	2.48	2.89	1.44	0.90
10 B 11½	10 × 4	9.87	11.5	3.39	3.950	0.204	0.180	51.9	10.5	3.92	2.01	1.02	0.77
8 B 10	8 × 4	7.90	10.0	2.95	3.940	0.204	0.170	30.8	7.79	3.23	1.99	1.01	0.82
10 M 25	10 × 5¾	9.90	25.0	7.35	5.86		0.35	117.0	23.6	3.99	9.84	3.36	1.16
8 M 28	8 × 6½	8.00	28.0	8.23	6.65		0.39	90.1	22.5	3.31	17.73	5.33	1.47
6 M 20	6 × 6	6.00	20.0	5.88	5.938		0.250	38.8	12.9	2.57	11.4	3.8	1.39
12 Jr 11.8	12 × 3	12.00	11.8	3.45	3.063		0.175	72.2	12.0	4.57	0.98	0.64	0.53

TABLE B-3

ELEMENTS OF AMERICAN STANDARD I-BEAM SECTIONS

Section Index	Depth of Beam In.	Weight per Foot Lb	Area of Section In.²	Width of Flange In.	Web Thickness In.	Axis 1-1			Axis 2-2		
						I In.⁴	Z In.³	r In.	I In.⁴	Z In.³	r In.
24 I 120	24.00	120.0	35.13	8.048	.798	3,010.8	250.9	9.26	84.9	21.1	1.56
24 I 105.9	24.00	105.9	30.98	7.875	.625	2,811.5	234.3	9.53	78.9	20.0	1.60
24 I 100	24.00	100.0	29.25	7.247	.747	2,371.8	197.6	9.05	48.4	13.4	1.29
24 I 90	24.00	90.0	26.30	7.124	.624	2,230.1	185.8	9.21	45.5	12.8	1.32
24 I 79.9	24.00	79.9	23.33	7.000	.500	2,087.2	173.9	9.46	42.9	12.2	1.36
20 I 95	20.00	95.0	27.74	7.200	.800	1,599.7	160.0	7.59	50.5	14.0	1.35
20 I 85	20.00	85.0	24.80	7.053	.653	1,501.7	150.2	7.78	47.0	13.3	1.38
20 I 75	20.00	75.0	21.90	6.391	.641	1,263.5	126.3	7.60	30.1	9.4	1.17
20 I 65.4	20.00	65.4	19.08	6.250	.500	1,169.5	116.9	7.83	27.9	8.9	1.21
18 I 70	18.00	70.0	20.46	6.251	.711	917.5	101.9	6.70	24.5	7.8	1.09
18 I 54.7	18.00	54.7	15.94	6.000	.460	795.5	88.4	7.07	21.2	7.1	1.15
15 I 50	15.00	50.0	14.59	5.640	.550	481.1	64.2	5.74	16.0	5.7	1.05
15 I 42.9	15.00	42.9	12.49	5.500	.410	441.8	58.9	5.95	14.6	5.3	1.08

12 I 50	12.00	50.0	14.57	5.477	.687	301.6	50.3	4.55	16.0	5.8	1.05
12 I 40.8	12.00	40.8	11.84	5.250	.460	268.9	44.8	4.77	13.8	5.3	1.08
12 I 35	12.00	35.0	10.20	5.078	.428	227.0	37.8	4.72	10.0	3.9	.99
12 I 31.8	12.00	31.8	9.26	5.000	.350	215.8	36.0	4.83	9.5	3.8	1.01
10 I 35	10.00	35.0	10.22	4.944	.594	145.8	29.2	3.78	8.5	3.4	.91
10 I 25.4	10.00	25.4	7.38	4.660	.310	122.1	24.4	4.07	6.9	3.0	.97
8 I 23	8.00	23.0	6.71	4.171	.441	64.2	16.0	3.09	4.4	2.1	.81
8 I 18.4	8.00	18.4	5.34	4.000	.270	56.9	14.2	3.26	3.8	1.9	.84
7 I 20	7.00	20.0	5.83	3.860	.450	41.9	12.0	2.68	3.1	1.6	.74
7 I 15.3	7.00	15.3	4.43	3.660	.250	36.2	10.4	2.86	2.7	1.5	.78
6 I 17¼	6.00	17.25	5.02	3.565	.465	26.0	8.7	2.28	2.3	1.3	.68
6 I 12½	6.00	12.5	3.61	3.330	.230	21.8	7.3	2.46	1.8	1.1	.72
5 I 14¾	5.00	14.75	4.29	3.284	.494	15.0	6.0	1.87	1.7	1.0	.63
5 I 10	5.00	10.0	2.87	3.000	.210	12.1	4.8	2.05	1.2	.82	.65
4 I 9.5	4.00	9.5	2.76	2.796	.326	6.7	3.3	1.56	.91	.65	.58
4 I 7.7	4.00	7.7	2.21	2.660	.190	6.0	3.0	1.64	.77	.58	.59
3 I 7.5	3.00	7.5	2.17	2.509	.349	2.9	1.9	1.15	.59	.47	.52
3 I 5.7	3.00	5.7	1.64	2.330	.170	2.5	1.7	1.23	.46	.40	.53

TABLE B-4

Elements of American Standard Channel Sections

Section Index	Depth of Channel In.	Weight per Foot Lb	Area of Section In.²	Width of Flange In.	Web Thickness In.	Axis 1-1			Axis 2-2			
						I In.⁴	Z In.³	r In.	I In.⁴	Z In.³	r In.	$\bar{y}$ In.
18 ⌶ 58	18.00	58.0	16.98	4.200	.700	670.7	74.5	6.29	18.5	5.6	1.04	.88
18 ⌶ 51.9	18.00	51.9	15.18	4.100	.600	622.1	69.1	6.40	17.1	5.3	1.06	.87
18 ⌶ 45.8	18.00	45.8	13.38	4.000	.500	573.5	63.7	6.55	15.8	5.1	1.09	.89
18 ⌶ 42.7	18.00	42.7	12.48	3.950	.450	549.2	61.0	6.64	15.0	4.9	1.10	.90
15 ⌶ 50	15.00	50.0	14.64	3.716	.716	401.4	53.6	5.24	11.2	3.8	.87	.80
15 ⌶ 40	15.00	40.0	11.70	3.520	.520	346.3	46.2	5.44	9.3	3.4	.89	.78
15 ⌶ 33.9	15.00	33.9	9.90	3.400	.400	312.6	41.7	5.62	8.2	3.2	.91	.79
12 ⌶ 30	12.00	30.0	8.79	3.170	.510	161.2	26.9	4.28	5.2	2.1	.77	.68
12 ⌶ 25	12.00	25.0	7.32	3.047	.387	143.5	23.9	4.43	4.5	1.9	.79	.68
12 ⌶ 20.7	12.00	20.7	6.03	2.940	.280	128.1	21.4	4.61	3.9	1.7	.81	.70
10 ⌶ 30	10.00	30.0	8.80	3.033	.673	103.0	20.6	3.42	4.0	1.7	.67	.65
10 ⌶ 25	10.00	25.0	7.33	2.886	.526	90.7	18.1	3.52	3.4	1.5	.68	.62
10 ⌶ 20	10.00	20.0	5.86	2.739	.379	78.5	15.7	3.66	2.8	1.3	.70	.61
10 ⌶ 15.3	10.00	15.3	4.47	2.600	.240	66.9	13.4	3.87	2.3	1.2	.72	.64

9 ⌐ 20	9.00	20.0	5.86	2.648	.448	60.6	13.5	3.22	2.4	1.2	.65	.59
9 ⌐ 15	9.00	15.0	4.39	2.485	.285	50.7	11.3	3.40	1.9	1.0	.67	.59
9 ⌐ 13.4	9.00	13.4	3.89	2.430	.230	47.3	10.5	3.49	1.8	.97	.67	.61
8 ⌐ 18¾	8.00	18.75	5.49	2.527	.487	43.7	10.9	2.82	2.0	1.0	.60	.57
8 ⌐ 13¾	8.00	13.75	4.02	2.343	.303	35.8	9.0	2.99	1.5	.86	.62	.56
8 ⌐ 11½	8.00	11.5	3.36	2.260	.220	32.3	8.1	3.10	1.3	.79	.63	.58
7 ⌐ 14¾	7.00	14.75	4.32	2.299	.419	27.1	7.7	2.51	1.4	.79	.57	.53
7 ⌐ 12¼	7.00	12.25	3.58	2.194	.314	24.1	6.9	2.59	1.2	.71	.58	.53
7 ⌐ 9.8	7.00	9.8	2.85	2.090	.210	21.1	6.0	2.72	.98	.63	.59	.55
6 ⌐ 13	6.00	13.0	3.81	2.157	.437	17.3	5.8	2.13	1.1	.65	.53	.52
6 ⌐ 10½	6.00	10.5	3.07	2.034	.314	15.1	5.0	2.22	.87	.57	.53	.50
6 ⌐ 8.2	6.00	8.2	2.39	1.920	.200	13.0	4.3	2.34	.70	.50	.54	.52
5 ⌐ 9	5.00	9.0	2.63	1.885	.325	8.8	3.5	1.83	.64	.45	.49	.48
5 ⌐ 6.7	5.00	6.7	1.95	1.750	.190	7.4	3.0	1.95	.48	.38	.50	.49
4 ⌐ 7¼	4.00	7.25	2.12	1.720	.320	4.5	2.3	1.47	.44	.35	.46	.46
4 ⌐ 5.4	4.00	5.4	1.56	1.580	.180	3.8	1.9	1.56	.32	.29	.45	.46
3 ⌐ 6.0	3.00	6.0	1.75	1.596	.356	2.1	1.4	1.08	.31	.27	.42	.46
3 ⌐ 5.0	3.00	5.0	1.46	1.498	.258	1.8	1.2	1.12	.25	.24	.41	.44
3 ⌐ 4.1	3.00	4.1	1.19	1.410	.170	1.6	1.1	1.17	.20	.21	.41	.44

TABLE B-5

ELEMENTS OF EQUAL ANGLE SECTIONS

(Slightly Abridged List)

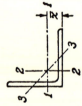

Size (In.)	Thickness (In.)	Weight per Foot (Lb)	Area of Section (In.²)	Axis 1-1 and Axis 2-2				Axis 3-3
				I (In.⁴)	Z (In.³)	r (In.)	$\bar{x}$ (In.)	r_{min} (In.)
8 × 8	1⅛	56.9	16.73	98.0	17.5	2.42	2.41	1.56
	1	51.0	15.00	89.0	15.8	2.44	2.37	1.56
	⅞	45.0	13.23	79.6	14.0	2.45	2.32	1.57
	¾	38.9	11.44	69.7	12.2	2.47	2.28	1.57
	⅝	32.7	9.61	59.4	10.3	2.49	2.23	1.58
	9/16	29.6	8.68	54.1	9.3	2.50	2.21	1.58
	½	26.4	7.75	48.6	8.4	2.50	2.19	1.59
6 × 6	1	37.4	11.00	35.5	8.6	1.80	1.86	1.17
	⅞	33.1	9.73	31.9	7.6	1.81	1.82	1.17
	¾	28.7	8.44	28.2	6.7	1.83	1.78	1.17
	⅝	24.2	7.11	24.2	5.7	1.84	1.73	1.18
	½	19.6	5.75	19.9	4.6	1.86	1.68	1.18
	7/16	17.2	5.06	17.7	4.1	1.87	1.66	1.19
	⅜	14.9	4.36	15.4	3.5	1.88	1.64	1.19
	5/16	12.5	3.66	13.0	3.0	1.89	1.61	1.19
5 × 5	⅞	27.2	7.98	17.8	5.2	1.49	1.57	.97
	¾	23.6	6.94	15.7	4.5	1.51	1.52	.97
	⅝	20.0	5.86	13.6	3.9	1.52	1.48	.98
	½	16.2	4.75	11.3	3.2	1.54	1.43	.98
	7/16	14.3	4.18	10.0	2.8	1.55	1.41	.98
	⅜	12.3	3.61	8.7	2.4	1.56	1.39	.99
	5/16	10.3	3.03	7.4	2.0	1.57	1.37	.99

Size	Thickness							
4 × 4	3/4	18.5	5.44	7.7	2.8	1.19	1.27	.78
	5/8	15.7	4.61	6.7	2.4	1.20	1.23	.78
	1/2	12.8	3.75	5.6	2.0	1.22	1.18	.78
	3/8	9.8	2.86	4.4	1.5	1.23	1.14	.79
	5/16	8.2	2.40	3.7	1.3	1.24	1.12	.79
	1/4	6.6	1.94	3.0	1.1	1.25	1.09	.80
3½ × 3½	1/2	11.1	3.25	3.6	1.5	1.06	1.06	.68
	3/8	8.5	2.48	2.9	1.2	1.07	1.01	.69
	5/16	7.2	2.09	2.5	.98	1.08	.99	.69
	1/4	5.8	1.69	2.0	.79	1.09	.97	.69
3 × 3	1/2	9.4	2.75	2.2	1.1	.90	.93	.58
	3/8	7.2	2.11	1.8	.83	.91	.89	.58
	5/16	6.1	1.78	1.5	.71	.92	.87	.59
	1/4	4.9	1.44	1.2	.58	.93	.84	.59
	3/16	3.71	1.09	.96	.44	.94	.82	.59
2½ × 2½	1/2	7.7	2.25	1.2	.72	.74	.81	.49
	3/8	5.9	1.73	.98	.57	.75	.76	.49
	1/4	4.1	1.19	.70	.39	.77	.72	.49
	3/16	3.07	.90	.55	.30	.78	.69	.49
2 × 2	3/8	4.7	1.36	.48	.35	.59	.64	.39
	1/4	3.19	.94	.35	.25	.61	.59	.39
	3/16	2.44	.71	.27	.19	.62	.57	.39
	1/8	1.65	.48	.19	.13	.63	.55	.40
1¾ × 1¾	1/4	2.77	.81	.23	.19	.53	.53	.34
	3/16	2.12	.62	.18	.14	.54	.51	.34
	1/8	1.44	.42	.13	.10	.55	.48	.35
1½ × 1½	1/4	2.34	.69	.14	.13	.45	.47	.29
	3/16	1.80	.53	.11	.10	.46	.44	.29
	1/8	1.23	.36	.08	.07	.47	.42	.30
1¼ × 1¼	1/4	1.92	.56	.08	.09	.37	.40	.24
	3/16	1.48	.43	.06	.07	.38	.38	.24
	1/8	1.01	.30	.04	.05	.38	.36	.25
1 × 1	1/4	1.49	.44	.04	.06	.29	.34	.20
	3/16	1.16	.34	.03	.04	.30	.32	.19
	1/8	.80	.23	.02	.03	.30	.30	.20

TABLE B-6
ELEMENTS OF UNEQUAL ANGLE SECTIONS
(Slightly Abridged List)

Size	Thickness	Weight per Foot	Area of Section	Axis 1-1				Axis 2-2				Axis 3-3	
				I	Z	r	$\bar{y}$	I	Z	r	$\bar{x}$	r_{min}	Angle with Vertical
In.	In.	Lb	In.²	In.⁴	In.³	In.	In.	In.⁴	In.³	In.	In.	In.	
9 × 4	1	40.8	12.00	97.0	17.6	2.84	3.50	12.0	4.0	1.00	1.00	.83	11.5°
	¾	31.3	9.19	76.1	13.6	2.88	3.41	9.6	3.1	1.02	.91	.84	12.0°
	½	21.3	6.25	53.2	9.3	2.92	3.31	6.9	2.2	1.05	.81	.85	12.4°
8 × 6	1	44.2	13.00	80.8	15.1	2.49	2.65	38.8	8.9	1.73	1.65	1.28	28.5°
	¾	33.8	9.94	63.4	11.7	2.53	2.56	30.7	6.9	1.76	1.56	1.29	28.9°
	⅝	28.5	8.36	54.1	9.9	2.54	2.52	26.3	5.9	1.77	1.52	1.29	29.0°
	½	23.0	6.75	44.3	8.0	2.56	2.47	21.7	4.8	1.79	1.47	1.30	29.2°
	7⁄16	20.2	5.93	39.2	7.1	2.57	2.45	19.3	4.2	1.80	1.45	1.31	29.3°
8 × 4	1	37.4	11.00	69.6	14.1	2.52	3.05	11.6	3.9	1.03	1.05	.85	13.9°
	⅞	33.1	9.73	62.5	12.5	2.53	3.00	10.5	3.5	1.04	1.00	.85	14.2°
	¾	28.7	8.44	54.9	10.9	2.55	2.95	9.4	3.1	1.05	.95	.85	14.5°
	⅝	24.2	7.11	46.9	9.2	2.57	2.91	8.1	2.6	1.07	.91	.86	14.7°
	½	19.6	5.75	38.5	7.5	2.59	2.86	6.7	2.2	1.08	.86	.86	14.9°
	7⁄16	17.2	5.06	34.1	6.6	2.60	2.83	6.0	1.9	1.09	.83	.87	15.1°

Size		Thickness												Angle
7	× 4	7/8	30.2	8.86	42.9	9.7	2.20	2.55	10.2	3.5	1.07	1.05	.86	17.6°
		3/4	26.2	7.69	37.8	8.4	2.22	2.51	9.1	3.0	1.09	1.01	.86	18.0°
		5/8	22.1	6.48	32.4	7.1	2.24	2.46	7.8	2.6	1.10	.96	.86	18.2°
		1/2	17.9	5.25	26.7	5.8	2.25	2.42	6.5	2.1	1.11	.92	.87	18.5°
		7/16	15.8	4.62	23.7	5.1	2.26	2.39	5.8	1.9	1.12	.89	.88	18.6°
		3/8	13.6	3.98	20.6	4.4	2.27	2.37	5.1	1.6	1.13	.87	.88	18.7°
6	× 4	7/8	27.2	7.98	27.7	7.2	1.86	2.12	9.8	3.4	1.11	1.12	.86	22.8°
		3/4	23.6	6.94	24.5	6.3	1.88	2.08	8.7	3.0	1.12	1.08	.86	23.2°
		5/8	20.0	5.86	21.1	5.3	1.90	2.03	7.5	2.5	1.13	1.03	.86	23.5°
		1/2	16.2	4.75	17.4	4.3	1.91	1.99	6.3	2.1	1.15	.99	.87	23.8°
		7/16	14.3	4.18	15.5	3.8	1.92	1.96	5.6	1.9	1.16	.96	.87	23.9°
		3/8	12.3	3.61	13.5	3.3	1.93	1.94	4.9	1.6	1.17	.94	.88	24.0°
		5/16	10.3	3.03	11.4	2.8	1.94	1.92	4.2	1.4	1.17	.92	.88	24.2°
6	× 3½	1/2	15.3	4.50	16.6	4.2	1.92	2.08	4.3	1.6	.97	.83	.76	19.0°
		3/8	11.7	3.42	12.9	3.3	1.94	2.04	3.3	1.2	.99	.79	.77	19.3°
		5/16	9.8	2.87	10.9	2.7	1.95	2.01	2.9	1.0	1.00	.76	.77	19.4°
		1/4	7.9	2.31	8.9	2.2	1.96	1.99	2.3	.85	1.01	.74	.78	19.5°
5	× 3½	3/4	19.8	5.81	13.9	4.3	1.55	1.75	5.6	2.2	.98	1.00	.75	24.9°
		5/8	16.8	4.92	12.0	3.7	1.56	1.70	4.8	1.9	.99	.95	.75	25.3°
		1/2	13.6	4.00	10.0	3.0	1.58	1.66	4.1	1.6	1.01	.91	.75	25.6°
		7/16	12.0	3.53	8.9	2.6	1.59	1.63	3.6	1.4	1.01	.88	.76	25.7°
		3/8	10.4	3.05	7.8	2.3	1.60	1.61	3.2	1.2	1.02	.86	.76	25.9°
		5/16	8.7	2.56	6.6	1.9	1.61	1.59	2.7	1.0	1.03	.84	.76	26.1°
		1/4	7.0	2.06	5.4	1.6	1.61	1.56	2.2	.83	1.04	.81	.76	26.2°
5	× 3	1/2	12.8	3.75	9.5	2.9	1.59	1.75	2.6	1.1	.83	.75	.65	19.6°
		7/16	11.3	3.31	8.4	2.6	1.60	1.73	2.3	1.0	.84	.73	.65	19.8°
		3/8	9.8	2.86	7.4	2.2	1.61	1.70	2.0	.89	.84	.70	.65	20.0°
		5/16	8.2	2.40	6.3	1.9	1.61	1.68	1.8	.75	.85	.68	.66	20.2°
		1/4	6.6	1.94	5.1	1.5	1.62	1.66	1.4	.61	.86	.66	.66	20.4°

TABLE **B**-6 (*Continued*)

ELEMENTS OF UNEQUAL ANGLE SECTIONS

(Slightly Abridged List)

Size	Thickness	Weight per Foot	Area of Section	Axis 1-1				Axis 2-2				Axis 3-3	
				I	Z	r	$\bar{y}$	I	Z	r	$\bar{x}$	r_{min}	Angle with Vertical
In.	In.	Lb	In.²	In.⁴	In.³	In.	In.	In.⁴	In.³	In.	In.	In.	
4 × 3½	5/8	14.7	4.30	6.4	2.4	1.22	1.29	4.5	1.8	1.03	1.04	.72	36.7°
	1/2	11.9	3.50	5.3	1.9	1.23	1.25	3.8	1.5	1.04	1.00	.72	36.9°
	7/16	10.6	3.09	4.8	1.7	1.24	1.23	3.4	1.4	1.05	.98	.72	37.0°
	3/8	9.1	2.67	4.2	1.5	1.25	1.21	3.0	1.2	1.06	.96	.73	37.1°
	5/16	7.7	2.25	3.6	1.3	1.26	1.18	2.6	1.0	1.07	.93	.73	37.1°
	1/4	6.2	1.81	2.9	1.0	1.27	1.16	2.1	.81	1.07	.91	.73	37.2°
4 × 3	5/8	13.6	3.98	6.0	2.3	1.23	1.37	2.9	1.4	.85	.87	.64	28.1°
	1/2	11.1	3.25	5.1	1.9	1.25	1.33	2.4	1.1	.86	.83	.64	28.5°
	3/8	8.5	2.48	4.0	1.5	1.26	1.28	1.9	.87	.88	.78	.64	28.9°
	5/16	7.2	2.09	3.4	1.2	1.27	1.26	1.7	.73	.89	.76	.65	29.0°
	1/4	5.8	1.69	2.8	1.0	1.28	1.24	1.4	.60	.90	.74	.65	29.2°
3½ × 3	1/2	10.2	3.00	3.5	1.5	1.07	1.13	2.3	1.1	.88	.88	.62	35.5°
	3/8	7.9	2.30	2.7	1.1	1.09	1.08	1.9	.85	.90	.83	.62	35.8°
	5/16	6.6	1.93	2.3	.95	1.10	1.06	1.6	.72	.90	.81	.63	35.9°
	1/4	5.4	1.56	1.9	.78	1.11	1.04	1.3	.59	.91	.79	.63	36.0°

Size	t												Angle
3½ × 2½	½	9.4	2.75	3.2	1.4	1.09	1.20	1.4	.76	.70	.70	.53	25.9°
	⅜	7.2	2.11	2.6	1.1	1.10	1.16	1.1	.59	.72	.66	.54	26.4°
	5/16	6.1	1.78	2.2	.93	1.11	1.14	.93	.50	.73	.64	.54	26.6°
	¼	4.9	1.44	1.8	.75	1.12	1.11	.75	.41	.74	.61	.54	26.8°
3 × 2½	½	8.5	2.50	2.1	1.0	.91	1.00	1.3	.74	.72	.75	.52	33.7°
	⅜	6.6	1.92	1.7	.81	.93	.96	1.0	.58	.74	.71	.52	34.1°
	5/16	5.6	1.62	1.4	.69	.94	.93	.90	.49	.74	.68	.53	34.2°
	¼	4.5	1.31	1.2	.56	.95	.91	.74	.40	.75	.66	.53	34.4°
3 × 2	½	7.7	2.25	1.9	1.0	.92	1.08	.67	.47	.55	.58	.43	22.5°
	⅜	5.9	1.73	1.5	.78	.94	1.04	.54	.37	.56	.54	.43	23.2°
	5/16	5.0	1.47	1.3	.66	.95	1.02	.47	.32	.57	.52	.43	23.5°
	¼	4.1	1.19	1.1	.54	.95	.99	.39	.26	.57	.49	.43	23.8°
	3/16	3.07	.90	.84	.41	.97	.97	.31	.20	.58	.47	.44	24.1°
2½ × 2	⅜	5.3	1.55	.91	.55	.77	.83	.51	.36	.58	.58	.42	31.6°
	5/16	4.5	1.31	.79	.47	.78	.81	.45	.31	.58	.56	.42	31.8°
	¼	3.62	1.06	.65	.38	.78	.79	.37	.25	.59	.54	.42	32.0°
	3/16	2.75	.81	.51	.29	.79	.76	.29	.20	.60	.51	.43	32.3°
2½ × 1½	⅜	4.7	1.36	.82	.52	.78	.92	.22	.20	.40	.42	.32	18.8°
	5/16	3.92	1.15	.71	.44	.79	.90	.19	.17	.41	.40	.32	19.2°
	¼	3.19	.94	.59	.36	.79	.88	.16	.14	.41	.38	.32	19.6°
	3/16	2.44	.72	.46	.28	.80	.85	.13	.11	.42	.35	.33	20.0°
2 × 1½	¼	2.77	.81	.32	.24	.62	.66	.15	.14	.43	.41	.32	28.5°
	3/16	2.12	.62	.25	.18	.63	.64	.12	.11	.44	.39	.32	28.9°
	⅛	1.44	.42	.17	.13	.64	.62	.09	.08	.45	.37	.33	29.2°
1¾ × 1¼	¼	2.34	.69	.20	.18	.54	.60	.09	.10	.35	.35	.27	25.9°
	3/16	1.80	.53	.16	.14	.55	.58	.07	.08	.36	.33	.27	26.4°
	⅛	1.23	.36	.11	.09	.56	.56	.05	.05	.37	.31	.27	26.8°

INDEX